I Heard There Was a Secret Chord

Also by Daniel J. Levitin

Successful Aging:
A Neuroscientist Explores the Power and Potential of Our Lives

A Field Guide to Lies:
Critical Thinking with Statistics and the Scientific Method

The Organized Mind:
Thinking Straight in the Age of Information Overload

The World in Six Songs:
How the Musical Brain Created Human Nature

This Is Your Brain on Music:
The Science of a Human Obsession

Foundations of Cognitive Psychology: Core Readings
(editor)

Foundations in Music Psychology: Theory and Research
(editor, with P. J. Rentfrow)

ALBUMS/MUSIC CDs:

Turnaround

sex & math

I Heard There Was a Secret Chord

Music as Medicine

Daniel J. Levitin

W. W. NORTON & COMPANY

Independent Publishers Since 1923

For information about permission to reproduce selections from this book,
write to Permissions, W. W. Norton & Company, Inc.,
500 Fifth Avenue, New York, NY 10110

For information about special discounts for bulk purchases, please contact
W. W. Norton Special Sales at specialsales@wwnorton.com or 800-233-4830

Manufacturing by Lakeside Book Company
Book design by Lovedog Studio
Production manager: Lauren Abbate

ISBN 978-1-324-03618-0

W. W. Norton & Company, Inc., 500 Fifth Avenue, New York, N.Y. 10110
www.wwnorton.com

W. W. Norton & Company Ltd., 15 Carlisle Street, London W1D 3BS

10 9 8 7 6 5 4 3 2 1

Contents

1.   A Musical Species   1

2.   If I Only Had a Brain
THE NEUROANATOMY OF MUSIC   15

3.   Oh, the Shark Bites
MUSICAL MEMORY   36

4.   Look at Me Now
ATTENTION   60

5.   Daydream Believer
THE BRAIN'S "DEFAULT MODE," INTROSPECTION,
AND MEDITATION   77

Interlude   95

6.   Music, Movement, and Movement Disorders   102

7.   Parkinson's Disease   126

8.   Trauma   139

9.   Mental Health   156

10. Memory Loss, Dementia, Alzheimer's Disease, and Stroke   168

11. Pain   187

12. Neurodevelopmental Disorders   200

13. Learning How to Fly   227

14. Music in Everyday Life   250

15. Fate Knocking on Your Door
PRÉCIS TO A THEORY OF MUSICAL MEANING   276

16. Music Medicine, Mystery, and Possibility   317

Acknowledgments   325

Appendix: Types of Music Therapy   331

Glossary   335

Notes   344

Index   384

I Heard There Was a Secret Chord

Chapter 1

A Musical Species

Science seeks to find truth in the natural world;
art seeks to find truth in the emotional world.

I'M WALKING ON REVERE BEACH OUTSIDE OF BOSTON where the low tide has pushed the ocean far away from the shore, and the wet sand is squishing up between my toes. I came here for no particular reason—just to clear my head after two frantic weeks in the city surrounded by noise, construction, and crowds. A briny, sweet smell hangs over everything in the thick humidity of September. Warm air and wafts of cold breeze intermingle—it is T-shirt and jacket weather both at the same time.

I sit in a hard plastic seat on the Blue Line as the car speeds smoothly forward. The Blue Line trains are some of the oldest; they feel like an anachronism. I'm counting the stops to Government Center, where I'll get off to pick up the Green Line, looking at the passengers. A couple of small children with plastic buckets taking sand souvenirs home. A college student lost in her book. A man with work boots and the dust of a day's work clinging to his clothes. As we get closer to town more people get on and I can no longer see the children or the young woman with the book. The man in the work boots gives me a nod and I nod back. The train lurches and we instinctively grab onto a pole to stabilize ourselves.

Before reaching my stop I find myself in a coffee shop, my

saxophone case under the table at my feet. I'm reading the morning newspaper, people-watching, enjoying a Danish and picking up bits of conversation around me. The place is jumping.

Different cities have distinct feels—the air smells dryer or damper, the sounds bounce off buildings and landscape in their own way. I walk through the redwood forests of Muir Woods in Marin County, California, and find a spot that is so quiet I lie down on my back, staring up at the tops of trees more than a thousand years old. The blue sky barely peeks through the dense greenery. I can't hear a single human-made sound, and at this time of day, the birds are quiet— I have to wait a few minutes to hear a song far off in the distance. I close my eyes. Hello darkness, my old friend. With no wind, even the trees are silent and I'm lulled into a state of pure calm, a stillness and majesty that are mind-altering. I'm hypnotized by the thick, dark red bark. These trees exist on their own time scale, so much longer than human life.

Someone at the bar drops a glass and it shatters. I open my eyes and realize I've been here all along, in this room. The Keystone Korner, a jazz club in my home town of San Francisco. Art Blakey and his Jazz Messengers have been playing, and I was lost in the music, here and yet not here, my mind taking me through different places and scenes all while sitting in my chair at this table. Wynton and Branford Marsalis have just taken solos on "In Walked Bud" and passed them on to Donald Brown, the young pianist. All of us were young then—still in our twenties, except for Art, who was my grandfather's age—beaming, moving things along, "directing traffic" as he says.

Until the glass crashed, I had been in what neuroscientist Richard Davidson from the University of Wisconsin calls *experiential fusion*— the state of being so absorbed that your consciousness itself becomes fused with what you are experiencing. During experiential fusion with music, you temporarily lose awareness of yourself as an individual entity, separate from the music; you and the music have become one. If someone touched your hand and asked you, "Are you aware that you're in a jazz club?" you'd almost certainly say, "Yes." But that awareness is born only in the moment of the interruption, as you get yanked out of your absorption, back into the mundane. In both moments you're attentive to the music, but only in the second moment do you have meta-awareness. If this sounds strange, compare it to sleep. If someone wakes you up and asks if you were sleeping, only retrospectively could you assess that sleeping was just taking place, and it was *you* who was doing the sleeping.

The band launches into "In a Sentimental Mood." Billy Pierce, who'd been my sax teacher just last year before Art picked him up midsemester, is also up on the bandstand. He looks at me and smiles. He starts to play, and I'm transported again to another time and place. I couldn't tell you where—but it is in turns thrilling, heartbreaking, bustling, radiant, and always, always moving forward.

Today, in my sixties, after a bad day at work, getting cut off in traffic, or just feeling blue and despondent for no discernable reason (that *is* a part of the human condition), there is refuge. Picking up my guitar, or sitting at the piano, it's as if I'm in a bubble—feeling safe, contented, and that all is right in the world. And when playing music with Victor Wooten, Rosanne Cash, or Carlos Reyes, on a good night we feel that bubble extend out into the back of the room and lift up everyone in it.

My father's father, Joseph, was a scientist and a medical doctor with an extensive collection of 78 rpm records of big bands, opera, and symphonies, and 33 1/3 rpm sing-alongs. As a physician, he lived in a world of evidence and scientific truths. As a humanist, he immersed himself in art; his home was filled with paintings, sculpture, literature, and music. We may think of science and art as standing in opposition to one another, but they are bound by a common objective. Science seeks to find truth in the natural world; art seeks to find truth in the emotional world. Medicine fits somewhere in between, bridging science, art, and the emotions that move us toward the will to survive, to heal, to take our medicine, to exercise, and to put in motion all those things that keep us healthy. It is no accident that the word "motion" is part of the word "emotion."

Both come from the Latin *emovere*: to move, move out, or move through. As music moves through us, it signals, exercises, and invokes emotions.

Prehistoric medicine relied on a collection of folk remedies and superstition, before evolving into what we do today: use evidence-based treatments to cure disease and promote lifelong wellness. For all of its seeming precision, the practice of medicine is both a science and an art. To an outsider, it may look like following a recipe, but as every bread baker knows, changes in humidity, age of the ingredients, and random factors affect the outcome. In medicine, every case is different; no two cases of a disease or injury present in exactly the same way. Good medicine relies on clinical judgment, refined through the same sort of trial-and-error and creative problem-solving that artists and scientists use. Both the master physician and master baker must improvise. (Although the thought of a brain surgeon "improvising" may fill you with terror, it's actually necessary, as neurosurgeon Theodore Schwartz explains. "Not only is the normal anatomy of every human variable, and unique, every tumor has its own configuration that distorts the landscape into which it has dug itself in a slightly different way. Inevitably, the reality we encounter differs from our expectations of what we thought we would find.")

The most important distinction, then, isn't in separating artists from scientists and doctors, but in separating creative thinkers from formulaic ones, separating those who can tolerate uncertainty from those who cannot. Art, science, and medicine trade in doubt, and in its remedy—improvisation. Moreover, to be effective, the musician, the therapist, the scientist, and the physician must establish a rapport and a relationship of trust with people they may never have met. They remain alert at all times while appearing relaxed. They monitor reactions unobtrusively, connecting and responding in real time to their behaviors, to the unexpected.

Music affects the biology of the brain, through its activation of specialized neural pathways, its synchronization of the firing patterns of neural assemblies, and its modulation of key neurotransmitters and hormones. Together, these drive a range of changes that are important to our survival and well-being. Music promotes relaxation when we're stressed; it can reduce blood pressure or make diabetes management easier; it soothes us when we're depressed and energizes us for exercise. Professional athletes listen to music to pump them up when they compete—often hip-hop, but not always (LeBron James listens to jazz and Beethoven). Patients with Parkinson's use it to help them walk; patients with Alzheimer's find that it reconnects them with themselves and improves memory. Music reduces pain, increases resilience and resolve, and can actually change our perception of time, such as when we're lying in a sickbed with nothing else to do, going on a long road trip, or immersed in a VR game.

Engaging with music, whether as a listener or a player, facilitates entry into the brain's Default Mode Network (DMN), a path to the subconscious that is instrumental to everything from problem-solving to relaxation, from creativity to immune system function. And for many, music can connect us to a sense of a higher power, of great and enduring beauty, and listening to or playing it can provide some of the most exhilarating and meaningful moments of our lives. Those moments may come from a concert by Wiz Khalifa, Beyoncé, or U2; we may experience them alone in the privacy of our earbuds, or when the stillness of a room suddenly comes alive as we play the first few notes of Beethoven's *Pathétique* Sonata on the piano.

Companies such as Spotify, Apple Music, YouTube, and Pandora hope to curate effective playlists that not only help with mood regulation or mood change, but target your individual tastes. Already, these music services are adaptable, learning what you skip, and what you listen to. Within the next five years they will do even better. Through data mining, the companies' software knows your age, gender, income, geographic location (often within a few meters), political views, what time of day it is, what things you've searched for. Through your smartwatch they log your heart rate and skin conductance. The algorithms register that you're driving your car across town on a Sunday at 4 pm, the GPS knows you're headed toward your parents' home and, based on previous months of data, that once you get there your blood pressure usually goes up (perhaps because they berate you for not making more out of your life). These algorithms have tracked what music you usually select for yourself on that drive. They also know from scanning your emails that your latest lab report showed you've got high blood pressure. Soon, personalized, custom playlists will be curated automatically and in the background, becoming essentially invisible to you—your car or smartphone will offer to play you what it thinks you need at this particular time, based on your individual tastes. If this sounds far-fetched, consider what existing smart devices already do. Smart thermostats in your home learn your comings and goings, which room of the house you're in at various times, what temperatures you like when asleep versus awake. If you have a smartphone or digital assistant, you may have found that when you say out loud to a friend that you need new walking shoes, a Tik-Tok or Instagram ad for shoes pops up in minutes.

To customize playlists to each person's tastes, most of the algorithms use collaborative filtering (such as: people who listened to Drake also listened to Doja Cat; people who listened to The Cure also listened to The Smiths). Elegant research by Cambridge University's Jason Rentfrow and his colleagues is offering a new and more precise approach to creating taxonomies of musical tastes and preferences, overlayed with demographic information, attitudes, and personality characteristics. This is the future of personalized playlists, and hence, the future of music therapy.

⌒

Beliefs about music's power to heal the mind, body, and spirit date back to the Upper Paleolithic era, around 20,000 years ago, when ancient shamans and other healers used drumming in the hopes of curing a wide range of maladies, from mental disorders to wounds and illnesses. Our word *shaman* comes from the Russian *shaman*, for a person of special status within a tribe who acts as an intermediary between the natural and the supernatural worlds, using magic to foretell the future, cure illnesses, and control spiritual forces. The term was originally applied to the Tungusic peoples of Siberia and northeast Asia. (It is often broadly applied to any ancient or modern person who "travels" in non-ordinary reality to gain information and possibly to heal spiritual, mental, or physical ills, such as the Inuit *angakok*, the Mentawai *sikerei*, the Korean *mu*, the Azande *boro ngua*, and the !Kung *n/um k"ausi*.)

The shaman, medicine man or medicine woman, medium, or psychic who is entrusted with healing powers is a cultural universal, recurring across human societies and nearly every hunter-gatherer tribe. The shamanistic tradition was begun by women, and contrary to popular misconception, it was far from a men-only profession. Shamanism is just one historic antecedent of twenty-first-century music therapy. During the 11th century BCE, near the end of his life, King Saul suffered from periodic depression and agitation. On such occasions, we're told, he would summon David, the same David who battled Goliath, and who was reputed to be among the greatest musicians in the kingdom.

> David would take the lyre and play it; Saul would find relief and feel better, and the evil spirit would leave him.
>
> *(1 Samuel 16:23 NIV)*

David could slay a giant with a rock, and depression with a lyre. The origins of music for healing are as old as our species and are still practiced by a majority of peoples in the world today; music's ability to heal, to provide a brighter day, to promote physical and mental health, knows no boundaries of language or culture. As Longfellow famously penned: *Music is the universal language of mankind.* Centuries earlier, a continent away, Confucius wrote: *Music produces a kind of pleasure which human nature cannot do without.*

It is a recent feature of Western society that we have separated these two, healing and music. We tend to see healing as the province of doctors, and music as entertainment. Perhaps it is time to reunite two of the most intimate parts of our lives. Scientific advances in the past ten years have provided a rational basis for this reunification, opening a dialogue between health care workers, health insurance companies, and all the rest of us. The research allows us to take what had been speculation, anecdote, and observations untethered from evidence, and join them in equal partnership with prescription drugs, surgeries, medical procedures, psychotherapy, and various forms of treatment that are mainstream and evidence-based. The advances were spurred by this 20,000-year history, as scientists tested the ideas in the laboratory and in clinics.

It has been said that "Music gives soul to the universe, wings to the mind, flight to the imagination, and charm and gaiety to life and to everything." Plato believed that music can bypass our rational mind: "More than anything else, rhythm and harmony find their way into the inmost soul and take strongest hold upon it." It is perhaps no surprise to learn, then, that the Greek god of medicine, Apollo, was also the god of music.

Hippocrates (5th century BCE) is known as the father of Western medicine, and he believed strongly that music could be used to treat a variety of physical and mental health conditions. Although no musical prescriptions have survived, we do know that specific scales were used to treat particular ailments, and Athenian physicians would prescribe particular tonalities to heal colds, aches, depression, or injuries. For example, Dorian mode was considered suitable for mourning; Phrygian mode was used to control digestive problems; Lydian imparted good cheer, optimism, friendliness, and a tendency to laughter, love, and song when it was well executed, but could lead to weeping and sadness when not. Seven hundred years later, Ptolemy, living in the Roman Empire, wrote his treatise *Harmonics*, in which he elaborated on the relationship between music, emotions, and therapy.

Today, some sound-healing therapists, meditation headphones, and relaxation music apps use tones of particular frequencies, or music that contains those frequencies, to achieve goals for healing and self-care. Although the ancient Greeks knew (or intuited) things that took years for science to catch up with, they were also wrong about many things, as science has borne out. For example, there is no scientific evidence that music's ability to heal, or to cause changes in mood, or any other cellular effects, derives primarily from the specific frequencies of tones used; this has been exhaustively studied. The notion also doesn't make logical sense, for much shamanistic healing was conducted by singers who would have had no way to calibrate their precise frequencies from one occasion to the next, or from one tribe to another. The power of music, then, probably comes not from specific frequencies, but from trance-inducing rhythms, or a combination of musical elements as they stand in relation to one another—elements including harmonic structure, melody, tonality (major or minor), rhythm, and tempo. In other words, whatever effect a piece of music or sequence of sounds has on you, it's unlikely that it would stop having that effect if it were shifted by a few hertz (Hz, cycles per second) in one direction or the other.

Another important attribute of music is timbre, the tonal color that distinguishes two instruments both playing the same note—or that distinguishes one person's voice from another's when they're saying the same thing. Ravel's *Bolero* in 1928 introduced timbre as a form-bearing medium—that is, an element of music that could impart both meaning and a unique identity to a musical piece. By the 1950s, timbre had become among *the* most important attributes of music. Timbre is what allows us to recognize John Coltrane's saxophone from that of Stan Getz, or Johnny Cash's voice from Elton John's. It is a unique sonic fingerprint. Even records have a timbre all their own—most of us can tell the difference between a recording made in, say, 1940, and one made in 1970 or 2022 by "the way it sounds"—just another way of describing timbre.

Some of the beneficial effects of music across history may have accrued by putting patients or entire tribes into a trance state, or by accessing the subconscious, thereby opening a route to healing. An important question is whether there's something intrinsic in music that promotes healing, or alternatively, if it's the trance state itself that facilitates healing. It may be that music is just one of several psychological enzymes—catalysts—that can lead to healing through trance. But if there are others, they have not revealed themselves to us as clearly as music has.

This book takes as its inspiration the many compelling testimonials of poets and philosophers. Consider the words of the poet Georg Philipp Friedrich Freiherr von Hardenberg (Novalis), ca. 1799: "Every illness is a musical problem—its cure a musical solution." And it was the philosopher and composer Friedrich Nietzsche who articulated what has become a north star for me in my own research, that music is a way to express thoughts and feelings that cannot be expressed in words. "Life without music," he wrote, "would be a mistake."

⌒

The neurologist Oliver Sacks loved playing the piano, especially Bach fugues, which he played with great joy and exuberance. Oliver felt a very personal and deep connection between music and the practice of medicine: "My medical sense is a musical one. I diagnose by the feeling of discordancy, or some peculiarity of harmony."

Oliver had witnessed musical cures in his practice, where he had a front row seat to the unexplainable and curious experiences of patients who came to see him. One patient, Tony Cicoria, a non-musician, began hearing piano music in his head after being struck by lightning. "In the third month after being struck by lightning, then," Oliver wrote, "Cicoria—once an easygoing, genial family man, almost indifferent to music—was inspired, even possessed, by music, and scarcely had time for anything else." Oliver met another patient, Henry Dryer, a 92-year-old living in a nursing home who spent most of his days in a near-catatonic state, "inert . . . unresponsive, and almost un-alive." When Oliver played recordings of music from Mr. Dryer's youth (as shown in the documentary film *Alive Inside*), the previously catatonic patient came alive, singing joyfully and reminiscing.

In 1996, with neuroplasticity expert Ursula Bellugi, Oliver and I met dozens of individuals with Williams syndrome (WS), a neuro-genetic disorder, who may not be able to get food from their forks to their mouths but can play musical instruments like clarinet and piano with no impairment. Some of them could not tell time on a clock but could keep musical time. They also were profoundly moved by music. In one study, we found that after hearing a sad song, they felt sadder and stayed sad longer than those without WS, so-called neurotypicals. The same was true of joyful music and scary music and restful music such as lullabies. They also loved dancing and could barely stop themselves from doing so. In short, their neurological impairment somehow rendered them more sensitive to music. And—perhaps relatedly—they were friendlier than the average person.

None of us scientists knew why a collection of sounds, config-ured *just so*, could move people to the highest peaks of emotion, help them sleep, bring them out of catatonia, and promote healing across a wide range of physical and mental ailments. This topic spurred sci-entists such as myself to try to explain the biological underpinnings of these fantastic stories in order to allow us to specify evidence-based treatment protocols, and to better understand humans—in Oliver's words, "a highly musical species." We know a lot more now than we did then.

How can we scientifically study something as magical, ineffable, and as spiritually moving as music? If we try to pin down the slippery thing that is art, will we demystify it or ruin it? When I'm listening to music with my eyes closed (the way I listen to most music), I'm letting it run free in my head and my heart; I give myself completely to it. I feel chills, and goosebumps; my mind wanders; I let the music wash over me and take me wherever it wants to go. In short, I treat music and my experiences of it as pure magic. However, when I go into the laboratory, I have to leave that belief in magic at the door and maintain scientific objectivity.

When people tell me about how profoundly moved they are by music, I understand because I've felt that way, too; I've come from those experiences thinking *this is what it's all about.* When somebody tells me that x, y, and z made them feel relief from pain for the first time in months, I don't question their experience. But when I walk through the door of the lab, I have to ask, *was it really* x, y, *and* z? *Would just* x *and* y *do it? For that matter, will* j, k, *or* l *do this too?*

My friend Howie Klein, while he was still president of Reprise Records, was given a diagnosis of aggressive prostate cancer and told he needed surgery immediately, and even then the outlook was uncertain. Howie, desperate, heard about Dr. Timothy Brantley, who drew blood and then did analyses that no one else did, formulating a special diet based on the results of that blood work. It took Howie a few hours a day to shop for and prepare the food for this special diet. That was 22 years ago. He's still alive, and healthier than ever. His doctors couldn't explain it. "We must have made a mistake," they said. Howie thinks the diet saved him. I think it's possible, but unlikely—for every statistic, there's an outlier. If that outlier wasn't Howie, it would have been someone else. On the other hand, x, y, and z may have worked for him, but that's *not* the same as saying it will work for other people, all told, when large numbers of people are treated. Howie is a vocal and living example of the special diet regimen, but cemeteries are full of people for whom such alternative treatments did not work.

If you're telling me that you like Genelec speakers better than KRKs, or that vinyl sounds better to you than CDs, or that listening to '70s rock makes everyone dance at family gatherings, who am I to argue? But if we're talking about medicine, and specific, measurable outcomes that affect health—mental or physical—then it's the job of people in my field to ask questions and to test those claims.

You may have read that listening to Mozart makes you smarter. If only that were true. Why stay up until 2 a.m. doing calculus problem sets when you could just be listening to *The Marriage of Figaro*? The study that started it all was published in the prestigious but not infallible journal *Nature* in 1993. The researchers were circumspect in their claims that listening to Mozart was associated with modest improvement in spatial reasoning tasks like folding paper and completing mazes (tasks that are part of standard IQ tests), and they called for work with additional composers, not just Mozart. A media frenzy quickly ensued and these caveats were lost in the story.

Their woefully incomplete study raised more questions than it answered. *Does it only work with Mozart?* No one who loves Mozart would argue that his music isn't sublime, but why not Bach, Beethoven, Bartók, or Berlioz? And if you hate Mozart's music, would it still make you smarter somehow? What about other forms of art or entertainment? The experiment lacked a rigorous control condition, as did so much of music and medicine research prior to that time. Their control group simply sat in a room, bored, with nothing to do, while the treatment group got to listen to Mozart.

Many dozens of studies failed to replicate the original result. It turned out that it wasn't that Mozart was making people smarter, it was that sitting in a dark room doing nothing made them dumber. Doing almost anything other than that was equivalent to listening to Mozart. That didn't stop an entire industry of Baby Einstein CDs from popping up, or the governor of Georgia requesting $105,000 of public funds to send Mozart tapes and CDs to the parents of every newborn Georgian. I'm all for people listening to Mozart, but the money would have been better spent on things that are shown to work, like Head Start programs, or improving the public school system.

If we're going to hand over our future health to music, we need to know that it works just as well as or better than other treatments.

If we want insurance companies to spend their money on music interventions, we need to convince them—with science—that their patients aren't going to get sicker and cost them more money in the long run.

At the same time, we must take care to protect the sovereign domain of our personal tastes and proclivities as well as our sense of wonder. Choosing the right music for pleasure or for healing is never going to be a one-size-fits-all affair. Even setting aside therapeutic uses, our tastes can change over the course of a life or even a day. If I've just heard my favorite song six times in a row, I may not want to hear it again. The right music is whatever music is right for us at any given time and place.

The scientists who are at the forefront of our field are learning to ask questions with an open mind, and to listen to our experimental participants to better inform the studies we conduct and frame the results we obtain from them. The interaction of music, mood, health, and the biology of our brain is yielding ever more clues about how it all works. *I Heard There Was a Secret Chord* will show you what we know, how it can be explained, and how we can harness the potential of music for healing and for staving off disease in the first place; for relieving pain; for helping us look forward and reimagine our lives.

Chapter 2

If I Only Had a Brain*
The Neuroanatomy of Music

If I only had a brain . . . a heart . . . the nerve.

THE BRAIN IS THE MOST COMPLEX BIOLOGICAL ASSEMBLY we know of. Its 80 billion neurons—nerve cells—communicate with each other to make trillions of connections, more than the number of particles in the known universe. These connections give rise to the sum of our experiences: all of our thoughts, desires, and beliefs; our emotions, changing moods, memory, even our heart rate and the contents of our dreams. How does this funny-looking, folded-in-on-itself, three-pound tangle of wires and blood vessels do all this, as well as orchestrate the complex interplay between emotion, memory, sound, and healing that comes from music? And how does it allow us to remember music we like, make playlists, and party like it's 1999?

Our six primary senses (touch, vision, hearing, taste, smell, and balance) create mental representations—sensory maps—of our environment (less well known are an additional 17 senses that keep track of our internal environment, things such as hunger, chronoception [time], and baroception [blood pressure]). Events occurring in the outside world—the rustle of leaves in the wind, a red-tailed hawk flying by, a sudden waft of skunk—are picked up by our sensory receptors. The receptors pass electrical signals on to various parts of our brain

---

* "If I Only Had a Brain," from *The Wizard of Oz*. Music by Harold Arlen, Lyrics by Yip Harburg. Recorded by Ray Bolger and Judy Garland, 1939.

where specialized neural circuits process and interpret features of the input. In vision, for example, one neural circuit detects the shape of an object, another its color, another its spatial location. Later, downstream, the output of these circuits is combined, and we end up with a unified, coherent representation of an object, say, an apple.

The brain also analyzes musical features separately. After sound hits our eardrums it makes its way to the brain stem, then climbs upwards to the cerebellum—a structure at the base of our head dating back to the reptilian brain. We used to think the job of the cerebellum was governing motor movement, keeping internal clocks running, and not much else. We now understand that it sends signals to and receives them from the cerebral cortex (the largest part of our brain, comprising the four lobes: occipital, frontal, temporal, and parietal) and is involved in emotion perception. From the cerebellum, sound—or more accurately, the neuronal, electrochemical representation of it—goes up to a subcortical (below the cortex) area called the inferior colliculus. The inferior colliculus helps with sound localization, pitch perception, and auditory attention, and is part of the startle reflex, an unconscious process that causes us to get out of the way if we hear a sudden, loud sound. This startle circuit connects sound arriving at the brain stem with movement centers in the cerebellum, thalamus, and motor cortex. Our conscious awareness of what's happening sonically only occurs after the signal makes its way to the cortex, starting in the temporal lobes at what we call the auditory cortex. (This is actually a term of convenience and is not entirely accurate; deaf people, for example, use their "auditory cortex" to process sign language. The brain remodels itself, through neuroplasticity, to use cortical real estate that is available.)

Once sound hits the auditory cortex, the signal is shipped off to a series of special-purpose circuits for pitch, duration, and loudness. Aggregation circuits then use the pitch information to construct higher-order features that depend on pitch: contour (the pattern of ups and downs, without regard to the *size* of the ups and downs), melody (the abstract pattern of intervals, without regard to the specific

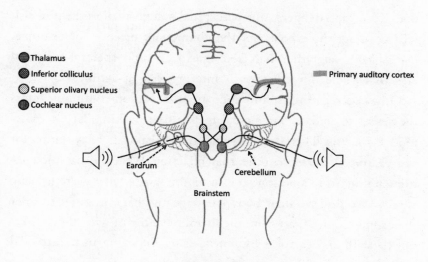

Auditory pathway.

pitches), and harmony (including tonality such as major-minor). Other circuits aggregate the duration and loudness information to compute meter, tempo, rhythm, and tactus (the point in a piece of music where you naturally want to tap your foot or snap your fingers). All three of these elemental features—pitch, duration, and loudness—are combined to compute more complex musical attributes such as timbre, spatial location, and reverberant environment. The different pieces of information come together later (where "later" is on the order of 40 milliseconds). It all happens so quickly that we don't realize our brains were analyzing the information piecemeal. Our subjective experience is that we simply hear the melody, we recognize it's being played by a trumpet, and if the trumpet has a very distinctive, identifiable sound, we even know who's playing it. The evidence that features are analyzed separately comes from neuroimaging and from people with brain lesions. We've seen patients who can recognize the pitches in a song but not the rhythm, and vice versa; patients who are "timbre deaf" and can no longer distinguish one musical instrument from another, yet still recognize a melody.

The coherence of the signal—the idea that we are hearing the song itself as it is played—is effectively an illusion, but an adaptive one. To escape dangers, our perceptual system is more helpful if it tells us about the whole, not the component pieces—for example, *that's a lion's roar*, rather than feeding us a stream of spectral-temporal cues that we need to consciously interpret one-at-a-time.

Other aggregating circuits compare information coming in across the senses to update our mental maps using multimodal, or cross-modal, analysis. This partly involves the thalamus, a relay station for aggregating all the senses other than olfaction. We use visual information coming from a person's lips to disambiguate what we're hearing aurally. We observe their body language and hand gestures to infer their mood, quite apart from the literal meaning of what they're saying (as with an eye roll or forehead slap). Engaging in multimodal activities such as playing an instrument, hiking in nature, or dancing appears to be neuroprotective and curative because large brain networks, spanning different regions of the brain, are unified, brought online to work harmoniously together toward a common goal.

Music is far more than just an auditory experience. Playing music activates the visual cortex as we read music, and look at other musicians as well as our own fingers and hands. Listening to music activates occipital lobe circuits related to visual imagery—most listeners report mental images of autobiographical memories or various mental images activated by the music—landscapes, activities, blue skies, or abstract shapes and colors. Whenever I hear a xylophone playing a staccato descending major scale, I get visions of Bugs Bunny tiptoeing down a staircase, thanks to Carl Stalling's evocative scores. The musician Sting volunteered to have his brain scanned while he listened to music. I was surprised to see that his visual cortex was extremely active. While he listened to James Brown, he reported, his closed eyes were flooded with images of James Brown dancing; while he composed a new song in the scanner, his brain was filled with images of rooms in a grand building, columns, buttresses. "I see music as structure," he explained, with the elements of songs supporting each other like elements of architecture.

Listening to music activates long-term memory as we recall events associated with a song we know well, and more implicitly, as our memory centers search their contents to find music that is similar. This is how you can recognize a tune you know in a version you've never heard before. It's how you can be surprised by something in a song, because your brain is comparing the ongoing development of this song with all the other songs you've heard before, registering what's similar and what's different. Music also activates centers for short-term memory, movement, emotion, prediction, reward, and many others. Music can lull us into a state of relaxation and invoke the daydreaming mode of imagination. It can activate private, happy memories that lift our mood, or old, traumatic memories, and offers us a chance to recontextualize them.

When we play an instrument (including singing), we are engaging more mental faculties than during almost any other activity: motor systems, motor planning, imagining, sensory feedback, coordination, emotions, auditory processing, and—if we're inspired—creativity, spirituality, pro-social feelings, and possibly a state of heightened awareness coupled with calm known as the *flow state*, a term coined by psychologist Mihaly Csikszentmihalyi. (Experiential fusion is a particular type of flow state.) Synthesizing all this information requires separate brain circuits for integrating and binding. The richly multimodal aspect of musical experience helps us to understand why it has the powerful therapeutic effects that it does.

Our brains are shaped not just by genetics, but by the environment and the culture in which we live, along with millions of random events. These give rise to robust individual differences: there is no one song that everybody likes, and no one song that everybody hates. On top of that, even if we have a favorite song, there are times of day and situations in which we really aren't in the mood for it. To be effective, music medicine must take into account individual taste and aesthetics, the subjective, individual, and idiosyncratic reactions we have to music. That individual differences are paramount here should not be surprising when we look at the history of pharmaceutical interventions. From an outsider's perspective, it seems simple: if you have ailment A you take a drug designed to treat that ailment, end of story. But the reality is that drug therapies are a bit hit-and-miss. A vice president from one of the big pharmaceutical companies told me that most of their medications only work in 50% of the people who take them, and then only 50% of the time. If you know anyone with major depressive disorder, you know that it can be a long, frustrating slog to find the antidepressant that works. This is changing with new advances in personal genetics, and in the next decade, we will see increased precision in prescribing drugs for particular illnesses (it appears that cancer patients will be among the first to benefit from this).

Even though prescription medication is wrapped in the aura of precision, it still comes down to individual differences in what will work for a particular patient at a particular point in time. The same is true of music. It's not the case that a music therapist will meet with a patient and say, "Depressed? Take two Joni Mitchell songs and call me in the morning." Music therapy, like any treatment, needs to be individualized. Anyone who has taken an aspirin for a headache, only to find it ineffectual, knows that this also happens with drugs, and is not peculiar to music.

Your inner metazoa: eukaryotes, fish, praying mantises and the origins of sound

All sound begins with some kind of motion, something that disturbs molecules in air, water, steel, wood, or other medium. Isaac Newton noted that light waves are colorless; the perception of color only occurs in the brain of a living organism. The same is true of sound. Thus perception of sound begins at our eardrums—before it hits our eardrums, it is nothing more than the disturbance of molecules. We can measure that disturbance with gauges and instruments, but until it hits a brain, it is not sound. You may have heard the old philosophical riddle, "If a tree falls in the forest and no one is there to hear it, does it make it sound?" No, because sound is a psychological concept, a product of brains. That tree will disturb molecules and make a mess, but if no creature hears it, no bird, fox, or human, it has not made sound.

The mystery of how we're able to hear begins about 800 million years ago with the first appearance of multicellular organisms that consume organic material, breathe oxygen, are able to move, can reproduce sexually, and go through a stage of embryonic development from simple cells to more complex cells—animals. The animals—also called metazoa—comprise pretty much every moving thing you can see (and many that you can't), including mammals, birds, fish, and insects. They include worms, spiders, stink bugs, snails, bedbugs, starfish, sponges, and lions and tigers and bears (oh my!). Zoologists have so far identified about 2 million different animal species, 1 million of which are insects.

All animal cells have membranes around them that protect the cell nucleus. The oldest animals on the phylogenetic tree, many of which are still with us, detect vibrations through their membranes by cell-to-cell communication, via a solid substance such as a leaf, a beehive, or the ground—not that different than you putting your hand on the top of a washing machine during the spin cycle. Over the following 600 million years or so, animals evolved sensory membranes that could detect movement at a distance through air and water.

Being able to detect movement in the air or water that surrounds us confers an evolutionary advantage, allowing us to perceive things we can't see, such as those around corners and in the dark. Hearing conveys the sound of footsteps, and if our senses are well attuned, we can tell whether those footsteps are the stomping of an angry neighbor (with the associated vibrational information received through the soles of our feet) or the tiptoeing of a loved one, whether the footsteps are approaching or departing. Each of our other senses has plusses and minuses. Vision only works when it's light. Smell can work in the dark, and at a distance, but only if the winds are right. Touch and taste both work in the dark and the light, but only when something is already right next to you—too late, in many cases, to help you avoid a danger. Ultimately the earliest use of hearing was probably for predation—to track something you want to eat, or escape behavior—to avoid being eaten.

Any medium can transmit sound, so long as it contains molecules that can be disturbed, which is why we can press our ears to the ground to sense approaching trains and buffalo herds, and why we, and fish, can hear under water. It's also why there is no sound in the vacuum of space: there are no air molecules to disturb.

Mammals have a thin piece of stretched skin, a membrane, that responds to these movements of molecules in air or water: the tympanic membrane or eardrum. We call it a drum because drums are made by stretching a piece of animal skin (or synthetic equivalent) across a rim. Suspended like that, the membrane vibrates, causing surrounding molecules to vibrate in proportion to how tightly stretched it is and how large it is—larger and looser skins make low notes, smaller and tighter skins make high notes. Drummers know that if they hit one drum that is near another, that second drum may start making noise because the air movements from the first drum cause some molecules to impinge on the second drum. In that case, the second drum is acting just like your own eardrum—it starts moving in response to vibrations in the environment.

When vibrating molecules hit our eardrum, they cause the eardrum to wiggle in and out. That sets off a chain of events of transduction of the signal into the electrochemical responses of neurons and neuronal assemblies.

Motion → Vibration → Transduction →
Neural Responses → Perception of Music

The wiggling in and out of our eardrums contains all the information of the sound. That information is decoded by specialized brain networks that first extract pitch, duration, and loudness, and then, through other specialized circuits, recreate additional features of that original sound, such as what direction it came from (requiring hearing in both ears), whether it is near or far, whether it's approaching or receding. The end result of this analysis lets you distinguish a flute from a trumpet, your mother's voice from another's voice, and whether a person is angry, happy, sad, or has a cold. It's a miraculous reconstruction.

Mammals aren't the only animals that hear. Insects also hear through vibrations of air using the equivalent of a tympanic membrane on their legs. The praying mantis has a single ear that detects the click pulses of bats that are its predators. A new paper in *Current Biology* found that the pitch range of spiders' hearing is remarkably similar to humans': 100 Hz to 10,000 Hz.

Birds, most reptiles, and amphibians have tympanic membranes. You can actually see a frog's because it's not protected by an external ear like mammals have—it's the oval membrane right behind the frog's eye. Fish detect the movement of water molecules using a row of pressure-sensitive hair cells called the lateral line, running along the side of their bodies from head to tail. The fish's lateral line senses disturbances of water molecules, helping it to navigate, find food, avoid predators, and seek mates.

The lateral line of a fish, used for hearing.

Animals with ear drums also have hair cells, like the fish, and they are tuned to specific frequencies. In humans, there are 15,000 or so of them. And instead of being arranged on the sides of our torsos, they're coiled up into an inner ear structure called the cochlea. If laid out flat, the hair cells would resemble a piano keyboard, with frequency selectivity going from low to high. When we hear the low notes of a piano, or the low rumble of a truck, the frequency-sensitive hair cells at one end of the cochlea send a signal to the brain about the frequency they encountered; when we hear the highest notes of the piano or the chirping of cicadas, hair cells at the other end of the cochlea send a signal to the brain. In this way, frequency information is available throughout all stages of processing.

Perception is a constructive process.

It is said that beauty is in the eye of the beholder. This is not quite correct—the eye is simply an organ of transmission of information. Beauty lies in the physical brain and the metaphorical heart of the beholder. This points to a central challenge for neurobiologists—how to account for the richness of our mental lives and experiences.

There is certainly a real world of clouds in the sky, flowers in the dirt, songs from birds, and the sound of babbling brooks. But we have no direct access to this world, nor to any of its properties.

Whatever we experience of the world has been mediated not just by our sense organs, but by a complex neurochemical chain of events that transform sensory inputs to neural inputs, and eventually to mental images or impressions of the world. Physically speaking, when you look at a tree, the sensory input is not the tree itself, but photons that began with the sun and then bounce off of it in particular ways, and that happen to reach the eye. These are focused by the lens and then fall on your retinae, where these photons begin a complex journey of being constructed into something we call seeing. The pattern in your cortex does not "look like" a tree in any sense. Special-purpose processing units in the occipital lobe detect edges, shading, shape, color, and movement, and then all of these come together a few tens of milliseconds later to present to you what you call a tree.

What begins as some kind of impingement on one or more sensory receptors ends up as an *experience* in our brains, constructed out of photons, or pressure waves, or interactions between certain molecules and chemicals in our nose, mouth, and tongue. A lemon is not "sour" or "yellow" in the real world; these are interpretations of the world. Musical instruments do not make music—musicians do, using their brains to guide their fingers, hands, and breath. A famous story illustrates this: A woman approached a musician after a concert (in some versions, it is Jascha Heifetz) and said, "Maestro, your violin sounds so wonderful." He smiled at her and held the violin up to his ear and said, "I don't hear a thing." The lady was perplexed. The musician smiled again. "Madame, the violin doesn't make the music. The music is being made by the man who is holding the violin."

Musicians* hold a mental image of what they want to hear and then make suitable gestures in order to approximate what they intended. I have never met a musician who said that what emerged from their instrument was precisely, exactly what they intended, though— there is always a bit of a discrepancy, again, because perception is

---

* Throughout this book, when I use the shorthand "musicians," I am including singers and conductors.

a constructive process, and also because our fingers and hands and vocal cords are imperfect—no matter how hard we train them.

Cognitive neuroscience is, in part, the study of how these objects-in-the-world become objects-in-our-heads, and how these objects-in-our-heads become things-in-the-world. It is the study of how these interactions give rise to emotions, actions, reactions, memories, and ultimately, consciousness.

Although many of us struggle with mathematics and statistics in school, our brains are sophisticated statistical processing machines. Through our statistical processing, we learn that certain combinations of letters are prohibited in English words (such as *rzhda*), and that certain word combinations almost always occur (the word *to* often follows the phrase *I'm going to give it . . .*). That implicit statistical processing (modeled by Bayesian statistics, and the basis for large language models—LLMs—such as ChatGPT and Gemini) allows even five-year-olds to recognize when a musical note is badly out of tune, or a chord is out of place.

The ability of the human brain to do many things implicitly that we cannot do explicitly is a great source of wonder and amazement. Noam Chomsky has noted that we are all expert speakers of our native language, and we don't need to know the rules of the language to speak it fluently. English has 16 tenses and we quite naturally use them all, but if I asked you to give me an example of the "second conditional" you'll likely be stumped, even though the phrase "if I asked you to give me an example" was just that.

Using these Bayesian statistics, and your knowledge of the world, you learned just by passive listening how music works. The process begins in the womb, and the auditory system is fully functional by 20 weeks of gestation. Music is filtered through the amniotic fluid, and the developing brain becomes accustomed to the rhythmic and melodic patterns it hears, constructing a representation of what is to be expected in music. As infants, and through childhood, we construct our mental model of music based on what we hear. If you grew up listening to Chinese opera, your brain became wired to recognize its typical patterns, as opposed to, say, Indian ragas or Western classical music.

Where does music happen in the brain?

The question posed here may not even be the right question to ask. In the late 1980s and early 1990s, new tools for taking pictures of the active, thinking brain became widely available to researchers. We spent much of the next 25 years mapping the brain, a pursuit that Mike Posner and I snarkily described in 2001 as mere neuro-cartography. Papers abounded in which scientists would breathlessly proclaim that mentally practicing your tennis serve happened in *this* part of the brain, and that memory happened in *that part over there*. Although we do still associate particular regions of the brain with specific mental operations, such as controlling the lips and tongue when we speak, or releasing dopamine when we experience pleasure, these descriptions are increasingly recognized as naïve oversimplifications. The reality is that the brain is vastly interconnected, and to speak of widely distributed networks is more accurate. Music, then, doesn't happen in one or even several parts of the brain. Music, like much of our cognition, recruits disparate parts of the brain in overwhelmingly complex, interconnected, and overlapping circuits. Indeed, music appears to recruit nearly every region of the brain that has so far been mapped.

We can localize some of the computational hubs that are integral to the perception of music, so long as we keep in mind that these are simplifications that gloss over how the regions connect to one another. A new technology called diffusion tensor imaging (DTI), a kind of connectivity analysis, has enabled us to track these connections.

Brodmann Area 47, a part of the prefrontal cortex, contains pattern detectors that look for patterns over time, and then try to predict what will come next. (Sensory systems have evolved to detect change—there is information in change, something that might help

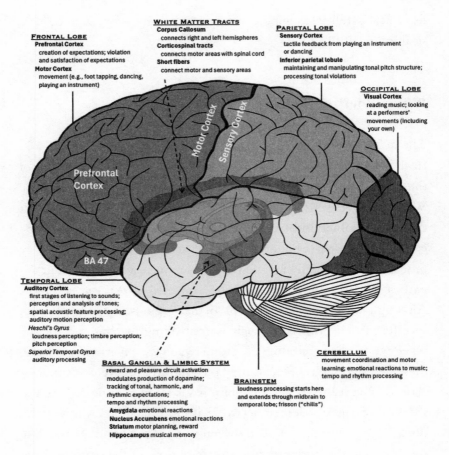

WHITE MATTER TRACTS
Corpus Callosum
connects right and left hemispheres
Corticospinal tracts
connects motor areas with spinal cord
Short fibers
connect motor and sensory areas

FRONTAL LOBE
Prefrontal Cortex
creation of expectations; violation
and satisfaction of expectations
Motor Cortex
movement (e.g., foot tapping, dancing,
playing an instrument)

PARIETAL LOBE
Sensory Cortex
tactile feedback from playing an instrument
or dancing
Inferior parietal lobule
maintaining and manipulating tonal pitch structure;
processing tonal violations

OCCIPITAL LOBE
Visual Cortex
reading music; looking
at a performers'
movements (including
your own)

TEMPORAL LOBE
Auditory Cortex
first stages of listening to sounds;
perception and analysis of tones;
spatial acoustic feature processing;
auditory motion perception
Heschl's Gyrus
loudness perception; timbre perception;
pitch perception
Superior Temporal Gyrus
auditory processing

BASAL GANGLIA & LIMBIC SYSTEM
reward and pleasure circuit activation
modulates production of dopamine;
tracking of tonal, harmonic, and
rhythmic expectations;
tempo and rhythm processing
Amygdala emotional reactions
Nucleus Accumbens emotional reactions
Striatum motor planning, reward
Hippocampus musical memory

BRAINSTEM
loudness processing starts here
and extends through midbrain to
temporal lobe; frisson ("chills")

CEREBELLUM
movement coordination and motor
learning; emotional reactions to music;
tempo and rhythm processing

Your brain on music.

us survive.) The nucleus accumbens, part of the limbic system with the amygdala, modulates dopamine levels during our experience of reward and pleasure.

Please don't be daunted if all these brain region names are unfamiliar. There's an unfortunate carryover from witchcraft and shamanism which is knowing the *secret name* of something. If you know the secret name, it gives you power over the phenomenon—or at least power over your colleagues who don't know the secret name (!). But the secret names themselves explain nothing—the explanatory power only comes when we can connect them to various phenomena. Someone who has a stroke in Brodmann 47, for example, might be expected to lose their ability to find patterns in music, something we have indeed seen in the clinic.

When we hear music that we don't like, especially when we can't get away from it (think of music in a public space), that activates the fear center in the amygdala. Many people list unwanted music in public spaces as a chief annoyance of modern life, and the U.S. Environmental Protection Agency (EPA) amended the Clean Air Act to include unwanted music under the heading of noise pollution. The brain stem is part of the auditory startle response and may be related to the feeling of frisson—the goosebumps or "chills" we sometimes experience in music of great beauty, pathos, or thrilling surprise. The basal ganglia are a group of subcortical nuclei responsible primarily for motor control, working in tandem with the cerebellum and the motor cortex. When the brain stem receives an unexpected loud sound, it recruits these motor areas to get you out of the way. In other cases, these motor areas help us to move more deliberately, as when we play an instrument or dance.

Neurons communicate with one another through electrical signals that must pass through a chemical gate known as a synapse. As you know, electrical wires in your house have a nonconductive insulating material around them—typically rubber or plastic. The purpose of the insulation is to protect the electrical transmission from interference, or from touching a surface that could cause it to spark or short-circuit. Evolution developed an insulating sheath for axons (a part of the neuron), called myelin. Myelin is a fatty white substance that your body is constantly replenishing. Diseases such as multiple sclerosis (MS) that cause the breakdown of myelin, or interfere with myelin production, can have disastrous effects on neuronal functioning. Because myelin appears white, we call the bundles of axonal tracts, insulated by myelin, white matter. The neuronal cell bodies are grayish, and so we call them gray matter. A simple description is that gray matter represents the computational hubs in your brain, the billions of miniature CPUs, whereas white matter represents the wiring between them, the transmission lines. Of course, synapses are colorless because they are just tiny gaps, 20–40 nanometers wide. How small is that? You could fit about 20,000 synapses in the thickness of your fingernail.

One study of white matter tracts by Nandini Singh at the National Brain Research Center in Gurgaon Haryana, India, found that people with increased musical ability (not necessarily as performers but as keen listeners) showed greater connectivity across a large set of brain structures that reads like an inventory of nearly every important computational and emotional center of the brain: superior frontal, rostral middle frontal, posterior cingulate, pars opercularis, caudate, putamen, insula, hippocampus, nucleus accumbens, and more. These networks tie together perception of low-level features such as pitch and duration to higher-level features such as

- tracking melodic and rhythmic structure,
- forming predictions about what will come next,
- categorizing musical elements and entire songs,
- comparing what is being heard at present to what we've heard before (musical memory),
- activation of emotion and reward circuits, and
- preparatory motor activity (the participants didn't move, but their brains seemed to be preparing them to—for foot tapping, hand clapping, finger snapping, or dancing).

Moreover, the corpus callosum—the tract of fibers connecting the two cerebral hemispheres, left and right—contained greater connectivity for white matter tracts with increasing musical perception skill. Musical experience, even in the absence of formal training, changes the very structure and wiring of the brain. The changes not only facilitate the flow of information across the left and right hemispheres, they also connect the frontal lobes, the seat of higher thought, with the motor cortex. Peter Vuust at Aarhus University in Denmark went on to show that professional musicians display enhanced neuroplasticity in connections between auditory and visual areas. Musical perception is truly multimodal, and can lead to lifelong improvements in brain function and connectivity.

Neurochemicals are necessary to activate the connections across white matter tracts. They are like a key in a lock that opens up neural connectivity under particular circumstances. They do different things in different regions of the brain, and we don't have a very detailed understanding of how they work in the musical brain. Added to that is the problem that while there may be as many as 100 different neurochemicals, we have the tools to track only about ten of them in the human brain, and so we're putting a lot of weight on the shoulders of a few. (Almost all can be measured in cerebral spinal fluid, but this is not practical as a research tool.) In all cases, their influence, like that of dopamine, is likely to be region- and circuit-specific.

Why we like the music we like

Just as different cultures tend to develop their own languages, different musics emerge across the world as well. All musical systems we know of have some sort of musical scale—a collection of notes that go together within that musical tradition. As far as we know, all societies in some way acknowledge the special status of the octave, which is produced by a ratio of 2:1 (in string length, the length of a pipe, and indeed, the ratio of vibrations). Pentatonic (five-note) scales are common, and our American blues scale is a pentatonic scale with an added "blue" note, the tritone. In our Western musical system, the octave is divided into 12 equal steps, and we tend to use only seven of them at a time; depending on which seven, and the order we play them in, we get a major or a minor scale. Music in Arab and Indian cultures uses microtones, notes between the notes on our piano keyboard; this is one of the reasons their music sounds exotic to Westerners. Some writers have claimed these constitute 48-note scales, but the reality is that most of the microtones are used as passing tones and ornaments, not components of a melody.

There also exists considerable variation in rhythms. Most cultures use rhythms with nominal low-integer ratios such as 2:1 (one note twice as long as the next) and 3:1 (a waltz or rocking rhythm), and many cultures use much more complex ratios. A common ratio in African music is 5:3, and some African musics use 13:11. The ratio 7:5 is found in Indian music, and 11:9 is used in Javanese music. We don't need to travel that far, however, to find such exotic ratios in music. Beethoven *notates* ratios of 5:4 and 7:4, but in performance, the actual ratios can create great complexity. Studies of actual performances show scientifically what any musician could have told you: that performed rhythms deviate in very complex ways that are impossible to notate. As first shown by Bruno Repp at Haskins Laboratories, and later by Anjali Bhatara in my laboratory, the rhythms played in expressive performances of Chopin and Beethoven piano music, although written in simple integer ratios, are performed very differently. Some eighth notes written on the score, for example, are played almost twice as long—or half as long—in actual practice creating ratios such as 20:9. This disparity between notation and performance is particularly apparent in the Western jazz tradition, primarily since the late 1920s and through to the present time.

To musicians, the word "swing" has a special meaning, referring to a feel, a momentum of the music, as exemplified in the music of Louis Armstrong, Benny Goodman, Mary Lou Williams, and Duke Ellington. There is even Western swing, heard in the music of Bob Wills, Asleep at the Wheel, and Brennen Leigh. When we talk about music that gets people moving, and wanting to dance, swing is among the most reliable. But what exactly is swing?

One well-known aspect of swing is the long-short pattern of notes—the tssss-dit-dit tssss-dit-dit-tssss of a hi-hat cymbal as on "Moanin'" by Art Blakey and the Jazz Messengers or Glenn Miller's "In the Mood." But how long is the "long" and how short is the "short"? Anders Friberg, from the Royal Institute of Technology in Stockholm, actually measured jazz beats and found ratios as complex as 7:2, and a subsequent study by Henkjan Honing of the University of

Amsterdam found ratios as complex as 12:5 for swing. A rhythm that is often taught to jazz players is "triple-feel," in which the beat itself, a quarter note, is divided into three parts, and the first two of those three parts become tied together to make the "long" part of swing. But Friberg found this was somewhat rare in actual practice, and that the precise formula for swing is a function of tempo and whether or not the player is attempting to play "behind the beat" or "on top of the beat." If all this sounds a bit like inside baseball, it is, but the point of it is that complex rhythmic ratios are found in the practice of Western music and are not the sole province of non-Western musics.

A related concept is that of polyrhythms, when two rhythms are played at the same time but are not synchronized with one another. This is one of the most noticeable features of non-Western musics, particularly African and Latin music (it's also something that the American musician Frank Zappa delighted in playing around with).

The brain processes notes, scales, and rhythms in dedicated circuits (as noted above). As we learn to label them with names, such as *C minor pentatonic scale*, or *slow blues shuffle beat*, the process of lexicalizing them causes a shift of processing to the left hemisphere (the language hemisphere for most right-handed people) and creates a binding of the name and the sound across the corpus callosum; this further enriches our mental representations of them. They become not just familiar but *named*.

Naming things adds to our appreciation of them. I no longer see just a *little brown bird* in a tree, I see a *house wren*. And knowing what a house wren looks like, I can now distinguish it from a house sparrow, a distinction that is harder to make if they're not lexicalized—if I don't have a name for them. Similarly, I'm not just hearing a catchy chord progression, I'm hearing a "one-six-two-five" (I–vi–ii–V) or a "1 Major, 2 minor, 3 minor" (I–ii–iii).

There are many variables that go into understanding musical preferences, and musical features are just one of them. My student Yuvika Dandiwal investigated personal and contextual factors that underlie music listening habits. Personal variables refer to things like age, gender, musical knowledge, personality type, temperament, sensitivity to music, autobiographical experiences, and whether or not a piece or style is familiar. Contextual variables refer to the *where* and *why* of your listening—are you at work, driving, doing this for leisure, or health; are you in the presence of others? In one paper, Yuvika studied the role of personal and contextual factors in music listening during the COVID-19 lockdown. She found that personal variables were more significantly associated with music listening for stress relief than were contextual variables—younger adults, women, and people who already showed a strong affinity for music were more likely to use music as a coping strategy, as were those who felt high levels of worry.

So with all of this as background, we are left with the question: Why do we like listening to music—is it the pitch, the rhythm, the combination? The answer is: it depends, because there are broad individual, cultural, and situational differences. Recall that there is no one song that everybody likes, no one song that everybody hates, and that even your own likes and dislikes change over time: a song you used to love can become tiring or even aversive, and you can grow to love songs you once hated.

Within these individual differences, some people are drawn to rhythm, others to melody, and others to lyrics. Some people know every nuance of a drum part but not the subtle filigrees and ornaments of a melody. My wife Heather can identify a song from the first few notes of the melody, or the chord progression, but has difficulty recalling the lyrics. Many performing musicians can play *thousands* of songs from memory, but need to have the lyrics in front of them on an iPad—even when they wrote the lyrics themselves.

Sometimes it is the *timbre* of a piece of music that grabs us—the overall sound quality. Rodney Crowell captured this beautifully in his song "I Walk the Line (Revisited)," in which he describes the first time he heard Johnny Cash, sitting in a 1949 Ford in 1956 as the song came on the radio. He describes it as sounding like it came from outer space, a feeling which parallels my own experience. When I first heard that song on the Zenith vacuum-tube radio by my bedside, it was like nothing I had ever heard before. Thankfully, I've had that experience of newness many times over, and still do: Johnny Cash, The Beatles, Jimi Hendrix, Stevie Wonder, Joni Mitchell, Rodney Crowell himself, Prince, and more recently, Clare and the Reasons, Daniel Knox, Victor Wooten, and Tom Brosseau. Their use of timbre, the sum total package of tonal color and resonances even independent of all the other musical elements, continues to delight and excite me. In the classical domain, I've become enchanted by the pianist Karin Kei Nagano. Even when she's playing the most complex and dramatic pieces, I hear an inner calm, an orderliness combined with great passion, that I find soothing and exhilarating at the same time, like a meditation.

Music has the ability to calm our brains, our hearts, our nerves. We tend to like music that reminds us of something we've heard before, but not too much. We like music that strikes the sweet spot between novelty and familiarity, simplicity and complexity, and between predictability and surprise. The job of the composer, and of the musicians who interpret the composition, is to hit these in just the right balance. The trick of it is that the sweet spot is not the same for all of us, and often not even the same from day-to-day. Loving music requires that we be receptive to it, that we make the mental space and time to allow ourselves to give into it, to be won over by it. If our defenses are up—as they can be in clinical, therapeutic environments—it may simply not work. Or it can catch us by surprise, evoking some of the deepest memories and deepest feelings of our lives, and in the process, help us through almost anything.

Chapter 3

Oh, the Shark Bites[*]
Musical Memory

IN 1960, ELLA FITZGERALD WAS AT THE HEIGHT OF HER career, widely regarded as among the best jazz singers in the world. An experienced performer by any measure, she would typically perform 45 weeks a year. On Saturday, February 13, 1960, Lady Ella took the stage at Deutschlandhalle in Berlin in front of 12,000 fans. After singing a repertoire of popular songs from memory, including "Misty," "The Lady Is a Tramp," and "Summertime," she and the band began "Mack the Knife," first made popular by her collaborator Louis Armstrong five years earlier. Although many might not have seen the show it came from, *The Threepenny Opera*, the audience knew the song well—it had been a number 1 hit for Bobby Darin just a year before—and they cheered as soon as they heard the first line, "Oh the shark has pearly teeth, dear . . ."

Then something extraordinary happened.

After the third chorus, about a minute and a half in, Ella forgets the words. Without losing a beat, or her composure, she continues to sing perfectly in time: "Ah, what's the next chorus, to this song now? This is the one now, I don't know . . ." She continues improvising lyrics, occasionally inserting a remembered word or two from the song. When the next chorus comes around, still stymied, she riffs on

---

[*] "The Ballad of Mack the Knife," from *Threepenny Opera*. Music by Kurt Weill, English lyrics by Marc Blitzstein (1952); original German version ("Die Moritat von Mackie Messer") with lyrics by Bertold Brecht (1928).

the song's history by referencing Darin's and Armstrong's versions, then self-deprecatingly sings that she's "making a wreck of 'Mack the Knife.'" The band keeps playing and after modulating up a half step, a new chorus starts. Having now made up words to two entire choruses on the spot while never losing the rhyme scheme, she begins to scat while imitating Louis Armstrong. The audience goes wild.

In the history of live performance, this stands as one of the most mesmerizing, thrilling demonstrations of mastery. And what is mastery if not the ability to deal with the unexpected as though it was expected? To take an error and turn it into something better than if there had been no error at all? This level of mastery requires deep memory for the tools of one's craft, arrived at through thousands of hours of practice, memorization of procedures, facts, and conventions, until one's art or craft reaches a state of automaticity, what some call the flow state.

We witnessed this type of mastery in a wholly different domain with "Sully" Sullenberger, the pilot who improvised a safe recovery of US Airways flight 1549 in 2009 after it lost power in both engines over New York. Sullenberger describes what went through his mind, and his words could equally apply to Ella's extraordinary performance. "Because I had learned my craft so well . . . I knew my profession so intimately, I could set clear priorities, and so I chose the highest priority items. And then I had the discipline to ignore everything I did not have time to do as being only distractions and potential detriments to performance." Landing the plane required that he not think deliberately so much as act out of training and instinct, on automatic pilot as it were—the flow state. Without memory, there can be no flow. But memory is not an all-or-nothing entity. It flows in bits and pieces; it stops and starts and sputters and spurts. Our left cerebral hemisphere then stitches the pieces together with what it thinks are plausible inferences to fill in the gaps.

Although there was less at stake in Berlin that day than on the Hudson, Ella exhibited the same kind of presence of mind and mastery as Sully did—she prioritized and didn't panic. Ella's highest priority was that the show must go on, and she ignored everything nonessential. Getting the words exactly right was far less important than keeping it swingin' with the familiar melody and rhythms.

When we listen to that magical recording, we hear that Ella has at her disposal a number of cues to the nature of the song—her memory for "Mack the Knife" has not been completely slashed. Her memory for the melody and rhythm are intact and precise. We hear her playing with the beat, as she's done throughout the song so far, using the rhythmic technique of swing that propels each phrase into the next in a long-short, long-short rhythm; she creates groove, even when forgetting the lyrics. The next line is supposed to be "There's a tugboat, down by the river, don't you know?" Do you notice what Ella did? She remembered the last word of the line: *know.* She also remembered, or reconstructed, the internal rhyme of lyricist Marc Blitzstein, with the long ō sound of *don't* when she sings "I don't know." Ella improvises her way into a self-deprecating last line that ends with the name of the song, just as her previous chorus did. She then skips to her memory of the final chorus, where the lyrics name-check characters from the opera: Jenny Diver, Sukey Tawdry, and Lucy Brown (Louis Armstrong added Lotte Lenya, a nod to one of the original leads in the show in both Germany and the United States, and wife of composer Kurt Weill).

Ella's stunning ad hoc performance earned her a *Grammy* award and was so admired that when Frank Sinatra sang "Mack the Knife" with the Quincy Jones Big Band in 1984, he incorporated some of Ella's Berlin lyrics into his version, adding her name to Bobby Darin's and Louis Armstrong's in the list of singers who had previously covered the song. (Ella herself reprised pieces of the improvisation during a return to Berlin in 1962.)

As the seldom-sung final stanza of "Mack the Knife" goes,

There are some who are in darkness, and the others are in light
And you see the ones in brightness. Those in darkness drop from
  sight.

Ella was the bright light, and her performance of forgetting stands as one of the great monuments to memory and forgetting in the history of American music. The truth is that memory and forgetting are forever entwined. Let's unpack this. Ella retained an exquisite memory for the compositional structure, rhyme scheme, melody, the underlying harmony, accent structure, phrasing, and rhythms of the song. She had intact memory for the song's history, who had sung it before, and the particular timbre of Louis Armstrong's voice.

Neurobiologists learn a great deal about memory by studying forgetting. To forget something means we had to have known it at some point, and that's different than never having known it in the first place. And even when we think we know something, memory is fallible in two distinct ways. First, we can lose things in our memory banks, sometimes temporarily, sometimes for a lifetime. Second, when we do locate and retrieve a memory, it can be fantastically distorted without our realizing it.

The reality is we have false memories every day, lots of them. We just don't know it because we're not often challenged. Of course, the place with the highest stakes is the courtroom. Not all eyewitnesses are reliable, but you can't always know which ones are reliable and which aren't. That's why, generally speaking, you want three witnesses, but prosecutors often must base a case on one. And even with multiple witnesses, false memories can creep in for all of them. For example, most Americans falsely recall seeing footage of the first plane hitting the World Trade Center on September 11, 2001. In reality, no media outlets had footage of the first plane until the following day. The brain is not as interested in *how* it comes to acquire information as it is in interpreting that information in the most helpful way. And so most Americans' brains overwrote the true sequence of events and constructed a correct-to-history but not correct-to-personal-experience version. Millions of people will swear that on September 11 they saw the first plane hit. It's a collective illusion.

To hardcore Beatles fanatics, getting the story straight is nearly a life-or-death enterprise—alas, one that is doomed to be frustrating. Beatles producer George Martin discovered this firsthand. "When I made the film *The Making of 'Sgt. Pepper'* we had [archivist Mark Lewisohn] come in as a sort of consultant. And I had George and Paul and Ringo come 'round, and I interviewed them about the making of the album. The interesting thing was there were parts of it that all of us remembered differently. When I was interviewing Paul, I had to keep telling Mark Lewisohn not to correct Paul . . . Paul would recollect something and it would be wrong. For Lewisohn to say, 'Well, that's not right. According to these documents here and these logs, it was this way . . . ,' for Paul it would be rather humiliating. So he tells his story and that's how he remembers it and it should remain that way." Memory is fragile, labile, and unreliable.

What it *feels* like to have a memory system in our brains is a common experience. When we experience something it feels as though it gets stored in a room somewhere in our brains for us to (hopefully) retrieve when we need it. Once in a while, the door to the room is rusty; we can only open a crack and let out just a trickle of information.

But the metaphor doesn't hold up. Memory is not in a *place*. Rather, it's in a distributed network of neurons. It's like asking, "Where is the gravity in my house?" It's not in any one spot. It's distributed. Or "Where is the University of Oregon?" I can circle the campus on a map and point to it, but during the pandemic, nobody was there. Yet the "University of Oregon" still existed as a network of people. Is it the buildings? Well, yes, it's that too—the university owns the buildings and it's one of the things we mean by "Where is the university?" There is no *one* particular place. What's true for schools is generally true for governments, clubs, social groups, and extended families—they are distributed in space and time, as are memories in our brains.

Our memory is largely governed by a brain structure called the hippocampus, which is shaped like a seahorse. It's found on both halves of the brain, deep down inside, and it arranges the storage of short- and long-term memory, including the connection of certain sensations and emotions to those memories. I've come to believe that memories aren't actually *stored* in the hippocampus the way you might store a book on a shelf. In the library of the brain, the hippocampus is the card catalog that tells you what shelf the memory is on, in which wing of the library.

The hippocampus evolved in humans and other mammals to help us find things: Where's the water? Where's the food? Where's the shelter? Memory for where things are is extremely important to secure food and shelter. The hippocampus evolved for that, not for remembering things like your PIN number or the lyrics to a song, but we use it for those tasks anyway. In just the last five years or so we've learned that the best way to improve your general memory is to exercise your geonavigation skills because they make the hippocampus function better, regardless of what type of information you are wanting to store in it. Go for a walk in a place you've never been! Especially on an uneven surface, where you have to negotiate obstacles like tree roots or holes. Or drive to a place you haven't been before. If you're not in a hurry, turn off the GPS in your car. All of this strengthens the functioning of the hippocampus.

Memory serves many disparate functions. Although it feels as though it is a single, unified faculty, it is a basket full of separately evolved capabilities. It's certainly helping us with much more than spatial geonavigation. Memory tells us what foods have poisoned us or others, which people and animals are allies, foes, predators, or prey. Without memory, we could not have language, for we would not remember what words mean or how grammatical structure conveys their meaning.

The different kinds of memory are sensory memory, procedural memory, semantic memory, episodic memory, geographic/spatial memory, and autobiographical memory. Sensory memory includes the afterimage you may have beneath your eyelids after seeing something, or the echo in your mind's ear after you've heard something. Procedural memory, what we often call motor memory, keeps track of things like how to tie our shoes, how to use tools, how to drive a car, brush our teeth.

Memory for specific events, or episodes, is called episodic memory, and memory for facts and general knowledge is called semantic memory. A quick rule of thumb is that if you remember *where and when* you experienced something, that is episodic. If you only remember the thing itself, it is probably semantic. Our semantic memory system is the storehouse for remembering everyday objects, historical events, and cultural practices—things like *the capital of Canada is Ottawa*; *the square root of 9 is 3*; and *if there's an emergency, call 9-1-1*. Extending the rule of thumb, if you know something but you *don't* have a specific recollection of learning it, it is likely a semantic memory. Of course there are cases in which they blend: I remember learning my times tables by practicing them on my walk to school; I remember when I learned the name of the album *Magical Mystery Tour* because I was over at my friend Glen McClish's house and his older sister played the title cut on her JCPenney record player. The term "semantic" in this context comes from the Greek word "semantikos," which means "significant." The idea behind the term is that semantic memory stores the significant or meaningful information that we have acquired about the world around us, and it is thought to underlie our ability to reason, communicate, and understand language.

Autobiographical memory is the most intimate form of memory; it is necessary for our sense of self. *I played little league softball and I was the centerfielder. I honeymooned in Hawaii. I know how to drive a manual transmission.* These combine your personal, episodic history and other general self-knowledge: *Am I someone who likes vanilla ice cream or chocolate? Am I an early bird or a night owl? What do I like to do in my spare time? Who were the great loves of my life?* This autobiographical memory is separate from episodic memory. Like episodic memory, it indexes events that you were a part of, but it indexes much more: the memories of the things you did and thought, insights, places you've been, your political views, your values and morals—all the things that make you *you*, things that no one else knows.

These different memory systems (geographic/spatial, sensory, procedural, episodic, semantic, and autobiographical) are subject to loss individually as the result of injury, disease, or trauma. The nature of memory, however, is that the different systems often provide access points to one another, cross-referencing and cross-correlating. Even autobiographical memory, which draws on both episodic and semantic memory, can be separately impaired; that is, dissociated from the other forms of memory. President Reagan, before he was diagnosed with Alzheimer's, but probably suffering from it already, famously confounded an episode (a movie he acted in) with his autobiographical memory—he thought that the things that happened to the fictional character happened to the nonfictional him.

Memory is the heart of who we are, and the very private sense of what it is like to be us. Inextricably linked to memory formation are emotion and attention. Not everything we experience makes it into memory. If you are really paying attention to something, it's much more likely to be encoded. Often the unconscious is deciding: What does this experience mean to me? What might I learn from this? How can I use this later? Say I walk into a house that I'm thinking of buying: I'm going to notice different things than if I'm thinking of robbing it. As a buyer I don't care about the expensive coin collection in the glass case; I care about the leaky roof. We tend to remember those things that carry the biggest emotional wallop. This makes evolutionary sense: an emotional event, such as finding water when you're lost and thirsty, or seeing a family member eat a plant that poisons them—needs to be encoded for survival.

A popular and intuitive metaphor is that memory is like a digital recording of our lives—when we want to recall something, we find the file in our memory banks and hit play. And we can fast forward, rewind, jump ahead, play in slow motion—our memory is scannable, and its "playback" is under our control.

Ah, but this metaphor is not very good, because memories are subject to distortion or change over time, making it difficult to rely on what we pull out of our mental attic. In reality, memory is more like film that has been edited and doctored, with entirely new scenes inserted, old ones rewritten with new characters, backgrounds, and dialogue, and some distorting lenses applied. Our recollections can be influenced by a range of external factors such as context, mood, and expectations, causing them to become altered. This can result in different people having different memories of the same event. Over time, our own memories of an event might change, sometimes drastically, as we recontextualize the experiences—we do this without conscious awareness and can't help ourselves. These intruding distortions can also affect retrieval as we search our memory banks. If you're searching your memory for that time you had shrimp scampi with your old high school friend Jim Ferguson, you may not find it because you and Jim never *had* shrimp scampi—you only talked about it—and this detail is somehow lost, merged, or rewritten. The maddening thing about all of this is that we are *so certain* that the thing happened *just like we remember it*. This often is not true; psychologist Elizabeth Loftus has shown how easily and readily false memories can form.

Understanding the neuroanatomy of the different kinds of memory is crucial for working out how music therapy and music interventions can and do work. Memories of the emotional and autobiographical features of music—*Barber's* Adagio for Strings *always makes me sad; Van Morrison's* Days Like This *is the album she and I listened to in bed*—are prone to distortion like any memory. But the *perceptual* components of music hold a privileged position: they tend to be more accurate than other memories, even for nonmusicians. Memory for the absolute pitches of songs that we know well are preserved in memory with great precision; the same is true for tempo, timbre, melody, harmony, and spatial location (such as, *the cellos are supposed to be on my right*). Memory for the structure of the lyrics, if not each and every word, is also well preserved. We remember, with astonishing accuracy, the rhythms and notes that lyrics are attached to, the accent structure, the rhyme scheme, the pauses. Even if we only know a song fairly well, it's likely we've encoded enough of this componential information that we can take a good guess at what the missing words are.

Why this privileged position for music? In general, it is easier to remember information when there is some sort of organization scheme inherent in the material. The most successful mnemonic devices are ones that contain a structure predictable enough that we can learn it (ideally, just through passive exposure, as we do with our native language). Music—even more so than language—is a highly structured medium, meaning that the structure itself can help to scaffold one's memories. Because of this structure, we don't need to remember every single musical detail in order to reconstruct our musical memory. Any one of the individual component attributes of rhythm, melody, timbre, lyrics, and phrase structure can help us to remember others, allowing us to make plausible inferences about what comes next. Suppose I sing:

Somewhere over the rainbow, skies are blue
And the dreams that you dare to dream really do _____

You may have forgotten what comes next, or you may never have heard the song and so never knew. But the mutually reinforcing constraints of rhythm, accent structure, meter, melody, and rhyme (it probably rhymes with *blue*) constrain the possible completions. How would you finish this? *Really do **fit my shoe**, really do **go to the loo**, really do **make a stew**?* Those might work in a Weird Al Yankovic song, but here they seem out of place. Really the only thing that fits is what Yip Harburg wrote for that lyric: *really do come true.* The lyric is memorable because structure and semantics—logic—reinforce it.

Similarly, a clarinetist playing a glissando doesn't need to remember every single note—she can simply remember the first note and the last note and, using her knowledge of keys and scales, reconstruct what goes in between the two endpoints. In some experiments, we ask musicians if a certain note appears in a piece they know well. If the note is on an accented beat and begins or ends a phrase, it is more easily remembered than if it is jumbled in the middle of something. The same is true of lyrics—if I ask you to recall where the word "the" appears in "The Star-Spangled Banner" you probably can't answer directly, you have to quickly scan the song in your mind's ear. By contrast, if I asked you if the word "rockets" appears, you could answer immediately.

A related phenomenon is at work when audiences at a sporting event all begin singing a song together. Many of the people singing along would not be able to sing the song all the way through in its entirety, but if each audience member can recall a different part of the song—because it was somehow salient to them—that serves as enough of a cue to bring everyone together to create a crowd-sourced rendition.

Musical memory is unique in that it combines procedural memory (motor memory) with auditory memory and a special kind of semantic memory, memory for the structural rules that hold everything together. Because there are so many redundancies in musical memories, they are among the most robust in our lives, surviving even advanced Alzheimer's disease.

But musicianship requires much more precision; a multifactorial network of brain regions must cooperate. Ella was lucky in that she only lost the words. In other cases, the sounds can remain in our brains as auditory memory, but we have difficulty getting them out.

The New York Academy of Sciences offers a series of programs aimed to engage the public with science. In 2008, the Academy invited Rosanne Cash and me to perform music together and talk about what goes into creating music, both from the musician-composer side and from the scientific side. Rosanne had undergone brain surgery that affected her musicianship a year before.

Rosanne is a rare jewel among marquee performers. She is more focused on the experience of those around her than on her own experience—she directs her attention to the crew, the sound engineer, the stagehands, the other musicians. She is the anti-diva. Knowing that I'd be flying down from Montreal in the winter—not a good time to bring an instrument on a plane—Rosanne offered to bring me one of her guitars to play. We had a wonderful time. We had played together informally at her house, but this was the first time we had performed in public. It was so completely natural and comfortable, it felt like we'd been doing it for years. The event led to a series of similar engagements during which we talk about music and the brain, play a few songs, talk some more, and play a few more songs. It's a novel and intimate format; the audiences say that they feel like they're eavesdropping on a pair of friends having dinner at the next table in a restaurant.

Prior to her surgery, Rosanne had been experiencing headaches for as long as she could remember. As she got older, they became debilitating. "Once, I even dropped to my knees, the pain was so intense," she recalls. "But a lot of times, singing, playing music myself, I would move out of the headache. You know, it would just dissolve. That's an interesting thing about music. You know, people say it's very healing. It is very healing, literally." After ten years of misdiagnoses, she underwent an MRI that showed she had a "Chiari malformation," a rare abnormality I had to look up, that caused a bit of cerebellar tissue to slide down into the spinal cavity in the neck, pushing against the brain stem, "the geographical equivalent of starting in Vancouver and wandering down to Houston," Rosanne said. The only solution was neurosurgery. When Rosanne arrived in pre-op, the nurse asked her what she was there for. "I'm having a decompression craniectomy and laminectomy for Chiari 1 and syringomyelia," she said. Rosanne knew far more about it than I did.

The cerebellum is responsible for the exquisite orchestration of movement, whether it's lifting one's leg up a stair, opening a jar of jam, or playing a musical instrument. Any disturbance there, no matter how slight, can have outsized consequences for movements from large to small to minuscule.

Following the surgery, Rosanne tried to play the piano again. Her cerebellum, unceremoniously poked and prodded, was having none of it. She could move her fingers just fine, but all of those precise movements and sequences that had been so carefully rehearsed over a lifetime were now lost. Undaunted, she set off on a path of cognitive and musical rehabilitation, getting out her second grade piano books and teaching herself those pieces, one note at a time. At first it was frustrating. But gradually the dormant patterns came back to life. What had taken her decades to learn the first time around, she relearned within six months. It is an extraordinary property of memory that once you've learned something, even if you *think* you forgot it, you can retrain yourself and it comes back much more quickly than you think it will, given the fits and starts that characterize the early stages.

⌢

The intricate processing of music unfolds component by component, commencing from the very moment we perceive its resonating sounds. The waveform, whether it reaches our ears through live performances, speakers, or earbuds, undergoes a deconstruction process that gives rise to neural representations across three distinct dimensions: frequency, amplitude, and duration. These neural representations then traverse specialized pathways where they are transformed, gradually assuming the more familiar attributes of pitch, loudness, and rhythm. Through further specialized processing circuits, the rich world of music materializes into a mosaic of the nine dimensions that engage and shape our perception: melody, contour, harmony, timbre, meter, tempo, tactus, spatial location, and distance.

Accomplished musicians possess the capacity to construct higher-order representations encompassing chords (clusters of pitches), harmonic rhythm (progressions of chords), and phrase structure (such as the distinction between 8-bar and 12-bar blues). As these higher-order representations crystallize through repeated listenings or deliberate practice, they often coalesce into meaningful "chunks." Chunking reduces memory load, by transforming a collection of disparate parts into a cohesive entity.

Say you've got a favorite song on your playlist, and you've heard it hundreds or even thousands of times. Each time you hear it, it lays down a trace in your memory, and repeated hearings strengthen that trace. Some of the features of this song are invariant across multiple listenings—the melody, tempo, and rhythms, for example. Other features are variable, such as how loud it is, where it is coming from in space (near or far, left or right), and whether or not it is tangled up with other noises, such as the sound of your windshield wipers swishing, birds singing in your backyard, or a crowd of people chattering in a café. Each of these listenings forms a distinct trace that registers the unique features, and at the same time reinforces the previous traces that encoded the common features. Contrast this with listening to a live music performance, where the band may play the song differently each time—they may change up the rhythms, the melody, and all sorts of things. From an information processing perspective, these are distortions of the original melody. Song recognition remains accurate in the face of very large distortions, or alterations to the original way we heard it. Indeed, jazz improvisation relies on the idea that part of your brain is keeping the melody and chord changes in your head while the soloists riff on it, sometimes into a completely different melody or tonal space. Each such listening stores whatever that particular performance had in common with your previous listenings, as well as whatever is new.

Every time you experience something, whether it's going on in the world around you or it's a thought inside your head, it lays down a trace in your memory. When you experience something many, many times, the traces overlap and the memory becomes exceptionally strong. If each experience is slightly different, the traces still exist in a kind of package, like sticks of dried sage bound into a single stalk of incense. The traces themselves are instantiated in a pattern of synaptic connections. All that synaptic activity is governed by characteristic neurochemical activity, a balancing act across up to 100 different neurotransmitters. That high-fidelity, accurate, and absolute memory we have for the pitches, tempos, and timbres of well-known songs essentially falls out of this model for free: once you've heard those same compositions hundreds of times, each time writing a memory trace into your brain, you have essentially rehearsed the song, as a musician might, without even knowing it. This sort of massive rehearsal leads to overlearning, a mastery of something that causes it to become automatic—like Ella singing the melody and rhythm of "Mack the Knife" or Sully landing an Airbus 320 without having to think much about it.

Each experience, thought, or piece of information creates a unique synaptic network. When we hear a particular piece of music, some of the 80 billion or so neurons in our brains become connected in a unique way, specific to that song, and specific to that particular episode. Those neurons become members of a special subset of neurons that represent that song. Brain scans from our laboratory show that when people imagine listening to music versus actually hearing it, the activation patterns are nearly identical. That is, the act of remembering music causes activation of the same neural circuits that were active when we heard the music in the first place. That special, dedicated subset of neurons that were tied together in listening comes online again when remembering. What are otherwise disparate and isolated neurons spread throughout the brain re-form to once again become *members* of that original experience group; the neurons are *re-membered* onto their original formation. Remembering is re-membering.

In this respect, Ella's memory for music appears to be not a recording, but a series of recordings, each of a musical feature—like the multitrack recordings producers make that have drums on one track, bass on another, and so on. So here, imagine that one track got erased by mistake and you still have the others to work with. It is fairly easy to fake your way through. Or imagine that you're used to playing with a five-piece jazz combo and the piano player doesn't show up one night. Each musician modifies what they play to some extent, to account for the change of information the audience will hear. On-the-fly adaptations like this are possible because music contains such rich, mutually reinforcing and redundant structure.

Human memories are cross-indexed, with multiple, almost limitless entry points. Some entry points are universal ("name a song normally sung at a birthday party"), some are well known within a culture ("name a popular Beyoncé song"), some are very specific to you or your cohort of friends ("name the song we danced to over and over again that night on the beach"). In cognitive psychology, we call these entry points *cues.* The cues can be almost anything: sensory, emotional, autobiographical, factual, geographic, associational, a lyric, a piece of melody . . . the list goes on.

Try it yourself: think of a song that makes you sad; think of one that always makes you feel happy; think of a song that is about driving; a song that reminds you of your first kiss; one that you like to sing out loud; a song with the name "Jane" in the title. Each of these serves as an entry point to memory, a cue.

Memories are not static or passive, but rather dynamic and evolving, subject to constant revision and reinterpretation. The meaning of a song can't be summed up by an appeal to the notes that constitute it, and if an especially emotional memory becomes associated with a song at any point, you will probably never hear it the same way again. That song you loved when you were dating so-and-so might be difficult to listen to after an acrimonious breakup. Hearing it and experiencing those negative feelings will cause these new negative feelings to become attached to the memory and then stored along with it. As the most recent feelings stored, they may dominate the older ones so that it will be hard to recall a time when you *ever* enjoyed that song.

The fascinating phenomenon of state-dependent memory retrieval emerges as a consequential companion to these observations. When our spirits are high, we retrieve joyful memories effortlessly. Conversely, when we find ourselves submerged in sorrow, the retrieval of any gleeful recollections can prove arduous, leading us to conclude that our existence has perpetually been tinged with melancholy. This sets in motion a discouraging cycle of escalating despair, rendering it increasingly challenging to emancipate ourselves from the clutches of despondency. Trauma operates on a similar principle— whenever we recollect a traumatic encounter, entwined with all the adverse sentiments it entails, liberating ourselves from this perpetual cycle becomes a daunting endeavor: each triggering stimulus evokes a sense of panic, imprinting itself as an additional memory trace atop the accumulated others. Even music associated with a distressing period of one's life can summon forth fresh negative memories in individuals afflicted by depression, exacerbating their condition.

The multifeatured, multifaceted aspect of music, and the different ways it can be accessed, are what allows it to be so powerful with Alzheimer's, depression, PTSD. When nothing else gets through, a little snippet can shoehorn its way into consciousness, mood, and memory itself.

Neuroscientist Amy Belfi found that music-evoked autobiographical memories are more vivid than autobiographical memories evoked by other familiar cues, such as photographs. That's what caused the remarkable transformation in Henry Dryer; it's what allowed dancer Marta González to recall her Swan Lake choreography, and Glen Campbell to tour, while each was deep in the throes of Alzheimer's. Like Ella's, enough of their musical memory was preserved that everything necessary fell into place. (An exception is that in cases of behavioral variant frontal-temporal dementia, music doesn't get through as easily and may require multiple exposures because the brain's pathways are too compromised.)

Last month, my 86-year-old uncle lost his wife of 50 years. In his misery we started playing the Baroque music that he loved listening to when he was younger and his life was carefree. It didn't take his mind off his wife's recent death, but it did connect him with the feelings of his better, happier self, which gave him the emotional strength and resilience to move forward.

Our musical memories begin before we are even born. In utero, the developing fetus hears sounds through the amniotic fluid and uterine walls, much like we do when we're swimming underwater and there is disco music blaring poolside, and the fetus would hear mostly bass notes. What does it do with all this information? The fetus's brain is taking in all these experiences and using them to literally wire itself up to them. Babies recognize their mother's voice and scent right out of the womb, because this has been their primary sensory input. And, even after a year, babies show a preference for music they've heard in the womb.

Because fetuses are hearing low-pitched sounds, like bass notes and kick drums, they are learning about the rhythms and chord progressions of the music that surrounds them. Whether they are hearing jazz, country, classical, rock, hip-hop, Indian ragas, Chinese opera, or Tuvan throat singing doesn't matter—auditory input goes in, and the result is brain circuits that have an understanding of the rules and structure of that music. (If you want your child to have wide-ranging musical tastes, start playing a wide range of music before birth.)

The first eight years of a child's life are unlike any other period— a phase of explosive growth of neurons and neural connectivity. The primary mission of the brain during these years is to learn as much as it can about the world around it (historically, humans would begin their own lives and families around age 13, so they needed to be prepared for both the wonders and danger of a tough world). Any learning that occurs during these years is special and holds a privileged position in our brains. Starting around age 10–13, the primary mission of the brain shifts to pruning out unneeded connections. Like a tree, the brain grows many branches, some of which are nourished by sunlight and nutrients from the soil, some of which aren't, and then either break off or just take up space that would be better used for additional growth. Similarly, many of the capacities our brains begin with don't receive enough reinforcement for them to take up neural resources, and they are pruned.

This doesn't mean that you can't learn anything after age 13! It's just that the speed of learning begins to slow down, and over the next decade it takes increasing concentration and a deliberate effort to learn and for information to be retained. It's why children who learn a language or an instrument become so natural at it, and why older teens and young adults who learn later tend to speak with an accent. It's why Carlos Santana told me that I play guitar with an accent. (I should explain. I was producing a record at The Record Plant in Sausalito, and it occurred to us that it would be nice to have congas on it. Carlos Santana was recording in another room, and his conga player, Armando Peraza, was just sitting in the coffee room, and so I asked Carlos's permission to borrow him. Later, Carlos put his head in the room just as I was adding some electric guitar. Carlos asked, "What was your first instrument?" "Saxophone," I answered. "I figured it was something like that . . . you play guitar with an accent." "A saxophone accent?!?" "I don't know—but it's some kind of accent. Like Portuguese or something." I always thought that was a really interesting and insightful comparison. Carlos is a native speaker of Spanish. If you've ever heard a Portuguese person speaking Spanish, it's subtle—both are romance languages, and Spain borders Portugal on the east, and so many Portuguese speak Spanish. No one else has ever noticed that guitar was my "second language," but Carlos has a very sensitive ear.)

As our young brains are hearing music, they are figuring out its structure. Our brains are still new at this, so they need simple songs at first, songs with a regular, predictable beat and notes that make small, easy-to-sing melodies. It's no accident, then, that the songs we play for children—and that they gravitate toward—are simple: the Alphabet song (which, you may have noticed, is basically the same song as "Twinkle Twinkle Little Star" and "Baa, Baa, Black Sheep," with the words changed), "Frère Jacques/Are You Sleeping," Brahms's "Lullaby" ("Lullaby, and good night . . ."). Young brains can listen to these ad nauseam because these songs are still not entirely predictable to them; around every phrase turn, there is something new for them. They are learning, that is, memorizing the structure.

The accepted theory behind why we like the music we like is that it strikes just the right balance between predictability and surprise. Too predictable and we're bored. Too surprising and it puts us off balance because we can't get a foothold (or ear-hold?) on what comes next. That balance point is different for each of us, which is why there is no one song that everybody likes, and no one song that everybody hates (just when I thought everybody hated "Baby Shark," I ran into kids gleefully singing it—and for their own pleasure, not just to annoy adults).

As adults who have to listen to these over and over and over again, the songs become annoying because *we* are not learning *anything* from them. But kids love them because their brains are soaking up every nuance and creating new pathways for all the different elements—melody, rhythm, tempo, meter, harmony, accent structure—all the components we need as adults in order to enjoy new and old songs, to unlock the trove of songs stored in our memory banks when only a single feature is all we have to grab onto.

As kids get a little older, they seek and gravitate toward songs that are a bit more complicated, giving them the opportunity to challenge themselves and build up more complex conceptual representations of musical structure. For example, "Yankee Doodle" (when given new words, it's Barney the Dinosaur's theme song), or "The Hokey Pokey." Eventually we gravitate toward music with even more varied and complicated features—melodies that have larger and occasionally non-diatonic intervals; three-against-two rhythms (triplets); *rubato* tempos, in which the time breathes, speeding up and slowing down expressively; syncopated beats; timbres we've never heard before.

Personality plays a role in all of this. Some children are naturally more curious and exploratory, some more conservative and cautious. In all of these musical adventures, our brains are seeking similarity and patterns. Our brains are giant prediction machines, trying to figure out what will come next to us in the world. To do that, we need to extract patterns, commonalities, and create groupings of objects that are similar. We make sense of orchestras because we don't hear the 100 or so instruments as separate—our ears naturally group similar sounds together. All the violins automatically get merged in our head into a section (or two or three when they are playing different parts). Similarity can also apply to melodies as they are handed off from one instrument to another. Our brains apply certain rules that are inborn, to define similarity in several ways. An example is that notes that are close to one another in pitch space tend to get grouped together, and they separate—pop out—from other note clusters that are far apart from them. This is what allows us to hear a bass line as distinct from a guitar or clarinet line. There are over a half dozen rules for detecting similarity, most of them worked out by the Gestalt psychologists back in the late 1800s; these "configurationists" rejected the dominant notion that one could understand a complex phenomenon by studying the parts one at a time.

This penchant for similarity draws us toward music that has something in common with music we've heard before, the strongest example being different versions of the same song, followed by songs that are based on the same structure (from 12-bar blues up to sonatas and symphonies). When we've learned some of these fundamental laws that govern the music we listen to—that is, when we've built up dedicated brain circuits to do the work—our brains take over and handle the ongoing tracking of melody, rhythm, and other musical features without our conscious control. Our brains notice small and large deviations from what they expect, and those get encoded as separate memory traces, allowing us to build up a larger vocabulary of patterns, enriching our experience of music.

All of these musical elements are being stored along with extra-musical and contextual memories—like those contextual variables that my student Yuvika Dandiwal was studying—smells, tastes, feelings you're experiencing, the place you are, the people you're with. Each of these independently serves as a retrieval cue. This is why a smell can trigger a memory, and particularly if it is an odor that you don't experience all the time, and so is uniquely associated to a very specific memory. That is the key, really—the more specific a memory cue, the easier it is to retrieve what you're looking for. If I ask you, "Do you remember that shower you took three Wednesdays ago?" you're unlikely to have a specific memory for it—it's blended in with all the other showers you've taken recently. But if I ask, "Do you remember that shower you took recently when the water suddenly turned ice cold?" that, presumably, only happened once in the last few weeks, and so the uniqueness triggers the memory.

All these processes can be tied together: unique memory cues, multiple trace memories, similarity, and memory for individual features of music. Say I hear a new Regina Spektor song for the first time, and I haven't been told what it is. The various qualities of a sound—loudness, pitch, timbre, duration—are each handled in the area of the brain most suited to them, then brought together to create what I now recognize as a Regina Spektor song. I can do that, even though I've never heard this particular tune before, because there are also circuits connecting the experience to my memory of other Regina Spektor songs, comparing what's similar and what's different, and connecting Regina's voice to how she makes me feel. I've got a memory of sitting at home with friends in 2001 doing a record pull—everyone brings their favorite new song—and my friend Morgan played "Love Affair" for us, and I became a fan. (I love the line "He was perfect, except for the fact that he was an engineer.")

When I simply *imagine* a Regina Spektor song without hearing it, that hippocampal index tells me, *These are the circuits that were involved, the neurons that fired, the firing rates of those different neurons*—all of that info in the neural library card catalog I described. But I don't get it back perfectly, which is how I know it's a recollection and not the actual thing—unless I'm hallucinating to the point where I can't tell the difference between reality and imagination. This is another point about memory—neurochemical tags tell us when something we are thinking about is a memory, versus something that we dreamt or hallucinated. That's a good idea, evolutionarily. You don't want to pick a fight with Caveman Og just because you *dreamt* that he stole your food. But this system of neurochemical tags sometimes goes awry. When that happens we can have experiences such as déjà vu, the feeling that we've been somewhere or done something before when we really haven't; as you are actively encoding the experience and putting it *into* memory, something goes wrong and it gets spit back out as though it is coming *from* memory, rather like a kitchen sink backing up. Jamais vu occurs when you're doing something you've done many times, but it feels like the first—as though you've never (*jamais*) done it. Jamais vu is a state that every musician and actor seeks to attain when they are performing. The late Burt Bacharach probably sang "Alfie" more than a thousand times. But as pianist and producer Shelly Berg said, "Every time he sings it, it sounds— and feels—like he is singing it for the first time ever. And that's what audiences connect with." Psychotherapy can often trigger déjà vu and jamais vu because it aims to recontextualize memories.

We also connect anything we're hearing with what we've heard before, and that deeply enriches the listening experience. Our musical memories begin before birth and evolve as we grow. As long as we are listening, we keep adding to those memories, reinterpreting the new in light of the old, and in the process, recontextualizing the old. Your brain is changing all the time, and so are your emotional states, and so the "music medicine" you receive is essentially a brand-new medicine each time. The brain that hears that favorite song today is different from the brain that heard it last month. The journey through music is a never-ending one, guiding us through moments of joy and sorrow, discovery and nostalgia, a faithful friend that is always there to lift us up or help us through rough times. As long as we keep listening, we move forward, one note at a time.

Chapter 4

Look at Me Now[*]
Attention

HAVE YOU EVER NOTICED THAT SOMETIMES, WHEN singing "Happy Birthday" at a birthday party, not everyone is singing in the same key? Singing and recognizing a song in different keys comes naturally to us because it is the relationship between notes—the melodic space between them—that defines the melody. We take our human ability to transpose for granted; transposition is one of the devices that Lady Ella used in "Mack the Knife," ratcheting the key up about every 25 seconds to add excitement to the song. Without transposition perception, we'd think that each verse was an entirely different song. It turns out that melodies can retain their identity across a wide range of transformations, slowing them down, speeding them up, changing their pitches (transposition), rhythms, timbres—even changing some of the intervals (the distance between pitches). This makes not just for a very rich compositional sandbox for composers to play in, but an expansive one for improvisers, too.

In 1960, MIT professor Benjamin White conducted a classic study to demonstrate this. It goes back to a puzzle posed by Christian von Ehrenfels, one of the founders of the Gestalt movement, in 1890. Von Ehrenfels asked: How is it that you can change *all the notes* in a song, such that the new song has no notes in common with the

---

* The chapter's title is from the song "Against All Odds," words and music by Phil Collins (1983). But the song title is not the part that people remember—they remember the refrain "Take a look at me now."

old one, yet we still recognize it as the same song? The first line of "Happy Birthday" can consist of G–G–A–G–C–B, or it can be F♯–F♯–G♯–F♯–B–A♯, and it retains its identity. Similarly, I can sing the song slower or faster, and it is still recognizable. (There are limits, of course; if I play the melody at a pitch that is outside the range of human hearing, or too soft, or at a tempo that is lugubriously slow— say one note per month—that may stretch your ability to recognize the song.) This led von Ehrenfels to propose that a song is an *auditory object*, just as a chair is a material object, or the idea of "justice" is a conceptual object.

Ben White adjusted the pitch intervals of well-known songs up and down and in various ways, both forwards and backwards. He played the songs with only quarter notes, removing the rhythmic information while retaining the pitches; he played the rhythms of the song all on a single note, removing pitch information and leaving the rhythms unchanged. Linear transformations that involved adding, subtracting, or multiplying intervals by a constant were easy to identify. Nonlinear transformations, including time reversal and elimination of all melodic information, were more challenging, but people still recognized the songs. Even though he had altered the song, participants found some familiar aspect to pay attention to. Their brains then performed what computer scientists call a pattern match: with only a piece of the melody at their disposal, they'd search their memory banks for the closest match and then retrieve the memory associated with it. If I take your favorite chair and paint the legs, or cut off the arm rests, or put it in very dark or dim light, you will probably still recognize it.

Fast forward 36 years to 1996. I was a postdoctoral fellow at Stanford, and Lee Ross, with whom I had studied social psychology as an undergraduate, invited me to lunch at the Faculty Club. It was a warm, cloudless day. Red-winged blackbirds and European starlings chirped in nearby trees, and we could hear the drumming and gargling, trilling call of acorn woodpeckers—"wacka-wacka"—as they studiously drilled holes in trees to stash food for later.

"We had a graduate student a few years back," Lee said, "Elizabeth

Newton. For her thesis, she had people think of a song, and then tap out the rhythm on a picnic table to see if other people could identify it from rhythms alone. I don't think it was ever published." We played the game together. Lee rapped out a few easy ones, like the theme from *The Lone Ranger* (properly called the *William Tell* Overture by Rossini), and the Queen song "We Will Rock You." I tapped out "Baa, Baa, Black Sheep" and "Deck the Halls." We stumped each other on Beethoven's "Ode to Joy" (from his Symphony No. 9) and "Frère Jacques." Lee (one of the fathers of situationist social psychology) speculated that the context or situation in which one was tested would affect the results. If it was Christmas time, people would be primed for Christmas songs; if the testing took place in a disco, they'd be primed for other sorts of songs. The trick, he said, would be to figure out a way to keep the situation neutral across different participants.

When we were through with lunch, we got a couple of coffees to go. We walked through campus and talked some more—Lee taught me that one often does the best thinking while walking in nature. The conversation drifted back to the rhythm study; Lee conceded that it wasn't just the situation that mattered, but also individual differences—some people might have a natural disposition toward rhythms (drummers come to mind), some toward pitches (singers of Gregorian chants, for example).

In 2000, when I had my own lab at McGill University in Montreal, I set out to replicate this study with my students. How to keep the experiment neutral? One of the problems with much human behavioral research is that it tends to be conducted at research universities, using undergrads, and therefore ends up being biased toward individuals who come from backgrounds that are Western, Educated, Industrialized, Rich, and Democratic (called WEIRD). McGill is among the most diverse universities, with students coming from over 150 countries, many on scholarships, and many from Non-western, Agrarian, Poor, and Authoritarian backgrounds (NAPA). Of course, students who make their way to university tend to be well educated before arrival, but many are first-generation college students who come from families and communities that are relatively uneducated and nonliterate. Surrounding McGill, the city of Montreal is similarly diverse.

To conduct the experiment, we would need to determine *which songs* we would use. To minimize bias (our own, or the bias of the season, or whatever), we distributed a survey to 600 people who were from the same participant pool we would later use for the experiment: students and community members. We asked each to name 10 songs they "knew by heart" and "could recognize immediately" if they heard them.

From that list, we selected the top 50 vote getters; these included children's songs like "Three Blind Mice," holiday songs like "Deck the Halls," folk songs, and other popular songs such as "Here Comes the Bride," "Happy Birthday," and "Theme from *The Addams Family*."

Next, rather than just rapping out the rhythms on a picnic table and risking that each version might be slightly different, we created the rhythms on a drum machine using a single snare drum sound (we retained the natural accent structure of the rhythms). While we were at it, we created a separate condition with a synthesizer that kept the original pitches but played all of them as quarter notes, keeping the rhythm constant—and this time with no accents. Now we were ready to investigate how many songs could be identified by rhythm alone, and how many by pitch alone. We also wondered what rules or principles might govern this. Did it have more to do with the songs themselves or the listener? Or both?

Identifying songs from rhythm alone might seem to be just a fun party game or a cute demonstration of musical ability. But it goes further than that, telling us about the intersection of attention and memory, two operations that are tightly interlinked, and necessary to understanding music as medicine. What features are we actually attending to, and how does this affect what eventually gets encoded in memory? If your brain is not paying attention to something, there is no mechanism by which your memory can register it. This experiment shows how even one piece of a stimulus, like a song or a chair-with-its-legs-missing, might provide an access point to your memory that allows you to fill in the rest. It has clinical implications for people who have lost one of the entry points into memory and might respond to another. The research also opens the door to explorations of individual differences in attention—the particular aspects of a stimulus each of us might naturally be drawn to—and how those inform our immediate perception of an object or event, and then our subsequent storage. Like the burglar and home-buyer attentional perspective from the last chapter, some of us may find that our brains give priority processing to lyrics versus melody, or rhythm versus timbre, and so on.

We presented our set of 50 songs to 460 new people, drawn from the same pool. They heard the patterns and simply had to name the songs. Our experiment found that about half the songs were reliably identified by rhythm alone, and another half by melody alone, with some overlap. Certain rhythmic and tonal patterns were high in cue validity, meaning they were distinctive, not easily confused with other memory entries.

As an analogy, imagine you're playing *Where's Waldo*, looking for a man in a white-and-red-striped shirt. Waldo's easy to spot against a black-and-white background, harder to spot if there are other red things and white things and other stripey things because your attentional filter has to do so much work. In the first case, a red-and-white-striped shirt has high cue validity because it is so distinctive against a black-and-white background; in the second case, it has low cue validity because other elements of the picture have similar features. Cue validity can be quantified, and is always a function of context, meaning that the attentional focus is part of the equation.

Our externally oriented attentional filter helps us to quickly take in a scene and segregate out what's distinctive, as it does in *Where's Waldo*. In effect, we "instruct" our visual system to ignore anything that is not red *and* white *and* striped as we perform a visual search. (When you think about it, one part of our brain is telling another part of our brain what to do. But which part of the brain made the first part of the brain the boss over the second? This quickly degenerates into a turtles-all-the-way-down, or infinite regress, problem.)

In parallel, we have an *internally* oriented attentional filter, tasked with trying to pick out a *specific* memory from a whole mess of competing ones. If I ask you what you ate for breakfast two Mondays ago, that's probably not enough information to go on if your breakfasts vary randomly and there was nothing distinctive about that particular breakfast—Monday breakfast is low cue validity. (Unless you're Sheldon Cooper, and your breakfast schedule specifies that Monday breakfast is oatmeal.) As another example, if I'm thinking of a robin, and I tell you only "it flies" or "it lays eggs" or "nests in trees," those are hints with very low cue validity because they don't uniquely distinguish robins from other birds. A high cue validity hint would be "has a red breast and is commonly found on lawns looking for worms." If I were trying to get you to think of a penguin, I might need to say nothing more than "a bird that looks like it is wearing a tuxedo." That is a probe with very high cue validity.

In music, some rhythms are unique to us. The galloping rhythm of the theme from *The Lone Ranger* (bada-DUMP bada-DUMP bada-DUMP-DUMP-DUMP) is so tightly coupled to that song— and no other—that when we played it for people in our experiment, they would often spontaneously burst out in laughter because it was so easily recognizable. Other songs had low cue validity for rhythm: the first three measures of "Ode to Joy" are all quarter notes, as are the first two measures of "Frère Jacques"; they are impossible to tell apart based on rhythm alone. The distinctive, differentiating rhythm only occurs in the next measure for each one, and even then, they are difficult to tell apart. Those same two songs had very high cue validity for pitch because the pitches are relatively distinctive.

Other songs were low in cue validity for both pitches and rhythms. What differentiates such songs tends to be timbre, instrumentation, or lyrics. An example (not in our set) is one of Bob Dylan's songs, "Tangled Up in Blue," which has an undistinctive rhythm and pitch, but as soon as you hear the guitar part and his iconic voice, your mental search set is now guided to the Bob Dylan catalogue. Playing even a snippet of the song, so short that only timbre is present—no melody or rhythm—allows someone who knows the song to identify it in less than half a second. The same can be said for hundreds of songs we know. Timbre turns out to be higher in cue validity than rhythm or pitch for popular music because most popular music of the last ninety years—the recording era—is highly differentiated by the sound of the singer's voice, whether it's Patsy Cline, James Brown, Audra McDonald, Robert Smith, or Adele.

Experiments like these help us to unravel the features of songs that enable people to identify them based on the single cue (of melody, rhythm, or timbre) alone; we were also able to rank the songs by how recognizable they were in these impoverished versions. We found stark individual differences—some people were more attuned to rhythms or to pitches, and some more attuned to timbre or to lyrics. In one case, after hearing the rhythm of the clavinet part to Stevie Wonder's "Superstition," a young man said, "That's the song about the broken mirror and the seven years' bad luck." He could neither identify the song by name, nor hum or sing the melody. This exemplifies the type of clinical applications for music therapy that can help people with memory impairment. There was no indication that this young man had permanent memory impairment—this was probably more of a temporary gap, a tip-of-the-tongue moment, or possibly he knew these features but not the title of the song.

Part of the richness of memory, as it interacts with the multitude of cues we can attend to in music, is that even nonmusical, extramusical cues can trigger a memory. Cues like these can recover memories of songs that might otherwise remain buried:

- Name songs about driving
- Name songs by Moby
- Name songs that relax you
- Remember as many songs as you can from when you were in high school

A near limitless number of cues can thus activate a mental representation of a particular song. Hearing the song can open a floodgate of memories about where you were, who you were with, or, for example, what it felt like to be 16 years old. This is an important cornerstone of how we can use music in cases of severe memory loss, cognitive decline, or dementia.

The separability of musical components has been shown in patient case studies. A few months after my meeting with Lee Ross, cognitive psychologist Irv Rock took me to lunch on the Berkeley campus. (It seems much of my education has come from lunches with older mentors, a tradition I try to continue with my own students. Something about being outside of the lab, out in the open space, gives rise to more expansive and exploratory thinking.) Irv was the first person to interest me in the phenomenon of tone deafness. "I think the whole term is a misnomer," Irv said. He held a lumpy dill pickle in his mouth for a few seconds before taking a bite out of it. Irv liked to let ideas sit before expanding on them. "It seems to me that *tone* deafness has to be quite rare—people aren't deaf to *tones*, they are deaf to the melodies. It should be called *tune deafness*. They can recognize different tones or pitches. What they can't do is string them together into a representation of melody."

Tone deafness had first been formally described in 1895, but the term was used rather indiscriminately to define any deficit in musical ability, and nowadays the term *amusia* is used. We make a distinction between people who are born with a music processing deficit (congenital amusia) and those whose deficit is the result of injury or disease (acquired amusia). When this manifests as an inability to recognize or sing melodies, this can arise from poor pitch discrimination, poor pitch memory, or problems with discriminating and remembering rhythms. When the deficit is only in singing, it can result from poor control of the vocal cords, something that can be trained, provided that the singer can accurately self-assess when they are hitting the wrong notes (there are cases of individuals who can hear music just fine and identify mistakes made by others, but for reasons we don't entirely understand, have a "self-monitoring" deficit).

Beyond these common problems are more exotic cases of individuals who lost the ability to write music while still being able to read it (musical agraphia without alexia), and vice versa. Some people lose the ability to experience the emotional (hedonic) qualities of music, resulting in musical anhedonia. We've seen patients who lost one or more of the components of music perception while others remained intact (melody, rhythm, chords, harmonic structure, timbre, musical memory); and loss of musical tonal perception that extends to tones in spoken language—the prosody or inflection of language. The dissociation of musical abilities extends from perception to action: in other words, one can lose the ability to play something on an instrument while retaining the knowledge of how the fingers are supposed to move.

If you're born with one of these deficits, it's a bit like being colorblind: you don't realize you have a disability compared to others because it is all you've ever known. Everyone around you talks about seeing a quality that you don't, but your brain has adapted to navigating the world and identifying objects based on the information you do have. What most of us call "red" and "green" could potentially be differentiated based on differences in brightness and lightness without ever having the internal experience of "redness" and "greenness." If you acquire colorblindness, it's a different story; now you know what you're missing and it can be frustrating.

It is difficult to identify the precise neural deficits associated with congenital amusias because no two brains start out the same; they grow and remap in ways that are influenced by genetics, environment, culture, and random factors. The acquired amusias aren't all that much easier because damage to the brain is never exactly the same from one patient to the next—nature's experiments are not carefully controlled the way laboratory experiments are. But we have been able to draw out some patterns.

Matteo was a 20-year-old musician and law school student. Out of the blue one day, he got a blinding headache that would not go away.

It felt like it was coming from an area in his brain just underneath the space between his left ear and his left eye, the frontal-temporal juncture. He went to the emergency room and had a CT scan. The scan revealed he had suffered a mini-hemorrhagic (bleeding) stroke and was still losing blood. A surgeon inserted a tiny device, an embolic coil, to stop the blood flow. Afterwards, Matteo found himself unable to process pitch in music and speech. He couldn't distinguish a question from an exclamation, or sarcasm from sincerity. "I can't hear musicality," he said. "All the notes sound the same. . . . Sounds are empty and cold." And yet, Matteo could still discern other musical qualities. He could easily distinguish the properties of loudness, timbre, familiar rhythms (waltz, tango), and the directionality of notes as ascending or descending.

Imagine that you woke up one day to find that conversations that once flowed effortlessly, and that you readily understood, were now bathed in ambiguity. The rising inflection that signaled a query or the slight tonal shifts that conveyed a sly remark, a reproach, a plea, or a playful invitation run together like ink on a piece of wet paper. Language, once a trustworthy ally, was now a confusing morass of mixed signals. If you were lucky, your sensitivity to other cues, such as the timbre of a whisper, soft voice, or raspy yell could help you out. But still, previously normal human interactions would have become a world of hidden meanings, missed cues, and endless misunderstandings.

In another case, Ms. T., a 62-year-old piano teacher, suffered an ischemic (clotting) stroke in Wernicke's area, a region normally associated with understanding the meaning of words and their relation to one another. In classic textbook cases, damage to Wernicke's area (in Brodmann Area 22, in the superior temporal gyrus) causes fluent aphasia—an inability to understand what is said, coupled with speech that is fluid but that utterly lacks coherence or meaning— what is often called "word salad." Patients with Wernicke's aphasia are unaware that they cannot comprehend. (It's hard to know if this side effect is merciful or cruel.)

The stroke unexpectedly affected Ms. T.'s ability to read music in a very specific and peculiar way: she could still read musical pitches, but not musical rhythms. She could still *perform* musical rhythms on the piano, but only for songs she had previously memorized. Her reading and performance skills became disconnected like a train car with a broken hitch. Reading a musical rhythm requires making a temporal connection between one note and the next, just as reading a sentence requires making the connection between one word and the next. In effect, the meaning of a melody becomes lost when it has no rhythm to indicate how long a note plays, or when the next note starts. Wernicke's area, it turned out, was not a "language" area as we had thought for a century, but an association area, a region connecting meaning to the motor movements underlying communication—a conclusion further reinforced by Daphne Bavelier's finding that speakers of American Sign Language also rely on Wernicke's for attaching meaning to gesture.

In Italy, a 58-year-old man, Mr. P., suffered a mini-ischemic stroke in the left hemisphere and lost the ability to identify timbre, both in music and in nonmusical settings. He could identify environmental sounds like trains, typewriters, and doorbells by their rhythm, not their timbre; violins sounded like flutes or clarinets; the piano sounded "like a thunderstorm," the organ like "a combination of sounds . . . an orchestra?" he asked.

Oliver Sacks describes a time when he was driving and, as part of a pre-migraine aura, he experienced a complete loss of pitch and timbre, so that a favorite Chopin ballade sounded like "a sort of toneless banging . . . as if the ballade were being played by a hammer on a metal sheet." His rhythmic perception remained entirely intact, to the point that he could still recognize the ballade by its distinctive rhythmic structure.

University of Manitoba neuropsychologist Lorna Jakobson reports on an amateur musician, K.B., who suffered a stroke that was highly localized in the right fronto-parietal area. K.B. lost his ability to recognize and identify music, but in a very peculiar way. His pitch- and rhythm-processing abilities were both totally impaired and he could no longer recognize instrumental melodies, but he retained the ability to recognize familiar songs with lyrics, even when those songs were played *without* the lyrics. He recognized the pitch and rhythm components of melodies at a preconscious level sufficiently well to act as a memory cue for the lyrics, and thereby allowing him to make an association between the lyrics and the song.

This functional disconnect between conscious and unconscious perceptions is a mainstay of the attention and memory literature. Perhaps the most famous neuropsychological patient, Henry Molaison (patient H.M. in textbooks), lost his ability to form new explicit memories, but was able to improve his performance on puzzles using a form of residual procedural memory. When shown a puzzle, he would claim he had never seen it before, despite having worked with it for months. His improved performance over time could only be accounted for by an intact attentional system, leading to an intact memory trace, albeit one he had no awareness of. Henry's hippocampus and amygdala were damaged, and possibly other regions unrelated to implicit memory.

Sacks also writes of Clive Wearing, a musicologist and conductor who contracted encephalitis, profoundly damaging his hippocampus, after which he became amnesic. Like H.M., he was unable to form any new memories, but his impairment was far worse. Whereas H.M. retained intact memories from before the hippocampal-amygdala damage, Clive lost access to a lifetime of episodic and autobiographical memories; the only person he recognized was his wife. When shown musical scores sitting next to his piano or organ, he said he had never seen those pieces before, but when Oliver asked Clive to play them, Clive played them flawlessly, and as he was playing, said he remembered them—only to forget the pieces he had just played as soon as he finished. Something about the act of physically playing the music brought forth memories—it was one of the few activities he could perform that tapped into knowledge he had prior to his encephalitic infection.

Like memory, attention is not an all-or-nothing affair. It can function consciously or unconsciously, conspicuously or inconspicuously, and it allows us to focus on the big picture or on particular features. Attention refers to a variety of control processes. But defining it can be confusing—so confusing that even Immanuel Kant stumbled and fumbled his way through it:

> What is meant by the word attention is not clear.

Over a century later, Edmund Husserl hadn't gotten any further:

> The phenomenological analysis of the theme of attention is beset with difficulties and requires detailed clarification.

William James did better:

> It is the taking possession by the mind in clear and vivid form, of one out of what seem several simultaneously possible objects or trains of thought.

Martin Heidegger captured the phenomenology of attention existing in a strange netherworld of being neither entirely inside of us nor outside of us:

> Attention is a basic state of Dasein [being human], which is always already with us and about us.

I'm partial to the current scientific definition by Mike Posner:

> Attention is the mechanism of selection, enhancement, and integration of information.

Colloquially, we use the word *attention* as a synonym for concentration and focus. We say "pay attention." The person this is directed at might be looking in another direction or their mind is wandering. We are asking them to stop what they are doing and direct their minds to a particular thing—to take possession of it in their mind, as

James said, to will themselves or instruct themselves to select this one thing from the infinity of things they *could* be attending to. When it comes to paying attention, we sometimes feel like a squirrel in traffic—easily distracted and constantly dodging shiny objects that are coming from every which way.

Focus doesn't happen easily. The human brain can only pay attention to a limited amount of information at a time, typically around four or five things. Attention is a limited capacity resource; directing your attention to *this* means that you must stop paying attention to *that*. You might be thinking: I can do all sorts of things at once—it's called multitasking. Earl Miller, Ed Awh, and others have shown that multitasking is a myth; the brain just doesn't work that way. Instead, our attention flits around from one thing to the next, to the next, and then back again. It does this in such short time frames that we feel like we're doing all of the tasks at once. We might even say we're "juggling" all these things at once, without realizing how accurate a metaphor this is. In actual juggling, some of the objects are always in a state of suspended animation, in the air, and jugglers only hold them in their hands briefly until another ball, or burning torch, or chainsaw, lands.

All that attention switching comes at a neurobiological cost. *Paying* attention is an apt metaphor because we literally are paying for the process of both attending and switching; the currency is the oxygenated glucose in our bloodstream needed to provide fuel for neurons and neurochemicals that help us focus.

Attention has a zoom-in, zoom-out quality to it. I look up from my writing desk and on the wall in front of me I see an oil painting of the Québec countryside. At first glance, I take in the color of the green-yellow meadow, with the dark green trees at the top and amethyst-colored wildflowers in front. If I direct the spotlight of my attention in just the right place, I get a powerful sense of depth that otherwise eludes me. If I allow my eyes to scan the scene, to take it in as I might if I were there in person, I perceive a gentle flowing and moving, and can almost see the wildflowers swaying in a light breeze. I can zoom in to look at just one particular flower or a fence post, and fix my gaze deliberately; I can zoom out and let the painting manifest itself without my consciously directing attention. In those moments, the painting draws me in in wordless authority. As Shakespeare might have said, the art doth seize my soul, compelling gaze and heart to yield.

We do the same with music. Sitting in San Francisco Symphony Hall, listening to Michael Tilson-Thomas conduct Mahler's Second, I can just take it all in and let it envelop me. I can decide to focus on just the oboes, just the tympani, whatever I like. If I am having trouble distinguishing the violins from the violas auditorily, I can use the spotlight of my visual attention to help to disambiguate (so long as they are bowing differently).

What is happening when we zoom in and out? We're engaging in *selective attention*, a funneling and filtering of this limited-capacity resource. Our nervous systems are subjected to far more external stimulation than can be used. Our brains receive a massive amount of input from the world around us, and from our own bodies through our various external and visceral senses. In addition to that, we are stimulated by memories and ideas that arise internally, either in response to a stimulus (*mmm . . . that cupcake looks good, I think I want one*) or just in response to our own thoughts (*I haven't heard from my friend Eric lately—I should give him a call and see what he's up to*).

Attention can be either voluntary or involuntary. If I'm attending to the orchestra, and then hear a fire alarm, I don't voluntarily attend to the alarm—the loud, sudden, and shrill noise was designed to capture my attention. And yet, not everything that is coming from outside of the orchestra will capture my attention—people rustling in their seats, my own breathing. The fact that our attention can be captured and redirected without our conscious awareness tells us there is a filter monitoring everything at a preconscious level, so that we are only bothered if something important is going on. Voluntary attention is goal-driven—I am choosing to direct my attention to something in particular; involuntary attention tends to be an interruption that either pulls me out of my own head (from sleep or daydreaming) or that pulls me away from the thing I was voluntarily attending to a moment ago.

Scientists are still trying to work out the functional networks that serve these different kinds of attention. In the visual domain, there is evidence that a dorsal (top) pathway governs goal-directed attention (also called top-down, because we are instructing our brains what to pay attention to), and a ventral (bottom) pathway governs the "grabbing" part of attention (called bottom-up, because the brain interrupts what we are thinking about to reorient our attention).

Seeing and hearing are selective—much more so than touch, taste, smell, or balance. We register what is needed at the moment and unconsciously ignore most other input. It may seem that our eyes are like a camera and our ears are like microphones, objectively recording everything, but we know from fifty years of experiments that our senses are not at all like those devices. Put a microphone in a crowded restaurant and you get cacophony. Put human ears in there and you can selectively follow one of several different conversations, all while monitoring, preconsciously, many others. Not only that, if we examine how information makes its way from the world to our sensory receptors and then into our brain, we see that a lot gets distorted or lost. People don't like to believe this, but psychologists can easily demonstrate it. Intro psych classes in college are pretty much devoted to demonstrating that you don't know what you think you know. Our brain, without our conscious awareness, is constantly trying to find patterns, extract order from disorder (think finding bunny rabbits in cloud formations on a summer's day), and predict what might happen next. We've all had the experience of reading and getting to the end of a sentence and realizing it doesn't make sense. Then we go back and find the word we misread. Misreading or mishearing is often the result of unconscious expectations. We thought the sentence would say something different before we even read it.

Implicit in all of this is that attention and consciousness are related. In language, we often use the terms interchangeably: to say I'm conscious of something is to say that I am aware of it, that I have paid attention to it. Yet consciousness is something more than mere awareness—it's a feeling, a sense, that has its own ontological qualia (that is, its own subjective "feel" to it that is not easily described). It's your consciousness that tells you what it feels like to be you, what it feels like to be hungry, exhausted, amorous, or bored; you know what it feels like to experience altered states of consciousness, such as sleep, or perhaps being drunk or stoned or on a psychedelic trip. The state that you are hopefully in right now, dear reader, is one of goal-directed, focused attention on this book. But occasionally your attention may drift while you're reading (Darwin forbid!). Your eyes follow the words, but your mind is off somewhere else—daydreaming. You may also have experienced a third state that combines these two, a meditative state that can arise through the intentional practice of meditation, or through yoga, music listening, or just lying in bed before drifting off to sleep. Where does this state come from? Could that drifting off actually be a good thing?

Chapter 5

Daydream Believer[*]
The Brain's "Default Mode," Introspection, and Meditation

I TOOK MY FIRST COGNITIVE PSYCHOLOGY CLASS IN 1976. The field was so new, there were no textbooks for the course. *Cognition: An Introduction*, a short dense paperback by Michael Posner, had come out just two years earlier, and was on reserve in the library as supplemental reading. Our instructor, Susan Carey, a newly minted PhD from Harvard, spoke so admiringly of Posner that I stayed up late at the 24-hour campus library to read his book. (When I applied to graduate school, Susan and I were equally delighted that Mike agreed to be my thesis supervisor.) The new field of cognition was delineated as that portion of experimentally driven, scientific psychology charged with studying the information processing capacities of the human brain: perception, attention, memory, decision making, expertise, intelligence, and language use. The few people studying music cognition were considered "way out there." (No serious scientist was talking about something as squishy as "art," let alone "consciousness." That was reserved for people who gave up their academic appointments, moved out west, and took acid trips.)

[*] "Daydream Believer." Music and lyrics by John Stewart. Recorded by The Monkees, 1967.

After graduate school, Mike had studied human factors psychology, the psychology of how humans interact with tools and machines—what might now be called HCI or human-computer interaction. Early on, Mike became interested in the methods that cognitive psychologists had for peering into the working brain. He developed the idea that the amount of time it takes for the brain to do something is an index of how difficult the task is, known as mental chronometry. When he started publishing work on this, many in the field followed.

In those days, there were no neuroscience departments. Cognitive psychology had not yet morphed into cognitive science or cognitive neuroscience. There was just experimental psychology. But a funny thing happened on the way to the 1990s—new tools became available, and Mike Posner was at the forefront of a paradigm shift involving brain scanning.

In 1985, Mike set up a laboratory at Washington University in St. Louis, with radiologists Steve Petersen, Mark Mintun, and Marcus Raichle (a neurologist and violinist), and neurologist Peter Fox (it takes a village). Marc (Raichle) came up with the idea of using radio-tracers to track blood flow in the brain in real time, inside a PET (positron emission tomography) scanner. Blood carries oxygenated glucose, the fuel for neurons, and the neurons that are working the hardest call for more glucose. If we could trace blood flow, we could start to map which regions of the brain are involved in various cognitive operations. For example, I could put you in a scanner and ask you to mentally practice your tennis serve, solve arithmetic problems, remember your first kiss, or play jazz piano. The result of this was a landmark paper published in *Nature* by Peterson, Fox, Posner, Mintun, and Raichle. This 1988 study was the first to gaze into the human mind in the act of thinking. PET was a huge advance over the widely used *structural* imaging techniques that simply told us the size and shape of the various anatomical structures in the brain—nothing about what they actually *do*. Structural scans are like looking at an aerial view of a city and mapping all the buildings but not knowing what goes on in any of them. Some are probably apartments, some are offices, warehouses, homes, or small businesses, but from the air it's hard to tell. The only source we had for understanding brain functions was from postmortem case studies of people who experienced brain damage. This is like going into that same city after a natural disaster and seeing which buildings had been small businesses, and then trying to take what you saw on the ground and correlate it with those aerial photos. Everything changed with this one new technology, with this one paper. Now, scientists could catch people in the act of thinking, imagining, dreaming, solving problems. We could study brain *function* and map the regions (and eventually pathways) involved in a variety of perceptual, motor, and cognitive tasks (even people thinking up new brain scanning experiments).

As other labs began conducting PET studies, Raichle and his colleague Gordon Shulman started to notice a strange pattern emerging in the brain imaging data. While participants were in the scanner, performing tasks that required active, goal-directed attention, some areas of the brain *deactivated*; blood flow decreased and neural activity was suppressed. It makes sense that if one set of brain regions is working harder others will not, but the consistent deactivations across such a wide variety of experiments was difficult to explain.

Scientists delight in being wrong; the most exciting part of the job is not providing more evidence for an old theory, but finding evidence that refutes one. When that happens, we know we've misunderstood something about the world. An unexpected or weird result is a gift that tells us we didn't know what we thought we did. And then the fun begins—of trying to design the right experiments that will fill in our knowledge gap.

Do you remember when you were a young child and everything was new? Asking an endless series of *how* and *why* questions?

Child: Why do birds have feathers?
Parent: To keep them warm. Later they learned to use them
to fly.

Child: Why?

Parent: Feathers help birds to catch the air and stay up in the sky.

Child: Why?

Parent: It's like when you hold out your arms in the wind and feel the air pushing against them. When the air pushes against feathers, birds can fly.

Child: Why?

Parent: I don't know.

Child: Why?

Parent   *(exasperated):* Because I was too busy goofing off with my friends in the back of physics class and I didn't pay attention.

Child: Why?

Parent: I have an idea! Let's go get some ice cream!

Scientists are basically like that annoying four-year-old. We keep asking questions and we never get tired of the game. There are always unexpected developments to keep us going.

Marc and Gordon's finding was one of those exciting and unexpected developments. What to make of these brain regions that were *deactivated* across a variety of goal-directed tasks? Parts of the brain were going offline while other parts were coming online, in a synchronized dance, choreographed to some unknown orchestra, led by an unseen conductor. Maybe something could *activate* these regions—but what? What is the opposite of a goal-directed task? A resting state: mind-wandering. Raichle and colleagues looked for activation in participants awake in the scanner with their eyes closed. That is, participants were asked to simply do nothing as compared to the *something* the experimenters usually asked them to do. Their experiments confirmed that some regions that had a *lower* level of activation while engaged in goal-directed tasks indeed reached a *higher* level of activation during a resting state. Because people seemed to automatically slip into this mind-wandering mode when they were not doing something, Raichle dubbed it the default mode (DM).

Two thousand miles west, Vinod Menon and I were postdoctoral fellows together at Stanford Medical School, pursuing several different directions of research, answering a tall stack of "why" questions we had accumulated. Vinod had an active research program conducting fMRI studies of individuals with psychiatric disorders, while I was conducting experiments in psychophysics and developing new statistical methods for music research. (fMRI uses a different neuroimaging technology with greater spatial resolution than PET.)

Like Marc and Gordon, Vinod had noticed clusters of deactivations in his clinical populations, including individuals with schizophrenia. Vinod and his Stanford colleague Mike Greicius developed a technique called network analysis. (We were the first to apply network analysis to music in our 2007 study of the DM.) The idea behind network analysis is that rather than looking at which brain regions are involved in various mental operations, we identify the distributed network of brain circuits that come together to perform these operations. In other words, we look for those nodes that are more highly connected to one another than to other nodes in the brain, forming a "community structure," what Gordana Derado at Emory University dubbed a "functional neighborhood." This technique allowed Menon and Greicius to identify the components of the default mode, and map out in 2003 what we now call the Default Mode Network (DMN): medial prefrontal cortex, anterior and posterior cingulate cortex, precuneus, and angular gyrus.

The DMN is a collection of distributed, interconnected brain regions that are suppressed when a person is focused on something in the external environment. In the absence of attention to something, our brain switches—or defaults—to internally guided thoughts, such as self-reflection, recalling the past (autobiographical memory), and envisioning the future. This same network is active when our brains are at rest during quiet waking, or when our thoughts are unguided during daydreaming. The discovery of the

DMN has fundamentally transformed our understanding of attention and consciousness, for it can explain a number of everyday occurrences. If the DMN's neurochemical switch keeps toggling on and off, we experience successive lapses of attention and absent-mindedness; we find ourselves paying attention for a few seconds, then losing focus, then focusing again, until we are exhausted from all the task switching. If the DMN switch gets stuck in the OFF position, we can't disengage our focus from an absorbing task (for example, when we don't see the car that is about to pull out in front of us because we are talking on the phone). If it gets stuck in the ON position, we can't be easily pulled back into a task that needs to be done (such as when we're lying on the beach with one of those drinks with the little umbrellas in it).

In our 2007 study, Vinod and I demonstrated that listening to music can facilitate entering the DMN daydreaming state. The following year we showed that the neuroanatomical switch is in a region of the brain called the insula. The insula is one of my favorite brain parts because it is a bit of a dark horse. As recently as 2006, no one gave it much thought, nor understood what it did.

The insulae (there's one in each hemisphere) sit in a strategic place in the brain, kind of like crossing guards next to a school at 3 pm, getting to see traffic from all different directions and helping to slow or stop it when necessary. They are right where three of the brain's four lobes—temporal, parietal, and frontal—come together. They each have the shape of a pyramid, but you wouldn't know that by looking at the surface of the brain because they are buried deep inside, about 2 cm (3/4") behind your temples. It's the portion of insula that sits in the right hemisphere that acts as this all-important switch.

The DMN's activation of inward-directed thoughts provides privileged access to episodic and autobiographical memory—mental operations that require looking inward. Clive Wearing's condition, with his focus on the present moment and only the present moment, is just what you'd expect if his DMN was no longer accessible to him—his attention was only directed outward, all of the time. He could play piano pieces and enjoy them in the moment, but not recall having played them. Even more profound was the effect that the encephalitis had on Clive's consciousness. Most of us experience our consciousness as a more or less continuing, unfolding narrative, connecting the present moment to immediately prior ones, stretching on back a lifetime. Clive lost this temporal connectivity, and felt as though he were continuously waking up from a dreamless sleep. Open his journal to a random page and you'd read "I am finally awake now for the first time." Then, that would be crossed out and replaced with "Now I am really awake for the very first time." And that would be crossed out, and so on. The journal went on like this, heartbreakingly, for hundreds of pages. In contrast, Matteo, the 20-year-old law student, retained his autobiographical memory, and thus he recognized that music sounded different post-stroke than it did before.

Entry into the default mode can be triggered in several ways. Our brains can grow tired while engaged in a task that requires focused attention, our minds start to wander and, voilà, the DMN has kicked in. Or we can engage in tasks that facilitate entry to it, such as meditation, or low-demand tasks, such as driving on a long, straight stretch of highway—this is why many of us have our best ideas while driving, or walking, especially in nature.

The DMN can also be triggered by tasks that require self-consciousness. In one fMRI study, participants were presented with trait descriptors such as "timid," "polite," and "perfectionist," and asked, "Do these words describe you?" That self-reflection, or self-referential cognition, engages the DMN, particularly the prefrontal cortex (medial), cingulate cortex (posterior), and left angular gyrus. And these same regions, plus the precuneus, show up when people listen to music, as Vinod and I found across several papers.

The activation of the precuneus (spanning Brodmann Areas 7–31) has always been particularly interesting to me, because it has kept popping up in our studies of musical reward and pleasure. The precuneus mediates self-awareness; we've all had the experience of being pulled out of whatever we were thinking about (or daydreaming about) by some startling sound. When that happens, we immediately and instinctively become aware of our bodies, our position in space, as a precursor to a possible emergency that would require us to duck, jump, or move out of the way. This is partly what the precuneus is doing. But the precuneus is only connected to the rest of the DMN when we listen to music that we like. When we listen to music we dislike, the precuneus severs its ties with the DMN. It's as though the brain is saying, "This is not part of me." *That* explains the neuroscience underlying why people find unwanted, piped-in music in public places so aversive: they are actively rejecting it: "This is not me, this is not my auditory space."

When we listen to music—really listen, give in to it, allowing it to take us over—we often find our minds drift to thoughts of who we are, where we've been in our lives, and where we're going. When we are not task-oriented in listening to the music, instead just letting it wash over us and through us, this self-reflection naturally comes up. And if there are lyrics that evoke issues of relationships, desires, failures, frustrations, and hopes, this all the more so triggers cognitive appraisal of ourselves. Love songs, or love lost songs, almost automatically cause us to see ourselves in them. This is how music helps us enter the default, mind-wandering mode, and once there, to stay in it.

Why do we see ourselves in songs? Art, literature, and music afford us the opportunity to try on different situations and emotions without actually putting ourselves in harm's way. We can watch Tom Cruise jump out of an airplane and simultaneously hold the knowledge that it is not me, but . . . what if it *was* me? We can experience both true love and absolute heartbreak without having to live through the ups and downs of them. And we can develop empathy, vicariously seeing and hearing what others are going through and how they react.

The capacity to make the distinction between ourselves and others is obviously important; without it, we can't get our own needs met, and we risk invading the physical and emotional boundaries set by others. When everything is working properly, the posterior cingulate cortex assists in making this distinction. There are two primary ways in which this distinction can dissolve; one leads to the Zen state of unity and oneness with the world, the other to schizophrenia. Experienced meditators who reach a state of oneness with the universe—dissolution of the "I"—find a state of decreased self-referential thinking. This is accompanied by increased connectivity between the posterior cingulate cortex and the medial prefrontal cortex. By contrast, people with schizophrenia show *decreased* connectivity in this network. New studies show that DMN activity between two or more individuals becomes synchronized when they process shared narratives. Vinod and I showed that when people listen to the same music, even if not at the same time, there is neural synchrony in their brain waves: listening to music literally puts your brain on the same wavelength as other people's brains.

Marc Raichle and his colleagues' discovery of the default mode, followed by Vinod Menon's discovery of the Default Mode Network, fundamentally changed the way we think about attention and its corollary, consciousness. We use the terms almost interchangeably—to be paying attention to something means that we are conscious of it. To have not paid attention to something is to be unconscious of it. That's not the same as being in a *state* of unconsciousness such as after fainting or being in coma. And what of unconsciousness, subconsciousness, subliminal perception—how do these fit in? (And do they have any relevance to music therapy? Yes, they do; stay tuned.)

Perhaps the most familiar alternate state is sleep. At first glance, it may seem that being asleep and being awake are opposites, that you're either one or the other, but this is a false dichotomy. Neither sleep nor wakefulness is a pure state; they fall along a multidimensional continuum, just as consciousness does. Most physicians and researchers hold to the traditional conception that sleep is a global process, occurring uniformly across the whole brain. If you asked a sleep researcher to account for how we can be awakened by someone calling our name or shaking us, they would just wave their hands. Clearly, some part of the brain is awake, monitoring the external environment. This is the attentional filter mentioned in the previous chapter, capturing our attention. But that filter is not a unitary thing in a specific place—it, too, is a network of circuits, monitoring all our six external senses (sight, sound, touch, taste, smell, and balance) plus our twenty or so internal senses (such as knowing that you need to get up to pee).

Then there are cases of sleepwalkers who can be "awake" enough to drive a car; sleep misperception in which patients who report being awake a large portion of the night were actually asleep, according to polysomnographs; and lucid dreamers who are aware inside the dream that they are dreaming—sometimes being able to guide it—and so in some sense are awake inside their own sleeping dream.

The cognitive psychologist Roger Shepard once had a dream in which his legs had an Escher-like appearance, seemingly attached to his hips, but on closer inspection, they formed an impossible figure. He recalls that in the dream, he mused that this created a l'egs-istential quandary for him and he laughed at the joke. On awakening, Roger further reflected that in order for the joke to work, one part of his brain had to conceive of the joke and then temporarily withhold it from another part of the brain.

Roger's wry observation is another case of one part of our brain telling another part of our brain what to do, as we saw with solving *Where's Waldo* puzzles. Which part of the brain "decided" to play the joke, and how did it keep the other part of the brain in the dark?

So far, our inventory of attentional-conscious states includes:

- the central executive (goal- or task-directed mode);
- the default mode (mind-wandering and self-directed);
- sleep;
- two attentional filters, one voluntary (find Waldo) and one involuntary (wake up if someone calls your name).

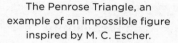

The Penrose Triangle, an example of an impossible figure inspired by M. C. Escher.

Roger Shepard's Impossible Elephant, entitled "L'egs-istential Quandary."

The elephant in the room, the existential quandary as it were, is what to make of different meditative states. Mindfulness meditation prompts its practitioners to dwell in the continuous present, gently observing their thoughts, emotions, and bodily sensations with curiosity, devoid of judgment or preconception. This is the state that most musicians aspire to when playing music. For both meditation and playing music, this means that the brain needs to deactivate circuits associated with self-consciousness and with judgment—the finger-wagging schoolteacher in your brain that is saying "that's terrible!" or "you're not good enough." That school marm is useful when you're editing, but anathema to creativity and spontaneity.

Neuroscientist Charles Limb of UCSF (then at Johns Hopkins) found this when he put jazz musicians in a brain scanner and asked them to improvise over difficult chord changes. During improvisation, he found deactivation in a region that is associated with effortful problem-solving and self-consciousness, the dorsolateral prefrontal cortex (DLPFC), as well as in our friend the precuneus. Together, these provide evidence that the musicians had entered the "flow" state. He found increased activation in the medial prefrontal cortex (MPFC), an area associated with accessing autobiographical memories; Charles's interpretation is that improvisation is a way of expressing one's own musical voice or story. He also found increased activation during improvisation in Wernicke's area, that portion of the brain that attaches meaning to sound, and that was damaged in the case study of Ms. T.

Different flavors of mindfulness meditation yield distinct effects on brain function and connectivity mediated by specific neural networks. *Focused attention meditation* anchors one's awareness on a singular point—our own breathing, a mantra, our own bodies. This practice increases neural connectivity in left hemisphere regions involved in attention and working memory—the left inferior frontal gyrus and the left inferior parietal lobule—and may enhance these cognitive processes. During *open monitoring meditation,* the practitioner brings attention to the present moment, allowing thoughts and sensations to arise and pass without judgment, while maintaining an open and curious attitude toward the experiences that arise. This style of meditation increases connectivity in the right hemisphere, specifically between the right superior temporal gyrus and the right insula—regions involved in interoception (the sense and perception of the internal state of the body) and body awareness. *Compassion,* or *loving-kindness meditation*, as practiced by the Dalai Lama, directs one's attention to feelings of love, empathy, and goodwill toward oneself, others, and, ultimately, all beings. Loving-kindness increases connectivity in both hemispheres, suggesting a more integrative pattern of neural coupling. Increased connectivity between the anterior cingulate cortex and the medial prefrontal cortex—involved in social cognition and emotion regulation—suggest loving-kindness meditation may enhance empathy and positive social emotions.

What all of these forms of meditation have in common is a certain degree of meta-awareness, and that distinguishes them from the mind-wandering of the DMN. This is the difference between being aware of something and *being aware* that you are aware. In mind-wandering, you're aware of what you're thinking about, but time and space seem to disappear—hence a friend snapping his fingers directly in front of your face and snarkily saying "Earth to Dan! Earth to Dan!"

Meta-awareness is a double-edged sword. It can help you understand consciousness and conscious processes; it is what allows the student violinist to be able to step outside the sheer joy of listening in order to study how to play the instrument. That same meta-awareness, however, can prevent us from becoming fully absorbed. When that happens, we hear only the flaws in our own performances; we sit at the concert and analyze what is happening, rather than synthesizing and experiencing it. Richard Davidson frames it this way: "Across a range of traditional and contemporary contemplative traditions, the absence of meta-awareness is viewed as an impediment to various forms of self-monitoring, self-regulation, and self-inquiry." Some meta-awareness is clearly necessary; too much of it robs us of truly becoming one with those things in life we care most about.

This idea of becoming one with the music, or the world, is famously part of Buddhist teaching. The Dalai Lama, for example, meditates four hours every day in order to strip away his sense of "I" and experience how interconnected all of us are. "I could sit here and think, 'I am the Dalai Lama.' There is only one Dalai Lama. Very lonely. I can think 'I am a Buddhist monk.' Thousands of monks, so less lonely. Instead I think 'I am only one of 8 billion people.' Not lonely." These aren't just words for him; he lives this, practices it, and finds great calm and compassion without sacrificing either awareness or meta-awareness. To pull off this trick of having both at the same time can take a lifetime of practice. Both meta-awareness and experiential fusion are important. Meta-awareness is required for us to self-assess and better ourselves. Experiential fusion is necessary for mental health, for hitting the reset button. Our ability to maintain a state, and not have our states switch willy-nilly without our intention to switch, is fundamental to cognitive control; and yet, our ability to have our attention reoriented when the situation warrants it is fundamental to our survival.

The ultimate form of musical engagement is experiential fusion while performing or listening. When a performer experiences it, it is more typically called *flow*. Musicians with deep musicality—Miles Davis, Victor Wooten, Ella Fitzgerald, Daniel Barenboim—can reach this when they perform. Audiences at Grateful Dead and Phish concerts felt a special connectedness to the music because those artists cultivated that sense of deep immersion as they played.

Flow is characterized by total absorption in the task, effortlessness, and ease (because you are not consciously directing the course of events), and a loss of meta-awareness alongside an almost paradoxical feeling of complete control over the task at hand. Flow is what takes over Steph Curry when he's "in the zone" on the basketball court, and what allowed Robin Williams to navigate so rapidly and seamlessly through an endless repertoire of voices, characters, and ideas during his stand-up comedy. Reaching flow, musicians shed their meta-awareness. The music carries them from note to note, and many say that they feel as if they are not playing music—music is playing them.

The flow state forces a reckoning of the way we think of the DMN as being in strict opposition, like a seesaw, to the central executive. It's as though the flow state allows us to rise up above both, to enter a state that borrows from both but is actually neither. Being in that zone, in the flow state, requires staying there; from neuroimaging studies we find just what we'd expect then: decreased activation of that neurochemical switch, the insula.

So are we conscious when we're in flow? Yes and no. We're involved in what we're doing and so can react, but we lack the explicit awareness that we're reacting. Until we're pulled out.

A peak musical state for the listener is to experience the chills—goosebumps—a unique and powerful part of flow's shift of time and space, when the music is taking you over, when time seems to stand still, and you experience an almost overwhelming burst of intense pleasure, a sense of awe, sometimes leading to laughter or tears.

Prediction and anticipation, alongside recognition of unfolding patterns, are at the heart of the musical experience. I asked Donald

Fagen, a writer far more elegant and articulate than I, how best to capture this important idea. Here's what he said:

> As one gets older, life loses its mystery. Good music puts the mystery back in life because you're always in a state of anticipation, wondering what's about to happen in the next few bars. No matter how many times you've heard, say, "Dewey Square," Bird's always gonna surprise you with exactly the way he pulls off the next phrase. It's a beautiful trip.

Typically in music, we know *when* something is going to happen, but we don't always know exactly *what*. The "what" can either be surprising or expected. The delicate interplay here appears to be what allows us to enter the default mode while listening to music: we are neither vigilantly holding our attention on a specific musical event, nor intentionally monitoring for an event that will capture our attention and snap us out of our reverie. The time-bound nature of music, the steady forward self-progression of it, is unique among the variety of things we experience in life.

The implication for music therapy will depend on one's therapeutic goals. If the therapist aims to facilitate entry to the default mode, they should select music that hits the sweet spot between surprising and predictable, especially when it falls slightly to the "predictable side" to calm us, but not so much that it puts us to sleep or annoys us with its repetitiveness. This kind of music will reinvigorate us, allowing us to hit the reset button in our brains, and to calm all the chatter and distraction that our attempts at multitasking create. We've all had the experience of having our thoughts spin out of control after we've been working hard among distractions, and we find it tough to slow down and focus on any one thing (this is why air traffic controllers must take mandatory breaks every 90 minutes or so).

That there is such a variety of attentional states and modes supports the idea that what we call "conscious awareness" is a rather shifty concept, and that it falls along a multidimensional continuum, not simply a line. The subjective feeling that consciousness is a unitary thing or process has been challenged by Daniel Dennett and Marcel Kinsbourne. They reject the "Cartesian Theater" model that posits a place or circuit in the brain where all the information we have "comes together" for some form of mental viewing on a virtual screen in the theater of the mind. If there is some conscious "me" sitting in my brain watching what's happening, where is the brain inside that conscious "me"? And what is *it* watching? Such a model leads to an infinite regress and makes no sense. It's more likely that unified consciousness is an illusion, and that we have lots of different consciousness awarenesses, monitoring systems, and narratives all taking place at once.

Vinod Menon believes the DMN is necessary for such a human consciousness. It connects disparate parts of the brain to support self-directed thought, mind-wandering, and to unify and ultimately build an internal narrative and sense of consciousness. "These narratives shape our understanding of our individual, highly personal, experiences," Vinod says. We can consider a thought experiment on what human cognition might look like if the DMN stopped functioning. "This would effectively end the narratives we tell ourselves. Our brains would not bind experiences to create a coherent internal narrative in the context of our experiences and shared social interactions. This narrative, part monologue and part dialogue, is central to construction of the epistemic self, component processes of which include episodic memories, semantic knowledge of facts about one's life and the world, representations of individual values and beliefs, and the ability to experience and produce evaluative directions to our perceptions, actions, and reasoning." The DMN is instrumental in providing an experience of subjective continuity, even in the face of inattentiveness, "brain freezes," and other gaps in our perceptions and cognitions. It is no accident, then, that all these brain circuits are also implicated in music listening and performance, one of the most absorbing and pleasurable experiences.

For decades, it was believed that music therapy was effective simply because it was pleasurable, or distracting, taking our minds off our pain, both bodily and psychic. We now understand that music is one of the few things that is present across all these different modes of attention (even sleep—many people hear music in their dreams). Music can then help to serve as a unifying source, a glue that connects our different modes of awareness with our internal narrative, our sense of self, where we've been, and, perhaps most important, where we want to go.

Interlude*

SEEING, HEARING, REMEMBERING, WANTING, LIKING, and feeling are complex states, built up of component parts. All are acts of construction. They make more or less use of incoming stimulus information depending on our experiential history—genetics, environment, family, culture, random occurrences, and current circumstances. The cognitive neuroscience perspective is that there exist identifiable neurophysiological processes underlying such states. These physiological processes are anatomical, electrical, and chemical, giving rise to synaptic pathways, which in turn bind into networks, or circuits. These circuits tend to be distributed throughout the brain and are not easily observed nor easily accessible to medicine. The *Music as Medicine* perspective is that for some of these constructive processes, music can find a privileged entry point, owing to its own multifactorial qualities, and the multifactorial ways that we initially take it in.

Forty years before Daniel Kahneman elaborated on the idea in his book *Thinking, Fast and Slow*, Ulric Neisser wrote:

> The constructive process of perception is assumed to have two stages, one of which is fast, crude, wholistic, and parallel, while the second is deliberate, attentive, detailed, and sequential.

Kahneman's landmark book applies this principle to the largely

---

* "Interlude" aka "A Night in Tunisia," by Dizzy Gillespie (recorded 1944).

irrational way that we tend to make judgments and decisions. System 1 makes rapid, intuitive decisions based on first impressions (the topic of Malcolm Gladwell's *Blink*); those decisions are sometimes right but often are not, particularly when they involve medical or financial decision-making. System 2 is slower, more deliberate, and analytic. It would be nice if we could map these onto the default mode and the central executive mode. We can . . . sort of. The mapping isn't neat and tidy, because these processes are not always mutually exclusive, and they often interact dynamically. A rule of thumb is that System 1 thinking (intuitive, *Blink*-like) maps loosely to the default mode. System 2 thinking (slow, analytic) maps loosely onto the central executive. Given that the DM is the mode in which we often find solutions to problems we couldn't otherwise solve, this may seem contradictory. But the kinds of problems that System 1 solves are not usually the ones that require rational, mathematically or statistically based solutions, but ones that require creativity, nonlinear and divergent thinking.

Music listening, performing, and writing also involve these two systems. Musical behaviors that are intuitive map to System 1, and those that are analytic map to System 2. System 1 is the place we aspire to be when we're performing, and where the greatest gains from music therapy can occur. When the music carries us away, we're in System 1. When someone behind us knocks our seat and pulls us out of our reverie, we get kicked back to System 2.

Up until now, everything I've written doesn't so much explain how we recognize songs as assume it. I did some hand-waving earlier about components making a pattern match in our memory. How do you know "Summertime" when you hear it? The obvious answer is that you know that it is "Summertime" because it sounds like "Summertime," but this is not very helpful. We could argue the opposite—that it only sounds like "Summertime" because you know what it is. And the problem is that not all "Summertimes" are alike. There are loud ones and soft ones, raucous ones and respectful ones, perfectly pitched ones and cacophonous ones, ones in a language you don't speak. There is Ella Fitzgerald and Louis Armstrong's slow and moody rendition, backed by the lush Russell Garcia Orchestra; Janis Joplin's earthy and raw interpretation backed by Big Brother and the Holding Company; Miles Davis's cool, spacious, and atmospheric version, instrumental only, marked by his signature muted trumpet sound; Willie Nelson's take on it that combines jazz with Tex-Mex; and Jacob Koller's piano hip-hop arrangement. All of these are "Summertime."

If all of these are "Summertimes" because they sound alike, then what process creates that similarity? The process of stimulus generalization, the same process that allows us to recognize a friend's voice whether they are near or far, angry or joyful, whether they have a cold or are panting and out of breath. The pattern of stimulation on the eardrums may change, but it's an important evolutionary adaptation that we recognize these divergent sensory inputs as coming from the same object. The same is true with music. Every experience you've had lays down a trace in your brain, a memory. The evolutionarily newer layers of the cortex, increasingly sophisticated as we move from vertebrates to mammals, work in the background to allow for increasing levels of variability, farther and farther away from the prototypes, such that you can recognize the face of a friend you haven't seen since childhood.

In music, we have more patterns and components to work with than in other mediums. At the most basic level, we recognize that we're hearing a song, not a voice, or a cough or a sneeze. Any of the perceptual attributes of music—pitch, melody, harmony, note duration, rhythm, meter, tempo, loudness, timbre (and sometimes even reverberant sound such as in Depeche Mode's "Personal Jesus" or The Weeknd's "Save Your Tears")—can trigger our brain's pattern-matching circuits to locate a stored memory trace, matching elements of the song. And in combination, these attributes are even more powerful and flexible. Our brain's concept of "Summertime" isn't indexed by a single feature—not by the rhythm, or the melody, or the words. That is, our musical brain doesn't work according to Aristotelian rules but with family resemblance. To Aristotle, all things could be defined with rules such as "A triangle is a closed, three-sided figure in two dimensions, the sum of whose angles is exactly 180 degrees." None of us looks at a triangle and thinks those things. We look at a triangle and recognize it as one because it bears a family resemblance to other triangles, even ones we've never seen before. As Wittgenstein argued, the categories of things are defined by how much they resemble other things, damned be rules.

The easy case for the brain, of course, is when we hear a song in a particular version that we've heard before—then the brain simply has to make a pattern match as a computer would. That's the underlying process that Shazam and other "music identifiers" use. To Shazam, it is only "Summertime" if some person has previously labeled it that. (As of today, you can sing "Summertime" to Shazam, or play it on the piano, and it goes unrecognized—this should change as machine recognition improves.) But even when we hear a particular version that we've heard before, the pattern of sound waves (pressure waves) impinging on your eardrum are unlikely to be identical with every listening—you're in a different room, or there are people talking, or your earbuds aren't fitting exactly the same way they did before, and yet your brain still finds the pattern match.

The next easiest case is if you're hearing a version of "Summertime" that has some features *nearly* identical to the ones you've heard before—maybe it's still Ella singing it but with a different backup band, or she's singing it faster, or in a different key. The pattern detectors in your cortex are hardwired to notice essential similarities, allowing you to quickly and effortlessly identify all these as the same song. We can get farther and farther away from our prototype or template for the song as we change these parameters, until all of them have changed at once—such as in the radical experimental version by Sun Ra and his Arkestra. Your brain may not recognize it instantly, but the essence of the song emerges from variations in these parameters. One of my favorite examples of a wholesale change is the Austin Lounge Lizards' bluegrass performance of Pink Floyd's "Brain Damage." The interesting thing about this is that the first time you hear it, it may take a while to put it all together, but once you've heard it, your brain remembers it and you'll recognize it much more quickly the next time around.

Listening to music is also a constructive process, not merely in the low-level piecing together of acoustic elements, but in the larger phenomenological experience of whether we like a song, whether we feel like hearing it right now, how it makes us feel, and what memories, introspections, hopes and desires it may unlock. It's because our brains process the different features of music separately (rhythm, tonality, timbre, pitch, harmony, etc.) before they become integrated that music therapy can work across a wide range of applications. Rhythmic therapies unlock one set of circuits, melodic-based therapies another. Passive listening with guided or unguided imagery unlocks different circuits than performing or dancing and the attendant motor movements.

No two of us, not even identical twins, share the same experiences. No two musical tastes are exactly alike. It's quite miraculous that there are songs almost everyone likes. Statistically speaking, a song so popular that most people like it is very, very rare. Apple Music and Spotify each boast 100 million songs in their catalogues, and by the time of publication of this book, they project they will hold 200 million songs. The number of songs that people like, and play over and over, is a tiny fraction of that number—at most, 0.002%. (That's one out of 50,000.) A song that relaxes you may agitate me. A song that usually relaxes you becomes irksome if you've heard it too many times, or just aren't in the mood. A song that you used to love may give you the willies now because it was the song you listened to with your ex before that anything-but-amicable separation. No two illnesses, injuries, or depressions are exactly alike either, because we all have different brains. Recall the words of the poet, that every illness is a musical problem, and every treatment or cure has a musical solution. The skill lies in finding that solution, that alchemy, that combination of tones, durations, and volume that will help the body and brain along in their natural, evolutionarily developed, healing processes.

We are all budding scientists, asking questions about the world, a process that started in early childhood, but that many of us unfortunately had trained out of us by well-intentioned but impatient parents, teachers, and older siblings. We are all budding music therapists, too. Most people in the world today use music in some medicinal fashion, and most people know what music to reach for when they want to maintain or alter their mood state (which is, neuroscientifically speaking, a brain state). We know what combination of sound waves will change our brain waves.

Our individual differences, however, cannot be ignored. There is no such thing as "relaxing" music, there is only music that *you* find relaxing. There is no such thing as a one-size-fits-all playlist for "focus music" or "serotonin music" or "exercise music," but there do exist some general principles. In general, for instance, music that makes us want to exercise vigorously has a predictable and easily discerned beat (the "tactus"), low bass, and groove to keep us engaged. In general, relaxing music is slower in tempo and has fewer surprises in it, whether those surprises are timbral, harmonic, rhythmic, or— as in the slow movement of Haydn's "Surprise" Symphony—volume. And our individual differences dictate what we will find surprising both in general, and in any given moment. If you put this book down right now and listen again to the last song you heard, it will affect your brain differently now because your brain has changed since the last time. It was changed to some degree by that last listening, even if you've heard the song a thousand times. It is changed by the mood you're in right now, whether you're hungry, or distracted, whether you had a good night's sleep. The playlists that people post on social media or on Spotify and other streaming services are *their* playlists. They might work for you a bit, but they are never going to be as good as the playlist you come up with, either on your own or with the help of a licensed music therapist.

As we approach treating diseases such as Parkinson's disease, Alzheimer's disease, PTSD, and depression, among others, we would do well to bear in mind that there is never going to be a musical prescription when it comes to the particular musical pieces we use. And that whatever music works for treatment for a few weeks or months may need to be adjusted. If it's beginning to sound as though music therapy is a guessing game, with treatments that may or may not work, and that may work but only for a while, ask a doctor how often they have to switch up a prescription drug for their patients— statins, blood pressure medications, antidepressants, even antibiotics for infections like strep, pneumonia, or sepsis. It's not that a particular drug doesn't work, it's that our bodies, our lives, our minds, our intentions, and the sum total of our experiences are continually changing us. We don't always realize it because "consciousness" sews it all together so seamlessly. The music that has helped you up to now may not always be effective. With more than 100 million songs available today on streaming platforms, there is surely a new batch of songs out there waiting for you to discover them. The very act of discovery is healthy, releasing dopamine and endogenous opioids, and boosting the immune system. And it's fun to find a new favorite and play it for your friends, passing it forward. That discovery helps keep all of us young(er) and mentally nimble.

Music treatment begins as an invitation to start exploring new music, and feeling the joy of discovery.

Chapter 6

Music, Movement, and Movement Disorders

M OVEMENT DISORDERS AND NEUROLOGICAL DISEASES were described by the Babylonians as long ago as the second century BCE, and although they knew little about the brain, their descriptions are strikingly similar to what you'd read in a twenty-first-century neurology textbook, as are the descriptions of motor system (movement) disorders that appear in the Bible. In the story of the young David slaying the giant Goliath (I Samuel 17), we learn that Goliath was "ponderous and moved slowly." Neurologic historians have concluded that his giantism, acromegaly, was likely the result of a rare large, growth-hormone-secreting pituitary tumor. In its later stages, the disease causes weakness, numbness, and pain, to the degree that Goliath might not have been able to dodge David's slingshot. He may not even have seen it—pituitary tumors can cause tunnel vision due to their pressing on the optic nerve. In Kings, and then, centuries later, in Matthew, Mark, and John, there are over a dozen mentions of movement disorders and paralysis.

Movement is *the* fundamental attribute that separates animals from rocks and minerals, and although plants move, they do so very slowly. In varying degrees, from the sloth to the roadrunner, and the ant to the elephant, every animal has the power of movement, driven by what is at least a rudimentary motor system in the brain. That system doesn't always work, and movement disorders can occur as a result of genetics, incomplete brain development, disease, injury, or environmental toxins.

Scott Grafton is a clinical neurologist and researcher studying the neuroscience of movement (and he is a piano, banjo, and bass player; by the time you read this, for all I know, he will have mastered the tuba). Scott has shown that a huge part of what we think of as intelligence is the result of moving about, touching things, and getting feedback from these actions. The idea is that all cognition is "embodied" or derived from physical engagement and interacting with the world.

Learning through active movement is entirely different from learning by passively thinking, seeing, or hearing the world. Touching, feeling, manipulating, traversing, breaking, grabbing, pounding, caressing, climbing (and falling) are just some of the ways that action yields knowledge of a wholly different kind than we can get by reading, listening to a lecture, or watching a video. Scott explains that physical intelligence "is profoundly different from, say, the instantaneous remembrance of a face, name or phone number." He reflects on the evolution of such skills, formed tens of thousands of years ago by humans and proto-humans, trying to survive in the wild. That world was "physically challenging and complex, characterized by palpable tension arising from an inability to predict what might happen and few means for maintaining control. Here were perfect conditions for improvising, inventing, and enduring some of the most rigorous demands of the wilderness. . . . Although the wild is uncontrolled, physical intelligence provides the means to establish a sense of control."

The mental capacities and neural circuits we use when we move, and when we learn from that movement, are, more than anything, different kinds of learning machines in the brain. What this means, practically speaking, is that if we don't regularly exercise this capacity, not only will it weaken, but we will be weakening a large portion of our brains. All these different things we do with our bodies create cognitive reserve, and bolster even non-movement capacities, such as general memory and problem-solving skills.

Simply moving, and interacting with the environment, strengthens the amygdala-hippocampal circuit, even for learning and memory that are not associated with movement. Walk on a rough trail and your memory for names may get better. New VR ("virtual realities") technologies offer adventures and challenges, but they have yet to replace real-world embodied experience.

Scott's ideas are echoed by Daniel Dennett—that for a robot or any other computer to be considered truly intelligent, to possess *knowledge*, it must interact with, manipulate, and experience its environment. This is not to say that quadriplegics and paralyzed people can't gain intelligence—it's just much more difficult. Far from a strictly physical limitation, then, movement disorders unfailingly affect our cognition, reasoning, memory, and thought processes. Indeed, it is known within senior living facilities that the patients who move the least are at the greatest risk for cognitive decline.

Movement disorders then don't just affect our locomotion—they affect our cognition. Older adults who find themselves slipping into cognitive decline and have a hip or a knee replaced discover that they were not on the brink of dementia—they just weren't moving around enough to keep their brains sharp. Part of the benefit of simple movement is healthy oxygenation of the blood and brain, but more important, it feeds cognitive capacity.

Of all the uses of music as medicine, none is more closely connected to biology than the treatment of movement disorders. Even without us wanting or intending them to, the motor and movement pathways in our brain are activated by music, they synchronize to it, and our limbic system signals pleasure when they do. Five of the major movement disorders have been shown to be responsive to music therapy: stuttering, Tourette syndrome, Huntington's disease, multiple sclerosis, and Parkinson's disease. For three other major movement disorders, ALS (amyotrophic lateral sclerosis or Lou Gehrig's disease), focal dystonia, and essential tremor, we still lack evidence for the effectiveness of music therapy. That doesn't mean music plays no role in treating these; music therapies have been shown to decrease anxiety and depression in all three, to stabilize or lift mood, and yield improvements in self-control, in compliance with medical protocols, and in quality of life.

Stuttering, and a diversion into how we learn to speak and play music fluently

Americans born in the 1950s and '60s may remember watching country singer Mel Tillis on TV and being fascinated by a paradox: Tillis stuttered when he spoke, but not when he sang. Tillis's 1980 performance of "Stay a Little Longer" is a rapid-fire tongue-twister of lyrics that any singer would have difficulty with, but he sang it flawlessly. Indeed, the *Washington Post* ran his obituary with the headline "Mel Tillis, Stuttering Country Star Whose Music Spoke Pristinely." Other singers who stutter when speaking but not when singing include Elvis Presley, Carly Simon, John Lee Hooker, and Ozzy Osbourne. In a related phenomenon, actor James Earl Jones stutters when he is speaking, but not when he is acting. How could this be?

Stuttering (also called stammering) is a speech disorder caused by neurological dysfunction, although all of us may occasionally stutter when stressed. The NIH defines stuttering as the unintentional "repetition of sounds, syllables, or words; prolongation of sounds; and interruptions in speech, known as speech blocks." Stuttering is most common among children between ages 2 and 6, when the brain regions that support spoken language are still developing, still being wired up. Most childhood stutterers grow out of it, but stuttering remains in one out of 100 adults, and it shows a gender bias—more males than females stutter. Stuttering is not confined to spoken language, either—some sign language users also stutter, telling us that it is a *language* disorder, not strictly a speech disorder, but either way, a problem in getting the motor system to work properly.

Consider the complexity of speech, of creating 44 distinct speech sounds (in English), fashioning sentences out of thin air on the spur of the moment, and coordinating millisecond-level timing of our larynx (including the vocal cords), pharynx, tongue, lips, and jaw. It is astonishing that speech disfluencies aren't the norm. Communicating fluently in speech or sign relies on making the required motor movements quickly and smoothly, in a particular order. Our brains need to devise a plan, and stay a few steps ahead of what's happening in the present moment, for all this to work out, while simultaneously monitoring whether what we *actually* did is what we *intended* to do.

At the sentence level, we have a plan for what we want to say, even if we are to some extent making it up as we go. To pronounce the words, we need to break them up into their constituent phonetic components, and there are complex rules about how phonemes are strung together. For example, a typical native English speaker who says "I'm going to the store" elides the word endings and beginnings to pronounce it like "I'm going tothustore" with no discernable break between /to/ and /th/ or between /ə/ and /st/ (a similar joining happens in French and many other languages). The same sort of elision happens when musicians play fast passages or play legato—they join together the end of one note to the beginning of the next.

If our brains needed to analyze auditory feedback before moving on, they would have to do it very quickly. Brains are fast: you can process a maximum of nine syllables per second—but that's if you're hearing someone *else* say the syllables. The time it would take to get feedback from your own speech, decide if you said the right thing, and only *then* move on to the next sentence would require faster auditory processing and analysis than our brains can achieve. This means that our brains are planning waaay ahead of what we're actually doing; the order of events—whether it's tongues, lips, and jaws moving for speech, or hands, wrists, and fingers moving for sign language and musical instruments—needs to be well specified before we even start.

As an analogy from a completely different domain, a related fact that runs counter to intuition comes from professional baseball. If you played softball as a kid, you may have been taught to "keep your eye on the ball." Children's softball pitches move slowly enough that you could decide when to swing based on where the ball was right before it reached you. But in Major League Baseball, the average speed of a four-seam fastball is now 93.9 mph and the distance between the mound and home plate is 60.5 feet. Many players decide when to swing before the ball even leaves the mound, and the last possible moment to swing is when the ball is a quarter of a second away—about 25 to 30 feet. The processing speed of the visual system is too slow for an MLB batter to see the ball as it gets anywhere near home plate; MLB hitters need to anticipate the ball's path before it gets anywhere near them.

Accurate motor action plans require that a series of movements be performed in a particular serial order. This applies to speaking, performing sign language, touch-typing, singing, or playing a musical instrument. This serial order is implemented in the brain as a set of hierarchical plans. For example, we don't plan to articulate each syllable of the word "neuropsychology," one at a time; instead, we plan to say the word, and that sets off a pre-stored motor action plan. If you play an instrument, you've probably learned a new piece by practicing chunks at a time and then stitching them together—these "stitching points" constitute entry points and transition points, high-level nodes in the hierarchy. In a difficult piece, musicians often can't begin at just any random note, but have to go back to the beginning of one of the chunks or, in some cases, even the beginning of the entire piece.

Additional evidence for this hierarchy comes in the kinds of errors that musicians make. Performance mistakes almost always occur at the stitching points, not in between them where the sequence is well established. Also common is to miss a note by only a semitone—a finger slip—rather than by a large interval—and these finger slips also suggest that the motor hierarchy is intact. Musical mistakes almost always preserve the tempo and meter of a piece—I've never heard a musician, in the middle of a difficult passage, suddenly substitute a flurry of wrong notes at twice the tempo the piece was in, or substitute 3/4 time in the middle of a 4/4 metered piece. There are parallels to this in language: speech errors tend to preserve the grammatical class, and something of the semantic relation. Nouns get substituted for nouns, verbs for verbs, and so forth. So instead of saying "after three blocks, take a left turn at the *lamppost*" we might say "after three blocks, take a left turn at the *fire hydrant*." Still a noun, still a fixed landmark, just the wrong word floating happily within the grammatical hierarchy of the sentence.

When we look at slow-motion video of people moving their fingers while touch-typing, or playing a fast musical piece like Charlie Parker's *Bird Gets the Worm* or Rimsky-Korsakov's *Flight of the Bumblebee*, we see that whatever their fingers are doing *now*, their hands and fingers begin to twist, curl, stretch, or contract, as they prepare to move into the next position that will be required. Their muscles seem to know just what to do, independent of their brains.

What we call *muscle memory* is not literally in our muscles, but in the part of our brain that controls those muscles, united with circuits that store sequences and patterns. Stuttering could occur if something goes wrong with any of these three phases of motor control: the storage of the sequence, the retrieval of it, or its implementation. That is, our brains may be perfectly aware of the instructions we need to send to our muscles, but somehow those instructions don't get carried out. This happened to me after a concussion that affected my cerebellum. For several months I couldn't play the guitar, an instrument I'd played for 45 years. I sensed that my brain was sending the correct signals, but that the fingers weren't doing what they were told. In effect, I had a musical stutter.

Another requirement for fluent speech or music is timing. Even with a perfectly well-functioning motor planning and implementation system, one that instructs us in the correct order, neural systems that control the precise timing of actions can go awry— kind of a local, muscle version of what happens when we are jet-lagged and feel sleepy, awake, and hungry at the wrong times of the day. Oration and poetry clearly have a rhythm to them, and ordinary speech has an intrinsic rhythm, too, an alternation of stressed and unstressed syllables, accents on certain words to disambiguate an utterance ("No, you misheard me. I said the *cat* chased the *dog*!"). That rhythm, and the stringing together of syllables or notes or finger movements when you type, requires predictive timing, the ability to predict precisely when the next motor action will be required. Predictive timing is also needed to synchronize movements when dancing or clapping your hands to music. In laboratory experiments, 90% of adults who stutter have difficulty synchronizing finger taps to a metronome, suggesting that their predictive timing mechanisms are impaired even when they're not trying to speak.

A third thing that needs to happen for us to effectively make motor movements is to get some sort of feedback from our bodies (the proprioceptive sense) telling us that our bodies are doing more or less what we—or our motor action plans—instructed them to do.

Why do people stutter? One theory holds that stutterers are expecting auditory feedback sooner than they actually receive it, causing them to pause the motor action plan in the middle, rather than continuing on; they get stuck, waiting to confirm that what they did was the correct movement. That auditory feedback has to be coordinated with knowing the location of our body parts such as tongues, lips, jaws, and fingers through sensory-motor integration. Some people whose memory for sequences is intact (as far as we can tell) exhibit impairments in the networks underpinning sensory-motor integration and timing.

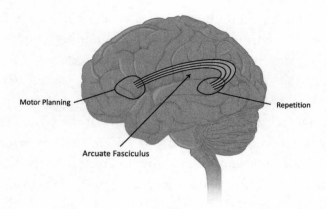

All these systems are interdependent—if any one of them doesn't function properly, we end up with halting, tentative, or frozen movements, getting stuck at some point in the hierarchical plan—stuttering. Hierarchical motor action plans use dedicated circuits in the motor cortex, pre-motor cortex, planning centers in the prefrontal cortex, and spatial memory hubs in the hippocampus. We've identified a bundle of nerve fibers—white matter tracts—that connects key language circuits involved in repetition and in motor planning, spanning three different lobes—the parietal, temporal, and frontal—and coordinates. Scott Grafton studied this bundle, the arcuate fasciculus (*ark*-yoo-at fa-*shick*-yoo-lus), and found it to be deficient in people who stutter—it simply wasn't making the streamlined connections among critical areas that it does for people who don't stutter. Even a minor interruption or decrement in motor and timing connections can interfere with fluency.

Parts of the right inferior frontal gyrus (IFG) are particularly active when we stop a repetitive action to inhibit further movement. One theory is that if this region is overactive, it fails to inhibit other brain areas that are required for the initiation and termination of movements. (Impairment to the IFG is also observed in patients with Parkinson's disease, another movement disorder.)

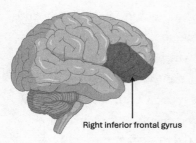

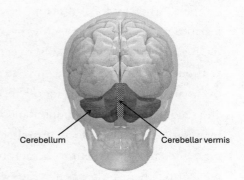

Right inferior frontal gyrus   Cerebellum   Cerebellar vermis

Stutterers tend to show impairment of the cerebellum and basal ganglia, regions responsible (as they are in Parkinson's) for motor control, specifically hyperactivation of the cerebellar vermis, as well as impairment of the left ventral prefrontal cortex (a hub for planning). The vermis is a tract of tissue that connects the two hemispheres of the cerebellum and helps coordinate the movements of the central body—trunk, head, shoulders, and hips (when Elvis was swinging his pelvis on TV, causing censors to run for their smelling salts, blame it on his vermis). My colleague Vinod Menon and I have been trying to sort out its role in music cognition for 25 years. Our lab has shown that it is necessary for processing musical structure, for being sensitive to the alternation of repetition and new material in a song, and for processing expectations of what will come next (in coordination with frontal lobe regions such as Brodmann Area 47). Because it is somehow involved in structuring information spread out over time, one can imagine it might play a role in disruptions in processing temporal structure—but so far this is just speculation.

Fluid speech, therefore, requires the integration of a great many pathways that are spatially distributed throughout the brain. When any one of these pathways becomes damaged, or simply doesn't run as efficiently as it could, the result can be stuttering. So how can we account for the situationally dependent improvement in Mel Tillis, James Earl Jones, and the rest?

Speech therapy for stuttering closely resembles the way musicians learn to play a new piece—slow everything down, work slowly and deliberately, and then gradually, over time, speed things back up. Even without explicit speech therapy, stuttering can be tamed through overlearning—making an action become automatized, unconscious, and second nature. Unlike spontaneous speech, which is only loosely planned, rehearsed performances become deeply entrenched in neural tissue, allowing them to be automatized—a fancy way of saying we can do them best when we're not thinking about them. So the likely reason Mel Tillis and James Earl Jones don't stutter when they perform is that they are performing sequences that they have practiced extensively—overlearned—and they know exactly what they're going to sing, or say, ahead of time. Once they get going on a well-practiced motor action plan, the plan itself takes over and generally runs smoothly. *Spontaneous* singing and speech put an extra load on a system that is somewhat rusty.

Do instrumental musicians ever stutter on their instruments? The great saxophonist Wayne Shorter was slightly disfluent in speech, an almost-stutterer. His playing has been described as a series of stuttering repeated notes—one of many examples of an artist turning their limitations into a feature, a signature of their artistic identity.

Stress is also a factor, and most stutterers report that the condition is exacerbated when they are tense, whether it's speaking or playing music. Fritz Kasten, an accomplished drummer who played with Vince Guaraldi and Joy of Cooking, notes:

A part of learning the instrument—not exclusive, certainly, to drums—is concentrating on relaxing. Assuming a normal neurology, invariably it's tension that creates the problems. . . . Most of us have experienced some degree of disturbed coordination in several musical situations such as learning new material. The more experienced—and confident—the player, the less the problem. Most musicians—if they're honest—will admit to tension in various situations, tension that affects their playing.

Part and parcel, then, of performing even an overlearned sequence is to learn to relax. Even the most accomplished musicians (or speakers) who have overlearned their parts can fall apart under stress. As any student knows, playing a piece perfectly when you're in a room by yourself is not the same as playing it in front of other people, when your cortisol levels are spiking, your heart is racing, fearing that you will be judged. And fearing that you will make a mistake in an overlearned sequence ironically causes you to focus so much on the details of the sequence that you pop right out of the motor action plan you've practiced so well, and suddenly it's as though nothing works. Once you get going and the music takes over, you can relax into it. That's because singing and playing an instrument invoke neurochemical changes that reduce overall stress levels, and decrease cortisol (the stress hormone) levels.

Individuals who stutter find prolonged episodes of fluency while singing that they rarely achieve during normal conversation. Singing thus helps to reinforce the normal activation patterns required for speech. Because singing relies more heavily on the right hemisphere than spontaneous speech does, this right-to-left hemisphere transfer is a way to retrain language circuits. Additionally, vowel sounds (as well as liquid consonants, such as /l/, /m/, and /r/) tend to be elongated in singing, compared to speech, thus the temporal planning and execution of the sounds themselves changes when they are sung. Successfully singing in this way is effectively like practicing a musical piece slowly—we have the time to learn good habits, and those good habits in turn transfer to the speech domain.

Many singers who stuttered as children, including Ed Sheeran, Kendrick Lamar, Ann Wilson, Shane Yellowbird, and Marc Anthony, credit playing music with helping them to overcome their stutter, or to dramatically minimize it. Just using the motor-articulatory circuits in synchrony with the pulse of music seems to have strengthened the pathways to improve their spontaneous speech. In the laboratory, stuttering is reduced when the speaker is paced with an isochronous tone (like a metronome) because the speaker knows exactly when the next "beat" is and can time their syllables with it, provided that their predictive timing circuits are somewhat functional.

The kinds of music that can help relieve stuttering are those with a steady beat, allowing the brain to leverage neural synchronization and planning systems to help keep it from getting stuck. Slow tempos are better, and songs with repetition—pop standards, with an AAB (verse, verse, chorus) form, for example. Most important are songs we enjoy and feel a connection to; otherwise, the stress induced by a song we don't care for could exacerbate the stuttering. Psychologist Frank Russo at Toronto Metropolitan University had 20 adult stutterers perform the lyrics of "Love Me Tender" by Elvis Presley, either with or without a rhythmic cue. Without the rhythmic cue, performance was highly variable, and there were many instances of stuttering. With the rhythmic cue, the participants displayed a high degree of temporal synchrony and almost no instances of stuttering.

Tourette syndrome (Tourette's disease)

The opposite case of a motor action plan getting stuck is a plan being triggered when we don't want it to be, and this can result in unwanted movements, or tics. Tourette syndrome (named after neurologist Georges Gilles de la Tourette, who first described it in 1885) is defined as having at least two different motor tics and at least one vocal tic; not everyone who has a tic has Tourette's, but everyone with Tourette's has tics. These can be controlled somewhat by medication that targets certain neurotransmitter systems in the brain (such as dopamine, GABA, and adrenaline). The tics themselves can be any of a nearly unlimited variety of abrupt movements or sounds. Individuals with Tourette's describe it as exhausting.

Singer Esha Alwani explains, "All of us experience something like it [Tourette's] when we get the hiccups: they're involuntary, and when we try to stop, we can't." Grammy winner Billie Eilish describes her Tourette's: "The main tics that I do constantly, all day long, are like, I wiggle my ear back and forth, and raise my eyebrow, and click my jaw . . . and flex my arm here and flex this arm, flex these muscles. These are things you would never notice if you're just having a conversation with me, but for me, they're very exhausting." She added that she almost never has them when she's playing music. Sound designer Jamie Grace turned her own Tourette's tics into music, creating a song with the drumbeats representing the tics. "I wanted to be as vulnerable as I could in my song . . . to be *me*," she said.

Tourette's tics often disappear during activities that require focused concentration, repetitive movements, and sensorimotor interaction, like riding a bicycle, playing ping-pong, dancing, or playing a musical instrument. Playing music may be especially effective because it entails precise planning, fine motor control, and more complex (and hierarchical) movement. To a lesser extent, simply *listening* to music can cause a reduction in tics, perhaps because it engages many of the same neural pathways as performing music. While music has not been shown to *cure* the ticcing of Tourette's, some musicians with Tourette's say that the tics can disappear completely while they're playing. Others report major tic urges right after a solo at a jazz concert, or between movements of classical pieces—for these musicians, the tic urges seem to become only temporarily suppressed, pent up, awaiting release.

The pathology of Tourette's disease is still poorly understood. Aberrant neural oscillations in the striatum (the largest nucleus within the basal ganglia) and the thalamus have been observed, and these abnormal oscillations might be responsible for the generation of tics and could lead to dysrhythmic activity in cortical regions. This partly accounts for why playing an instrument and listening to rhythmical musical stimuli can cause tic reduction: the music restores normal oscillations through neural entrainment.

The aberrant oscillations may be related to dysfunction in the cortico-striato-thalamo-cortical circuit (CSTC). One theory suggests that tics may be related to excessive cortical input into the basal ganglia, coupled with insufficient neural inhibition, leading to direct and unplanned motor output. Imaging studies have revealed weaker activity in the putamen, a hub for controlling motor movements, and in the caudate nucleus (another component of the basal ganglia), a circuit for planning motor activity. The caudate has a particularly high density of GABA neurons, which play a role in inhibition of behavior. Weaker activity there means that those neurons aren't able to serve that inhibitory function. Just why Tourette's may impair function in these circuits is still unknown, but we believe that it arises from a combination of genetic factors and bacterial infection.

Multiple sclerosis

Punk iconoclast and icon Exene Cervenka was diagnosed with multiple sclerosis in 2009. Her diagnosis, like many with MS, took a long time.

Multiple sclerosis is a "diagnosis of exclusion." There is no definitive test for it, and the diagnosis is only arrived at after other diseases have been ruled out, through a process called differential diagnosis. This takes a frustratingly long amount of time. (Other diseases with diagnoses of exclusion include Bell's palsy, schizophrenia, chronic fatigue syndrome, and long COVID.) MS is an autoimmune and neurodegenerative disorder characterized by destruction of myelin sheath in the central nervous system (CNS). Immune cells attack this protective lining that serves as an electrical insulator, surrounding nerve cells and nerve fibers in the brain and spinal cord. The word *sclerosis* means an abnormal hardening of organic tissue—plaques are developed along the neuron's axon, the white matter of the brain, causing communication problems from one neuron to another, and from the brain to the rest of the body. The loss of myelin causes reduced speed of information processing, and movement difficulties, that get worse over time.

The underlying cause of MS rests on a complex interplay among genetic and environmental factors, and recent evidence is that it is linked to a very common virus, Epstein-Barr (EBV), a type of herpes (HHV4). You probably have EBV antibodies—95% of adults do. In younger adults it can cause mononucleosis, but doesn't always. After initial infection, EBV remains dormant, sometimes for years; for reasons we don't understand, it can reactivate and cause all sorts of trouble. You might be thinking, "Why don't they just prescribe antivirals then?" A reasonable question. No antivirals have proven effective against this particular virus, and even if they did, in many cases, by the time symptoms appear, permanent damage to the body has occurred.

MS is diagnosed mostly in young adults, with disease onset occurring in most cases between the ages of 20 and 40 years. Exene was unusual in not being diagnosed until she was 53. MS can affect any area of the central nervous system, leading to visual, motor, and sensory deficits, as well as speech disturbances, sphincter disorders, cognitive impairment, sexual problems, and fatigue. The most common form of the disease—the form Exene was diagnosed with—is relapse and remitting multiple sclerosis (RRMS). Patients cycle between attacks (relapses) with new or worsening neurological symptoms followed by periods of recovery (remissions) where symptoms lessen or even disappear, but only for a while. The goal of pharmacological treatments for RRMS is to reduce relapse severity and frequency, but even with medication, the disease can have a severe impact on productivity, social participation, self-esteem, motivation, and quality of life.

Age aside, Exene's path to a diagnosis is typical, with her and her doctors trying to sort through a range of strange and seemingly disconnected symptoms, trying to exclude diseases they could test for. When she was 44, she developed blurred vision in one eye. Six weeks later it resolved. Over the following years, she had some episodes of confusion and severe headaches. Because her friend Victoria Williams had been diagnosed with MS 11 years before, it was on Exene's radar. One test that could provide some evidence was an MRI, which might detect the scleroses or lesions indicative of demyelination, although these could just as well arise from other causes. A positive test would be evidence that she had MS, but a negative test would not be evidence that she didn't, because a negative result could be explained by a number of things: the lesions could be too small; the MRI might not have been sensitive enough; motion artifacts or other unintended confounds.

Exene's MRI was inconclusive. The doctors just thought she was a hypochondriac. Her experience was typical. "A lot of people with MS are familiar with that diagnosis of 'there's nothing wrong with you—it's all in your head.'"

A few years later, her symptoms were undeniable and the doctors could no longer ignore them—or her. "I couldn't think, walk, talk, do anything. I was a passenger in a car, going down the highway, and suddenly there was two of everything for an hour and a half. I knew, as soon as I got home, that I had to go see the neurologist right away." She ended up in the hospital after a long bout of vomiting and intractable headaches. Her MRI was consistent with MS (we can't say "positive for MS" because many other things can cause the lesions observed), and Exene was diagnosed with the most common form, the cyclical on-again, off-again relapsing-remitting MS.

But Exene is indomitable. If your stereotype of punk rockers sees them as nihilists and fatalists, Exene is the opposite. (If you think about it, no one who is a fatalist could reach any high measure of success—it just takes too much dogged determination and hard work, whether you're John Lydon or John Doe.) Exene has the artist's drive and spirit, that nothing, neither rain, nor hail, nor gloom of night, nor neurological disease will stop her from continuing to create. She has continued to tour, writes books, exhibits her collages of found art, and makes spoken word recordings. The cycles of relapse are highly unpredictable, and so she has modified her lifestyle, eating more healthily, and doing yoga.

One of the hallmarks of MS is fatigue and malaise, to the extent that people frequently find themselves not having the energy to participate in just those activities that might make them feel better, like exercising, socializing, and listening to music. But once the music starts, the therapy kicks in, jump-starting motivation and dramatically lessening fatigue. The problem is having the motivation to get started. A practical solution is to program music to play at certain times of day—through the phone, a personal digital assistant, whatever . . . it's easier to turn the music off if you are in the middle of doing something else, harder sometimes to turn it on.

A promising technique for music-based movement therapies is rhythmic auditory stimulation (RAS), developed by Michael Thaut (now at the Rehabilitation Sciences Institute at the University of Toronto). Michael was born into a musical family in Austria, and excelled at the violin. After high school, he founded a professional folk ensemble, Fiedel Michel, specializing in the performance of authentic traditional German folksongs and dances from the past four centuries. His group performed 100 concerts a year throughout Europe and he simultaneously completed a degree in psychology and undertook advanced studies at the Mozarteum in Salzburg. "I fiddled my way through life for a few years," he says. After six years, he took stock. "I wasn't sure that I wanted to continue in music, playing in bands—unhealthy lifestyle. I had heard of music therapy but didn't really know what it was." Michael was awarded a Fulbright scholarship to study for an interdisciplinary PhD in music education, psychology, and music therapy at Michigan State. In short order he was directing a leading laboratory that studies the intersection between music, neuroscience, and therapy.

Michael Thaut's RAS method provides gait training. Patients practice walking while guided by an auditory rhythm, allowing them to synchronize gait movements to predictable time cues. The rhythm can come from anything—a metronome, a recording of footsteps, or music; emerging evidence is that music is the most effective of these, especially when the music is customized for each individual patient. Whatever the source, a tempo is selected so that the beats per minute of the auditory rhythm match the patient's normal walking speed, measured in steps per minute. The theory behind RAS therapy is that the body's internal clock (driven by the internal pacing circuits) has degraded and needs to be retrained, by way of guidance from an external source. RAS is successful in improving walking stability and balance, and in reducing falls, freezing episodes, and variability in walking speed. Originally developed for Parkinson's, RAS has also been similarly effective for rehabilitation in stroke patients, children with cerebral palsy, traumatic brain injury—and people with multiple sclerosis.

The ultimate goal of RAS is that the therapy can be discontinued; some studies show that once established, benefits can last for up to six months. One hypothesis about why the effect lasts is that RAS encourages the brain to bypass damaged circuits, relying instead on alternative routes in the cortical-thalamo pathway. The brain is, fortunately, marked by great redundancies—not because these redundancies were "planned" (evolution doesn't work that way), but because different pathways evolved independently for different purposes. It's kind of like having a Swiss army knife with both a screwdriver and a nail file. If you break the screwdriver, you can probably co-opt the end of the nail file to get the job done. fMRI studies show noticeable improvement in functional connectivity between brain regions crucial for timed, cognitive control after training. In particular, music strengthens the communication between the auditory cortex and the executive control network in the prefrontal cortex, and from the prefrontal cortex to the cerebellum. We are also finding evidence that rhythmic synchronization and entrainment build on central pattern generators in the spinal cord, leading to direct auditory-motor coupling without the involvement of cognitive or motor planning centers. RAS is not equally effective for everyone, however, and musical train-ing, proficiency, and sensitivity to "groove" (the ease with which a person can move their body to music) are all predictors of RAS effectiveness.

Rhythmic auditory stimulation for MS patients improves walking speed and stride length, as well as stability, reducing the amount of time a person needs external support while walking (from either a walker or another person)—after just two weeks. Even just imagining movements and repetitive sounds (motor imagery) yields improve-ment in walking, and patients report less fatigue and increased qual-ity of life.

A promising new direction employs musical keyboard training to improve the use of the hands and arms of MS patients. But why a keyboard? Wouldn't *any* fine motor hand exercises improve dexter-ity? You'd think so, but when Italian neuroscientist Roberto Gatti asked half of the participants in a study to play a keyboard with the audio on, and the other half with it off, those who received the auditory feedback from their movements showed substantially more improvement than those who didn't. What matters then, for a therapeutic outcome, is not simply movement, but the accompa-nying auditory feedback that allows the brain to strengthen already existing pathways that connect movement to sound. This finding is complemented by emerging evidence for the effectiveness of dance therapy; movement to music led by a music therapist can lead to reduced fatigue and improved cognitive capacity and coordination (including balance and walk time performance) for MS patients.

In 2011, at age 23, dancer Courtney Platt was performing on the Season 7 tour of *So You Think You Can Dance* when her legs sud-denly went numb. When she put her chin to her chest, she experi-enced what felt like electrical pulses shooting through her feet. "I thought I'd hurt my back and—as professional athletes sometimes do—I pushed through the discomfort. Yes, I couldn't feel my legs, but they still *moved*. I told myself, 'I can do this.' For months, I put up with not being able to feel anything from my waist down."

Soon, other symptoms set in—exhaustion and intense fatigue. To get through shows, Courtney relied on over-the-counter pain relievers and espresso shots. As her symptoms worsened, she began to wonder if something more serious was going on. Then her symptoms went away for a while; she figured it was just a typical injury, maybe a pinched nerve or slipped disc. Toward the end of 2011, her symptoms returned, worse than ever, even though she was no longer on tour and was working much less. While visiting with friends in January 2012, Courtney was so exhausted she began to slur her words. On February 1, 2012, she visited a neurologist at Weill Cornell Medical Center in New York. They found plaques on her spinal cord and lesions on her brain. She was diagnosed with multiple sclerosis.

Music had always served as a mood management tool for Courtney, "whether I need to be relaxed, I want to cry, I want to be happy, or I want to get amped in the gym. It's always been something that has helped me and something I turn to often to make me feel good, to get me to feel like I'm back where I belong." With the diagnosis, she explored music therapy, after initially being skeptical. "When I got to actually experience it—we did one where [the music therapist] played a song and she told me to just close my eyes and to dance. And she kind of led me through. It felt like more of a moving, musical meditation. And, like—I definitely don't meditate. I'm so not one of those people that just sits and breathes. That's not my thing. [Music therapists] really gave me tools to calm myself down, and channel my energy and my thoughts into a calm place using music."

She has also used music to help her as a mnemonic device for improving memory. "One of the symptoms of MS is memory loss, and forgetting things often, and I definitely experience that one. It's something as little as making a song about your grocery list, or something along those lines. It really gives you tools to be able to help manage a lot of the symptoms of MS."

Huntington's disease

Huntington's disease is a rare neurodegenerative disease, caused by a single defective gene on chromosome 4 that causes a progressive breakdown of neurons in the brain, and ultimately results in movement, thinking, and psychiatric disorders. The movement disorders include disruptions in both voluntary and involuntary movements; impaired gait, posture, and balance; difficulties with speech and swallowing; and dystonia and other muscle flexion and extension problems. The disease is sometimes called Huntington's chorea; chorea is the general medical term for any movement disorder that causes sudden, unintended, and uncontrollable jerky movements; it is found in many diseases and is caused by an excess of dopamine in the movement pathways of the brain.

Huntington's is inherited, with a prevalence of one person in 33,000 (making it 100 times less common than Parkinson's disease). There is no cure or treatment that can delay the onset or progression of it. Drugs are prescribed to treat the motor and psychiatric signs. Besides medication, non-pharmacological interventions, including music therapy, are used to improve motor and cognitive function, and quality of life.

The use of rhythmic auditory stimulation (e.g., a metronome or music) can improve gait in Huntington's, and these improvements remain for some time after treatment. However, more complex musical selections don't lead to gait improvement, and as Huntington's progresses, the tracking of musical rhythms declines more rapidly than the ability to track a metronome. This suggests that higher-order rhythmic function is more compromised in Huntington's than low-level timing is; higher-order processing deficits are likely due to the combination of hyperactivations of cerebellar and medial prefrontal structures, and are correlated with the severity of movement deterioration.

An emerging technique to improve speech articulation, clarity, fluency, tempo, and volume is Huntington speech music therapy (HSMT), based on the principles of speech music therapy for aphasia. In a joint session, a speech therapist instructs the patient to speak words and sentences accompanied by structured short melodies, played repeatedly on the piano by a music therapist. Continuously using the same melodic lines provides structure, which facilitates production and reproduction of language. In this way, as we see in stuttering and other disorders, patients are often able to sing words they have trouble articulating in speech. Unfortunately, though HSMT improves speech and communication during treatment, we still haven't figured out how to prolong the results.

Drummer Trey Gray spent his career providing the backbeat for superstars like Brooks & Dunn and Reba McEntire. After a seven-year run with Faith Hill, in 2001 he began playing with Jewel. Around that time, he started to notice some changes in his overall health. "I was having times when I would feel fatigue or off balance, for what seemed like no good reason. Sometimes my moods would change on a dime, or it was hard to leave my hotel room." In 2003, Trey was diagnosed with Huntington's disease and given 10 to 15 years to live. Twenty years later he continues to travel and play all over the world, all while producing and developing new artists. He's also the artist-in-residence at a private Christian school, teaching K-8 music and leading the choir and band programs. Trey believes that his drumming is helping him to beat (hahaha) the odds. "Drumming, I'm using all four limbs, getting blood and oxygen flowing, firing up the brain. Four limbs at a time [doing] different things. Imagine *that* in your brain—the neurons and stuff that are firing. It's fantastic."

Trey's experience using music to control his symptoms is paralleled in controlled studies. Huntington's patients were given bongo drums and instructed to play at home for 15 minutes a day. Their assigned exercises progressed with increasing complexity and speed. After two months, there were measurable improvements in working memory (perhaps spurred by embodied cognition), and evidence of changes in white matter microstructure in frontal–motor pathways and increased myelination. Drumming is a demanding task, even for people without a neurodegenerative disease. Unfortunately, drumming interventions have high dropout rates, suggesting that this kind of training may only be suitable for patients who have support and encouragement from their caregivers or family (who don't mind all the bongo playing coming from the other room), or it may be that (*gasp*) not everybody likes to channel their inner Alex Acuña or Sheila E.

Although our awareness of Huntington's is relatively new, it's probably been around for a long time, but hadn't been identified as such. For example, when Woody Guthrie began to show classic symptoms of Huntington's in the late 1940s, people (including Guthrie himself and his doctors) thought he was "just a stumbling drunk." In 1952 (upon release from the Brooklyn State Hospital), Woody told his pal Pete Seeger that he finally learned he was suffering from "the mental disease my mother had, Huntington's Chorea" and that the chorea kept him just as dizzy as whiskey, but was markedly cheaper.

Another movement disorder that is accompanied by memory loss is Parkinson's disease. From 1990 to 2015, the number of Americans with Parkinson's doubled to over 6 million, and that number is expected to double again, to 12 million, in the next 15 years. This makes Parkinson's the fastest growing neurological disorder in the world. These numbers are driven partly by aging—the biggest risk factor for Parkinson's, as with cancer and with dementia, is simply age. Environmental toxins, including pesticides, have also been implicated. The sheer quantity of people affected by it, either directly or indirectly, plus the proven use of music in its treatment, is a story unto itself.

Chapter 7

Parkinson's Disease

AFTER THE RELEASE OF MY FIRST BOOK, *THIS IS YOUR Brain on Music*, the PBS affiliate in New York City, WNET, got in touch with me to turn it into a documentary film. Their idea was to have me co-host the film with a musician who could add their own perspective. Whoever it was had to have unimpeachable musicianship, a sense of wonder, and be the sort of warm presence that would make people want to spend time with us for two hours. Bobby McFerrin was our first choice—an internationally known singer and one of the most accomplished improvisers in the world, whose music crosses every boundary. He had performed with Herbie Hancock, singing horn-line harmonies with the Marsalis brothers; he had conducted orchestras around the world, and had a song that reached number 1 on the *Billboard* Hot 100 chart ("Don't Worry, Be Happy"). Bobby is a bit reclusive, and so finding him took some time, but when we did, he gave us an enthusiastic "yes." He's a great mind who likes to probe how things work with a child's sense of wonderment and curiosity. Everything about the project appealed to him—and he to us.

Bobby and I met for the first time in New York. On the first day of filming, we hit it off, improvising, mugging for the cameras, cracking jokes, even co-writing some songs on the spot. Shortly after, we were invited to appear together at the World Science Festival in New York to talk about music and emotion with a panel of other scientists. I joined on saxophone to improvise a song with

Bobby, a sitar player, and a tabla player. After that, whenever Bobby came to Montreal on tour, he'd reach out, invite me to the show, and drop by to visit the lab to see what we were up to. Once when he came, he asked to be a subject in the experiments we were running. About a few of them, he said, "I see that you're doing it like *this*—wouldn't it be better if you did it like *this*?" and he suggested some brilliant modifications to our work that none of us on the "professional" experimenting side had thought of.

Whenever we weren't actively doing something—maybe while waiting for an experiment to be set up, or walking down stairs—Bobby would start singing, quietly. It was a delight to hear, never the same thing twice. He'd sing something he had just made up, maybe using the familiar as a launchpad. Just like his performances, a snippet of something here and there, from Scheherazade to Beyoncé, with some Thelonious Monk or Harold Arlen along the way. There is *always* music in his head, and when he performs, we all get to share in a bit of it.

As I began to write this book, I knew that I wanted to talk with Bobby. His music has been medicine to so many, and I wanted to ask if music had ever been medicine for *him*. When we spoke, he said he had recently been diagnosed with Parkinson's disease. It is one thing to read about the power of music in journal articles, to pore over the statistics, and to pick through the experimental design. It's another to see it for yourself, to be close enough to *feel* the change that can come over a person struggling with a disease when music comes into play. Our conversation changed forever the way I thought about music and medicine.

⌒

Like stuttering and Tourette's, Parkinson's is characterized by unintended or uncontrollable movements, although (unlike these others) it typically manifests as tremors, shaking, stiffness, and balance and coordination problems. Unlike stuttering and Tourette's, but similar to Huntington's, it is a progressive disease that affects the nervous system with increasing severity, gradually shutting it down entirely. Parkinson's degrades a host of different circuits that govern movement in the cortical spinal tract, affecting both involuntary and voluntary movements. Unfortunately, it also affects cognitive abilities as it progresses, leading to mild cognitive impairment, dementia, loss of decision making, planning ability, and memory. Parkinson's affects about three people per thousand, and 90% of diagnoses are in people over age 60; age appears to be a risk factor. Parkinson's reduces life expectancy by about a year when diagnosed at age 85, and by about seven years when diagnosed at 65.

Parkinson's occurs when critical neurons are lost in a part of the basal ganglia called the substantia nigra, leading to impaired dopamine signaling for movement. We still aren't certain what causes it. One afternoon, when visiting a clinic, Steve Keele, an expert on the neuroscience of timing and motor control, observed a patient who could walk normally but had difficulty starting—as though he was stuck. Others could begin walking and *then* got stuck. Some Parkinson's patients accelerate as they walk, sometimes to the point of losing control and walking too fast, a phenomenon called festinating. Because the basal ganglia are involved in interval timing, Steve wondered if the getting stuck part might have something to do with losing track of the beat, as it were: the timed synchronization of steps to an internal clock that is required to walk with a smooth and continuous gait. Perhaps some sort of visual cue could help Parkinson's patients get unstuck, given that the visual cortex in the occipital lobe has direct and fast connections to movement centers in the parietal cortex and the pons, an important relay station in the brain stem connecting the cortex with the cerebellum and spinal cord. Steve got a tape measure and set a row of coins on the floor about 30 inches apart. He brought in some Parkinson's patients who were in wheelchairs and helped them stand up, positioning their feet where the first coin was, and then instructed them to look down the line of coins before they started to walk. It worked—having the visual cue somehow triggered a motor action plan and . . . they walked. (Steve later called this the march of dimes.) Since then, it's been well established that playing music to people with Parkinson's is even more effective—for one thing, it's portable (they can have it on their cell phones), and for another, it works even when the disease impairs their vision. And it establishes an even stronger cue, connecting auditory input to a timed motor action plan.

Many people have Parkinson's for some time before they know it—they may notice their hands shaking, or a loss of balance, but a *lot* of things can cause those; only two people out of 10,000 are diagnosed each year, and so Parkinson's isn't always the first thing they or doctors think of. Bobby didn't know he had Parkinson's at first. He just felt himself getting more tired than usual, and in a sour mood more often. "One night I was in a hotel in Vermont the afternoon before a show and I felt I was really sick. I just didn't feel like doing the show. I thought about all the people who had bought tickets, how they had planned their evenings around this, that cancelling the show would disappoint them. I figured I could go for about 45 minutes, and although the audience probably expected twice as much, I could leave without creating too much disappointment.

"When I got on the stage and started singing, something miraculous happened. I was suddenly filled with energy. I played the whole show—two hours. And I felt great for the next few days. The music had somehow changed everything, and I felt 80 percent better after the show. Not long after that, though, I noticed I was having trouble getting my voice to do some of the things it usually did. I went to the doctor, and eventually, after a lot of tests, they diagnosed Parkinson's disease."

In 2022, he began performing his show Circlesongs every Monday at noon, at The Freight & Salvage, a concert venue in Berkeley, California. The audience is a wonderful mix of people of different ages and backgrounds—very Berkeley. The first day I went, there were aging hippies in tie-dye, millennials, Gen X-ers with their parents *and* their own children for an intergenerational family music outing. The energy in the room was electrifying. Four singers came out and sat down in the chairs on the ends, leaving one empty chair in the middle. The audience applauded. After the applause died down, there was a silent beat or two, and Bobby sauntered out. I hadn't seen him since the diagnosis; he looked thinner than usual, tired, a bit weary, but very present. He sat down in the middle chair and looked around the room, seemingly making eye contact with everyone in the room, one at a time, in no hurry to rush this personal connection. He was still handsome, and even from the fifth row, I could see the combination of impish mischief and loving-kindness in his eyes.

Bobby picked the microphone up off the stand. He closed his eyes, raised his head toward the ceiling, and the room was still, as quiet as a moonbeam. Without appearing to take a breath, Bobby started to sing. That sweet, emotive voice danced around a fresh, shiny new melody—sixteenth notes darted about, settling into quarter notes and breathless holds, then more notes dancing and swaying and holding their lover tight while spinning circles on the ballroom floor.

Bobby nodded almost imperceptibly at the fellow in the chair downstage right, who started to sing beats, a human percussion section, with the most stunning-sounding mouth snare drum I've ever heard. Bobby kept on going, then pointed to the gentleman between them, and soon the room was filled with the sound of walking bass, fretless slides and glissandi that married the rhythm to the melody. The two women to Bobby's left started to sing countermelodies, and we had a full orchestra.

The concert unfurled 90 minutes of resplendent improvisation and musical humor, and touched on too wide a variety of musical styles to list them all. In one moment, he was channeling Eartha Kitt or Bessie Smith, then quoting Aerosmith, Bananarama, or Thelonious Monk. He'd launch into microtonal Arabic mawwāl singing, or hold the microphone tight to the front of his neck and hit the subharmonics of Tuvan throat singing. He'd clap, beat his chest, play melodies on his cheeks; he'd alter his voice to sound like a saxophone, or Stevie Wonder, or Rex Harrison, or Judy Garland. But don't think that all this variety was in any way jarring or incoherent—everything followed what came before and led inevitably to what would come next, like a well-composed symphony, and each new vocal color served to support the music that was being created at any given moment.

Neither Bobby nor his singers knows in advance what he will do—he sits in a chair amid the other musicians, looks out at the audience (the lights are up), closes his eyes, takes a breath, and waits for some musical motif, some hook or idea to take hold.

We meet after the show, in his new neighborhood in San Francisco. It is quiet, residential; the neighbors all seem to know one another. We walk past verdant front yards that are lovingly tended. Stocky, 12-foot-tall palm trees sit above hydrangeas, ferns, and juniper bushes. Fir, spruce, and redwood trees, some 100 feet tall, stand guard like sentries protecting this quiet haven from the onslaughts of a large city. An Allen's hummingbird hovers in front of us before darting off to a nearby feeder, then speeds back, as if to say, "Thank you, humans—flower-food scarce this time of year." We see a tiny egg on the ground underneath one of the California live oaks. There is a nest in the crook of the branches, but if the egg fell, why wasn't it broken? It remains a mystery.

"Walking is good for me," Bobby says. "I try to walk a mile or two every day. It helps with my balance." He knows that the disease is progressive. When he received the diagnosis, he sold the home he loved, nestled in 300 acres of land in rural Pennsylvania, to be in a major city, near a top-ranked medical center so that he could get the medical attention he needs.

Having music in his life has helped him manage the psychological and physical challenges that the prescription medications don't. "Music helps a lot. Pilates and physical therapy help a lot. The music helps with rhythmic breathing. Look, when I got here I was depressed. Down in the dumps. I wasn't singing. Then I started singing and got even *more* depressed because I didn't like how my voice sounded. I found any kind of excuse not to do it. But as soon as I recognized that I might not have the chops I used to have, I realized I still have a creative mind that likes to play. It turns out that's the bigger part! *That's* the part you want to cultivate. Nobody cares about the quality of the voice if it don't got nothing to say.

"You asked earlier if music had ever been therapy for me," Bobby said.

"Yes—I can see that it is now."

"Yes, *now*. But I'm thinking of a really profound time. Did I ever tell you my Miles Davis story?"

"No."

"Well, when I was a month shy of 21, I took my date, Trudy Catania, to see Miles Davis. And we got there, and there was no way in the world we were going to get in. But I thought, 'We're here, where are we going to go? Let's just stay here for a while and see what happens.' After standing in line for maybe 10 minutes, a woman walks out of the front door, walks down the line, and stops at me. 'Would you like two? I got two extra tickets.'

"So we go inside with our tickets. The table that those tickets were for was right here [*gesturing*]. It's right there next to the piano. We can reach out and touch his back. Miles playing, sporting a black trumpet trimmed with gold lettering, a black shirt. And he would be out on stage for the first 10 minutes blowing and then he'd leave the stage and the rest of the band would just burn, *burn* for an hour playing the most incredible stuff. Up to that point, I'd never heard stuff like that. But I was so turned on. Literally, I felt like I had just died.

"I walked out of the club that night molecularly changed. I mean, I felt my whole cell system . . . my whole being was as if cells shoot and ping and move rapidly depending on the activity. I felt this real tingling in my . . . self. Whether I could feel it or not physically, I had a sense that something had changed. And what changed was from that moment on I devoted myself to improvisation. I wasn't going back. We sat there marveling at each other that half an hour earlier we were standing in a hopeless line. There's no way, no way are we going to get in here, but we waited. And then the rest of my life changed that night. It was in February, the eighteenth."

A week after our walk, Bobby texted me. "Singing is my mother tongue. Speech always felt like a foreign language to me." Many musicians have expressed the same idea, from guitarist Carlos Santana to saxophonist Billy Pierce. For them, it is the *music* that flows so effortlessly, natively, convincingly, gracefully.

Linda Ronstadt, another San Francisco musician, is also suffering from Parkinson's. She and I co-hosted a program on music education in Toronto in 2009. Everyone wanted her to sing, but she declined. She confided that she was starting to notice that the timbre of her voice was changing, and that her voice wasn't doing what she wanted it to do, but she couldn't figure out why. Three years later, she was diagnosed with Parkinson's, and it all started to make sense. Musicians are better able to describe neuroscientific concepts than the scientists themselves. "I'd . . . start to take the note and then it would stop," Linda said. "What you can't do with Parkinsonism is repetitive motions, and singing is a repetitive motion."

It's precisely because music is based on repetitive motions, either physically repetitive or conceptually or auditorily repetitive, that it has proven an effective adjunct therapy for Parkinson's. Normally, our basal ganglia are directly responsible for helping us construct a sense of internal timing, especially for repetitive movements, such as walking, clapping, swaying, doing jumping jacks, having sex. Activities like these require an internal sense of rhythm and timed motor control, abilities that degrade with the degrading basal ganglia in Parkinson's, causing individuals to lose their perception of internal timing. Police drummer Stewart Copeland explained that his job was to be a *rhythmatist*. A what? "Rhythmatism is the study of the patterns that weave the fabric of life. If you know where the beat is going to land, you can jump on it!" That is the reason that playing music, with the specialized and direct pathways between sound input and motor output, is so effective. Parkinson's patients hear the beat, and even if their sense of *internal* timing is impaired, another beat comes just when it should. Music compensates for the loss of control by the basal ganglia, and in addition, music enhances movement synchronization.

And yet, where Bobby McFerrin was able to channel his musical energy as a force for symptomatic release, Linda Ronstadt has been unable to do so. Every case is different. No two people experience *any* disease the same way, and Parkinson's is no exception. Like aspirin, music can treat the symptoms of some people in some situations, but not everyone in all situations.

The gait training provided by Michael Thaut's rhythmic auditory stimulation (RAS) that is effective for so many movement disorders is established for Parkinson's patients as well. Putting this into practice, some Parkinson's individuals or their therapists program music designed to meet the individual's musical preferences, and with a tempo consistent with the cadence of their normal, undisturbed walking. This has been found to significantly increase stability and speed, and reduce freezing. In one study, the participants reported a number of positive outcomes: "I walked with an increased pace, even after turning the music off"; "I stood taller and swung my arms more"; "I walked smoother, it was more even"; "I had an improved emotional state"; "exercising was less monotonous"; and "the music provided extra motivation to exercise."

In another study, a group of individuals with Parkinson's was randomly assigned to music therapy (MT) or physical therapy (PT). The MT sessions consisted of choral singing, voice exercises, rhythmic and free body movements, and active music-making involving collective invention. PT sessions included a series of passive stretching exercises, specific motor tasks, and strategies to improve balance and gait. The MT group had far fewer incidents of halting and slowing—and overall improvements in emotional functions, activities of daily living, and quality of life. By contrast, the PT session findings showed significant improvements only in flexibility. These interventions need to be performed regularly; in a follow-up, two months after the initial therapy, the advantages to both groups largely disappeared. Even video training is showing some promise, with games that require individuals to initiate virtual movements via an avatar, including picking up objects, flexing and extending one's knees to avoid obstacles, and rotating one's trunk to collect virtual coins. Following approximately two months of training, participants in one experiment showed improvements in gait and balance, and fewer falls. Unfortunately, that study had no control group, and so further research is needed.

When Parkinson's patients had their brains stimulated by transcranial alternating currents over the motor cortex using rhythmic pulses, an almost 50% tremor reduction was achieved. Another approach used a special music application for iPods. Here, music serves as a reward; the iPod starts playing music when patients walk using appropriate steps. Even stronger effects come from dancing to music, particularly tangos and foxtrot, which significantly improve overall mobility, balance, and backward stride length. iPods have already been replaced by newer technologies and advances in wearable sensors, treatment software, and delivery interfaces continually improve rehabilitation interventions that can be delivered in clinics, communities, or homes. Ambulosono, developed by Bin Hu and Taylor Chomiak at the University of Calgary, is one example of a wearable device that uses music in neurorehabilitation.

The conventional therapy for Parkinson's is levodopa (L–dopa), a substance that crosses the blood-brain barrier, and then is directly converted into dopamine through a chemical process in the brain. Because listening to music and performing can increase levels of endogenous dopamine, it is possible that patients who are regularly engaged with music can reduce their dosage of L–dopa, although there is not yet research to substantiate this claim. If true, this could save millions of patients the unpleasant side effects of the pharmaceutical, which include dizziness, loss of appetite, diarrhea, constipation (yes, both), forgetfulness, confusion, and—in higher doses—chorea, the uncontrollable and jerky movements that the drug is meant to alleviate.

So far, we've been looking at music-based treatments for movement, because Parkinson's is primarily a movement disorder. The body is not a set of isolated systems, of course, and a movement disorder impacts other abilities. A common and debilitating symptom of Parkinson's is impaired speech due to both decreased loudness and clarity, the result of weak air pressure generated during exhalation while speaking. Music therapist and neuroscientist Elizabeth Stegemöller at Iowa State University performed a study in which music therapists led a group of patients with Parkinson's in either once-weekly or twice-weekly singing. She found increases in phonation time, the longest time period a person can hold a single vowel sound, which serves as an objective measure of glottal efficiency. She also found improvements in respiratory pressure, and the participants reported increases in their overall quality of life.

Eun Young Han of Ewha Womans University in Seoul separated out any latent beneficial effects of being in a group (like the Stegemöller participants were) by studying the effects of an individual singing program. She replicated Stegemöller's findings and found significant improvements on the Geriatric Depression Scale. (It turns out that being able to speak clearly, hear yourself, and have conversations is important! Who knew?) Separate studies found that singing interventions had carryover effects in the form of reduction in body discomfort and improvement in overall motor function. An important area for future research is to study the effects of playing a musical instrument, particularly, but not exclusively, a wind instrument, because they require precise, bimanual movement. Other studies are showing preliminary evidence for carryover effects, where a variety of different music therapies can lead to significant improvement in cognitive flexibility, mental processing speed, attention, and memory. In one study, the beneficial effects were maintained for six months.

At the forefront of Parkinson's research is the new finding that a burst of sudden and intense emotion can temporarily improve

movement symptoms. Apart from difficulties stopping and starting, slowness of movement, called bradykinesia, is a hallmark of
Parkinson's (brady = slowing, as in bradycardia, the slowing of the
heart rate below 50 beats per minute, or bradylalia, abnormally
slow speech). Bradykinesia can be temporarily overcome in certain
cases. Paul, a grandfather with Parkinson's, was sitting in his chair
when—out of the corner of his eye—he saw his young grandson
Max heading for a steep flight of stairs. Paul had been living with
Parkinson's for over ten years and at this point, getting up and out of
a chair was an ordeal. But the fear that ran through him caused him
to jump up immediately and fluidly pick up Max to save him from
falling. This paradoxical kinesia (movement contradictory to one's
overall abilities or condition) is similar conceptually to the super-
human strength that people can demonstrate in an emergency—
such as being able to lift a car up off of a child, or break down a
solid door to rescue a baby from a fire. In the case of Parkinson's,
we are still trying to figure out what restores abilities that had long
ago been lost. Scott Grafton posits that this is accomplished through
an "open loop," a neurological circuit that is effectively a redundant, alternate pathway for motor movement that runs through the
amygdala, the cortical structure that regulates strong emotional
responses, especially fear.

Paradoxical kinesia is not specific to Parkinson's, but is a general property of the motor system. And it doesn't take an imminent
danger, it only takes emotion. Fluid movement, just like before the
onset of the disease, has been observed by Parkinson's individuals
engaging in activities that bring them great pleasure: playing tennis,
playing soccer, boxing, ice skating, and bicycling. In these cases, it's
not that the normal movement pathways suddenly work; it's that the
urgency or great emotional intensity of an experience triggers the
sorts of secondary or redundant pathways that are trained through
Thaut's rhythmic auditory stimulation (RAS). It seems logical to ask
whether music can also have this effect, especially music that the
patient finds thrilling or intensely emotional—whether happy, sad,

or bittersweet. Among professional musicians, it is well known that particular passages in certain songs create "chill moments" or frisson, the experience of hair standing on end. This could induce paradoxical kinesia as well, especially in Parkinson's patients who are sensitive to music.

Rapper and filmmaker Walter J. Archey III was diagnosed relatively young, at age 47. He strives to live longer and stay more active than the statistics predict. "I'm trying to be the Stephen Hawking of Parkinson's," he says. Having Parkinson's as a younger person has been challenging. "The first thing that a lot of people think when they see you is, 'Man, this dude is clearly on drugs.'" Rapping has helped his vocal cords, but he can't rap the way he used to because he runs out of breath. His goal is to overcome the public perception if not the disease. "When you see me, I'm going to have a gold chain and be with some fly ladies, making beats and rapping." His intention is that the audience won't even think about Parkinson's. They'll think about the music. And it will move them.

Chapter 8

Trauma

I MAGINE TWO GROUPS OF BABY RATS. ONE GROUP GETS lots of attention from their mothers, with regular licking and grooming; the other doesn't. McGill neuroscientist Michael Meaney asks a simple but groundbreaking question: Does it matter—does it in any way affect their lives as adults? The answer is a resounding yes. The loved rats grow up to be resilient adults, better able to handle the stresses of rat life. But here's the kicker: it's not just their behavior that's different; their brains develop differently and even their genes are expressed differently. Early experiences can affect which parts of your DNA get transcribed, and by how much—in effect, deciding which part of your genetic operating instructions get read. Meaney's work shatters the old nature-versus-nurture debate—like the butterfly's wings in California affecting the weather in Japan, Meaney's work shows that the experiences during infancy can echo throughout a lifetime.

Critically, through the work of Meaney and others, we've come to realize that the old distinction between a "psychological trauma" and a "biological trauma" (such as a bump on the head) is a myth. Every event we experience, positive or negative, traumatic or uplifting, changes the neurobiology of the brain, for better or for worse, even in small ways that aren't noticeable (due to redundant pathways and neuroplasticity). That includes talk therapies such as CBT (cognitive behavioral therapy) and psychoanalysis, drugs (prescription or recreational), and music therapy. In other words, an event that leaves no physical scar on the brain can still alter the connectivity profoundly, and therapies can effectively reverse those changes or rewire the brain to create beneficial change.

From this, you might think that it takes a lot of trauma to change the brain, a major emotional upheaval, but no. Our responses to adverse experiences are subjective. It is impossible to compare them, or rank one person's trauma as worse than another's. We differ so greatly in how sensitive we are, how resilient we are, what kinds of internal and external support systems we have; we differ in the strength of our memory.

When something traumatic, that is, subjectively traumatic, happens to us, a series of internal protectors are set into motion, each serving its own purpose. For some trauma patients, entire sections of their brains shut down, going dark on scans like a single house on a city block losing power. For others, defense mechanisms cause increased vigilance and hyperactivation of fear centers in the brain. It might be that the trauma occurred because the organism let its guard down; in that case, it's an adaptive response to be more aware of what's going on in the environment. But this evolutionary response can turn ordinary awareness into hyperawareness, increasing the response of the amygdala and the HPA (hypothalamic–pituitary–adrenal) system, leading to a condition of chronic stress. This in turn can shut down pathways to the hippocampus and prefrontal cortex and contribute to PTSD symptoms. These defense mechanisms may have been helpful responses at the time, but to free the mind from the trauma and return to the present moment, the defense mechanism must be slowly disarmed. Therapies for trauma range from standard talk therapy to pharmaceuticals (with MDMA, ketamine, and psilocybin on the horizon), and increasingly, music.

Playing or listening to music can soothe us through the release of prolactin, can reset our mood through changing serotonin levels, and can motivate us to seek pleasure through modulating dopamine activity. Simultaneously, music stimulates neurogenesis and neuroplasticity, enhancing brain recovery, and normalizing the stress response. A hub of the brain's pleasure circuit, the ventral tegmental area (VTA), becomes active when people listen to pleasurable music, and the VTA is instrumental in releasing dopamine, which can reduce anxiety, improve mood, and provide a sense of relief for individuals struggling with PTSD.

A new topic of research is music and epigenetics. Genes contain the instructions for making proteins, which are essential for many biological functions, including those that support neuronal health and neuronal plasticity. One lab found preliminary evidence that listening to music may modulate gene expression and activity relevant to reward pathways. This so-called upregulation of genes may account in part for why music therapies can have lasting effects in cases such as PTSD, rather than wearing off as soon as the music stops.

Most of us recover from minor traumas or losses within a month or so—meaning we can pick ourselves up out of the doldrums and get on with our lives, even if not as before. The trauma morphs into post-traumatic stress disorder (PTSD) when it interferes with one's daily life for more than a month, and in particular ways. These include being easily triggered by something related to the original trauma (a loud noise, the scene of the event); re-experiencing the event via intrusive memories and flashbacks; losing interest in everyday activities; and being unable to think clearly. The incidence of PTSD (number of new cases per year) in the United States is five out of every 100 adults, and the prevalence (the current total number of cases) is more than 13 million—about four out of every 10 (the prevalence rate is much lower than the incidence rate because, fortunately, many people get better). But these numbers represent only diagnosed cases; many more individuals don't see a doctor, and many others who don't meet the full medical criteria for PTSD still experience profoundly negative symptoms, such as reduced quality of life, sleep disturbance, and being short-tempered. All of these together—for both those diagnosed and undiagnosed—increase the risk for other mental health conditions such as major depression and substance abuse, as well as cardiovascular problems.

The key to overcoming traumatic memories, including those that lead to PTSD, is to employ treatments to recontextualize them in a more neutral, less fear-inducing light before they become stored again, a process called reconsolidation. Cognitive behavioral therapy (CBT) and exposure therapy are two of the most common treatments for PTSD, both of which involve working through traumatic memories with the help of a trained therapist. Medications such as antidepressants and antianxiety drugs may also be used in conjunction with these therapies to manage symptoms. But a great many cases are resistant to these treatments.

One treatment for PTSD and other forms of trauma is writing music, as so many professional songwriters and composers have modeled for us. But what if you lack musical experience and have no idea how to write a song?

Rodney Crowell runs a songwriting camp (I co-taught one of them, along with Lisa Loeb, Joe Henry, Allen Shamblin, Brennen Leigh, and Bernie Taupin). Attendees include people who have never written a song before or even picked up an instrument, to recording artists who are looking for new ideas and techniques. For several years, singer-songwriter Cris Williamson has led a songwriting workshop for beginners at Esalen Institute near Big Sur, California called "Catch & Release." One of Cris's exercises has the entire group write a song together as she plays guitar, each person contributing a line of lyrics and/or a bit of melody. Workshop attendees come to get expert coaching on songwriting, and often find that the songwriting also serves as therapy for them, even when they weren't seeking it—and many of them feel happier and more in touch with their emotions.

Collaborative songwriting improves veterans' PTSD symptoms of avoidance, depression, hyperarousal and hypervigilance, and overall coping skills. Often, individuals are encouraged to write something like a personal theme song, something that reminds them of who they were or who they want to be. This can take the form of an affirmation, a personal mantra, a song that makes the individual feel safe, or one that makes the individual feel power and agency. Comparable to prolonged exposure therapy, the reductions in PTSD symptoms may be due to the collaborative songwriting intervention gently and repeatedly exposing veterans to an artistic reinterpretation of their trauma. Listening to their song over and over may allow veterans to habituate to their trauma experience, dampening the associated heightened physiological response and avoidance.

Group drumming is also a powerful tool; because it bypasses language centers, it allows us to let go of the central executive network and more easily slip in the default mode in which our body takes over and guides movements without explicitly having to think of them or plan them. It allows us to experience great creativity and individual variation while still feeling part of a group, the collective beat continuing and pulling us along with it. Soldiers participating in group drumming sessions reported increased feelings of unity, togetherness, belonging, connectedness, and openness.

Other music therapies for trauma are centered around listening, and they are among the most effective and transformative treatments for PTSD available. Immersing themselves in the emotional experience of music allows people to experience deep emotions aesthetically where they are less likely to be overpowering. They can gain insight into their own psychological states, and hopefully resolve those that are troubling.

A cautionary note is that music can occasionally trigger PTSD. A number of veterans report that when they hear music they were listening to in combat prior to a traumatic event, it brings up the mood they were in during the trauma. This trigger can put them into a hypersensitive and hypervigilant state. It doesn't just happen to people who were in combat. There are pieces of music that remind us of relationships gone bad, and the mere hearing of them can send us into a tailspin.

The U.S. Departments of Defense and Veterans Affairs have

joined forces with the National Endowment for the Arts to establish Creative Forces®: NEA Military Healing Arts Network. This partnership places music and other creative arts therapies at the heart of patient-centered care in clinical locations across the country, as well as in telehealth services. Creative Forces® is enhancing the health, well-being, and quality of life for active military service members, veterans, and their families and caregivers. Among the most prevalent health issues faced by military personnel and treated using music are traumatic brain injury (TBI) and PTSD, with a striking 23% of recent-conflict United States veterans experiencing PTSD.

At Joint-Base Elmendorf Richardson (JBER) in Anchorage, Alaska, the Creative Forces Music Therapy program tackles the challenges faced by patients with mild TBI/concussion, PTSD, and psychological health symptoms using a three-phase interdisciplinary approach. The first phase consists of a six-week group music therapy program that incorporates music-based relaxation, music making, lyric analysis, and songwriting. The second phase emphasizes active music making, with jam groups providing a supportive environment where patients can express emotions, engage in positive social interaction, and find solace from life's stressors through group singing and playing. The third phase involves a gradual shift from the patient role in music therapy treatment to more active participation in community music engagement, culminating in planning and performing live music events both on and off the base.

Sam (a pseudonym) was 52 years old when he entered the program; he had served in the U.S. Air Force as a staff sergeant. "During music therapy I felt like something woke up in me that was gone. I found something in me that I had lost." As part of the protocol, he began playing his original songs during community performances at a local veteran's hospital. "When I played one of my songs . . . a patient there began to cry as he connected to my song. It was humbling and rewarding for me that I made an emotional impact on his life." Sam describes how his songs restored the bonds between him and his wife. "When I shared my song with her, she said, 'This was so good and now I better understand what you went through.' I could tell her in my songwriting what I wasn't able to tell her in words alone." This is a constant refrain: music helps us express and process emotions that we cannot express in words, to others or even to ourselves.

Major life events, like moving or getting married or divorced, are stressful. Even if a move is intentional and positive, the number of resulting changes in daily life—leaving behind family, friends, a job, your favorite coffee shop—cause stress and pose extraordinary challenges. Imagine the added stress of being forced to flee an unsafe home, facing a difficult journey, and arriving in an unknown country. As of May 2023, 100 million individuals were forcibly displaced worldwide. That's more than the entire population of Germany! Two thirds of people displaced across borders come from just five countries: Syria, Venezuela, Afghanistan, South Sudan, and Myanmar. Recently, the war in Ukraine has displaced 8 million people within the country and more than 6 million across borders.

Refugees and asylum seekers face extremely traumatic experiences in their home countries, and during dangerous escapes from them. Leaving home results in multiple losses—identity, culture, family, support networks, and more; and results in severe psychological effects—post-traumatic stress disorder (PTSD), emotional disorders, anxiety, and general grief. Music is one thing that can remind displaced persons of home and family and their old life. It is the one thing they can take with them.

In New South Wales, the Service for the Treatment and Rehabilitation of Torture and Trauma Survivors (STARTTS) offers music therapy to Mandaean and Assyrian communities. At STARTTS, women in group therapy often sing popular and traditional music from Iraq. This empowers them to use their individual voices and deepens their sense of a shared cultural identity. Singing also has physical, emotional, and social benefits. The action of singing strengthens the lungs, improves posture, and increases oxygen flow to the bloodstream and brain. Singing releases endorphins, which can help reduce feelings of depression and anxiety, making people feel uplifted. Singing slow and soothing tunes facilitates relaxation, slowing down respiration and heart rates. Singing in a group is a great icebreaker for strangers who are thrown together, and allows people to overcome shyness and nervous reactions. The women at STARTTS are also given opportunities to lead the group in singing or improvising with instruments. Being the "conductor" reinforces their capacity to lead, to be in control, and to participate in the group's creativity—it restores the sense of agency they have lost as refugees.

Luxembourg's Music Therapy Association implemented a project at Red Cross shelters for underage asylum seekers and refugees. The youths there are encouraged to play noisy instruments to blow off steam, alternating with calming, soft sounds for relaxation. One young participant, a boy from West Africa, described an activity playing Boomwhackers (hollow tubes used as musical instruments), and the catharsis of being able "to make as much noise as possible by banging on the chairs 'Bam Bam' . . . [a] way to let off steam [to] forget for a short minute the situation we were living in . . . to free oneself on the one hand from various evils and on the other hand a call, a call of hope." He describes other activities as more meditative and restorative, including the use of Tibetan singing bowls. "A half-metal, half-velvet sound carries you away with your eyes closed and empties your head."

⌒

Songwriters, either intentionally or unintentionally, often embed their fears, longings, pain, and trauma in their lyrics, sometimes composing happy, uplifting melodies as a kind of antidote or counterpoint. Think of Johnny Marr's jangly guitars in The Smiths, a band whose lyrics embody the morose and nihilistic, or the upbeat, jaunty music of a song about a serial killer, "Mack the Knife." The very structure of popular songs supports this duality, when, for example, a verse tells a sad story, with the bridge coming in to bring brightness and hope. Often the verse talks about a situation in the here-and-now, and the bridge pulls you out of that corporeal world, into a timeless interlude that gets you out of the current moment. Perhaps it's a nod toward something ethereal and heavenly, pointing to a larger, broader truth or a moment of wisdom bestowed on the listener by the writer; the bridge gives you a release from tension and an assurance that you're on the right path.

Tracy Chapman's breakout hit, "Fast Car," beautifully captures this dichotomy. Listeners are drawn to the catchy bridge, "So remember when we were driving, driving in your car." The driver's arm around her shoulder, she had a feeling like she "belonged"—to the driver, to the moment. But the tragic path of the narrator's life is another story. She is at wit's end. She works at a convenience store and has managed to save enough for gas money to get to the next city, where she imagines a better life. Her father's drinking has aged him beyond his years; he no longer works. His wife has left him. The narrator, now alone with her father, expresses a mix of compassion and frustration. Someone has to take care of him, she says, "so I quit school, that's what I did." Those last four words may seem like a small flourish, a little afterthought tacked on. "That's what I did" adds volumes to the story—the resignation, the matter-of-factness. We realize that any talk of free will or the power to be who you want to be are fantasies in a life guided not by her own will or wishes, but by the grinding, narrow dichotomy of choices presented to her at every turn (the type of situation rendered in Robert Sapolsky's book *Determined*). Do I stay in school and watch my father, the only remaining member of my family, wither away? Do I quit school and subvert my chances for a better life?

She meets someone with a fast car, and proposes a deal: if they team up they can get somewhere—anywhere—to get out of this place they are stuck in. She's got the resolve to leave now, and fears that if she doesn't, she never will. They leave and she sings that miraculous bridge of driving away, at speeds so fast she herself felt drunk. The bridge provides an explanation of the force that's driving her. She's telling us: *There was love there, and that's what I'm after.*

I had this feeling once—that's what life's all about. I just want to live in that place again; I want to ascend from this turmoil and live in that life. It's another version of "Love lift us up where we belong" (from the song written by Jack Nitzsche, Buffy Sainte-Marie, and Will Jennings, and recorded by Joe Cocker and Jennifer Warnes for the 1982 film *An Officer and a Gentleman*). Tracy is striving for the feeling of belonging—not just to another person, but the freedom of finally belonging to herself and to the world. As Joseph Campbell would say, "a glimpse of eternity."

They finally make it to the city, she finds a job as a checkout girl, and they have kids. Her partner never gets a job, instead spending time (and her money) at a bar. The song paints a vivid picture of generational trauma cycling—a woman falling in love with someone who has her father's worst flaws. Tracy simultaneously draws on the themes of Sophocles, Homer, Faulkner's *The Sound and the Fury*, Tolstoy's *Anna Karenina*, and Toni Morrison. The heartbreaking moment of the song occurs when she realizes, yet again, that she is stuck. She musters her courage to tell her lover to take that fast car and just "keep on driving."

I must have sung along with that song for a decade, mesmerized by Tracy's voice, the guitar work, the melody. When it came on the car radio one day in August 2023, I felt like I was hearing the story for the first time. It made me think of all the people I know whose hopes have been dashed, their dreams cut short, and the trauma inflicted by fathers who drink. But it was more than that. That same week five young people in the United States were shot, in four separate incidents, all because they made the kinds of mistakes we all do—turning into the wrong driveway, knocking on the wrong door, getting into the wrong car in a supermarket parking lot, going into a neighbor's yard to retrieve a ball. I could not find a tear for these kids when each of the stories came out because I was not yet ready; but when the song came on the radio, the sickening horror of it welled up in me, the cap coming off the bottle of all that fermenting emotion. That is the power of a song.

Many songs mean something different to the songwriter than to the audience, and that is as it should be. Donald Fagen and Walter Becker of Steely Dan intentionally kept their words opaque to allow each listener to draw their own personal meaning. Sung with the intense conviction of Fagen's delivery, the cryptic words invite you to interpret them in your own way. Bob Dylan, in his autobiography, explained that he tended to write his songs in a way that would allow people holding opposite views to feel that the song spoke to them. Stephen Stills had an alcoholic father and, like many children of alcoholics, he found it difficult to love himself, and to see his own worth. Of one of his most popular songs, "Love the One You're With," he said, "Most people misunderstand the lyric; it was about loving yourself, about me wanting to love myself. Wherever you go, the one you're with is you."

J. D. Buhl was the lead singer in two different bands in the San Francisco area, The Believers and The Jars, both darting around the edges of punk, new wave, and rock. They had a following, which is to say they could play good clubs on weekend nights (rather than getting stuck with Tuesdays when no one goes out), and their songs were played on local FM radio stations. In 1985, J.D. began recording more introspective, more personal songs. He sang with a voice that reminded me of Danny Kirwan, bittersweet tones that found solace in a sweet, longing, wistful, damaged, and tender voice, resonating with haunting beauty.

J.D.'s lyrics as a solo artist were more poetic, the arrangements built up around acoustic guitar and voice, rather than the guitar, bass, and drums of his rock bands. Like many artists, J.D. had experienced childhood trauma, and used music both as a salve for his wounds, and as a way to explore his emotions without having to directly confront them. His music shows how effectively metaphor can be used in songwriting-as-therapy. What you can't come right out and say is easier to say when hidden.

One of his songs, "Unity," begins with a high-strung, capoed acoustic guitar strum, with a second acoustic guitar interwoven with it.

We close our doors in motorland
Take the neighbors for a drive
Arm out the window to flick an ash
I catch it right in the eye
They say go on please—get married raise families
But don't complain when the pain breaks your knees

The very thought of unity, now that's really something
The very thought of unity, sends 'em running scared

Dads on the porches at evening time
Just sipping beer from the can
The sky is blue like the bruises when you hear
Those words reserved for a man
They say go on kid—just like your brother and sister did
But don't come cryin' when you find what we hid

The very thought of unity, now that's really something
The very thought of unity, sends 'em running scared

We'd find baby birds who'd burned in the streetlight
Bury them in the bushes
The Shriner's circus, the swing set in the yard
Why did they hurt us? What was so hard?

We close our doors in motorland
Taking turns at the wheel
Learn that town like the back of their hands
I'm sure you know how it feels

On the surface, these are innocent, everyday images—driving in a car, sitting on a porch drinking beer, baby birds—and if they tell a story, that story is buried underneath poetic imagery. The underlying meaning is there in plain sight, but only if you know what you're looking for, a masterful callout to sexual abuse, hinted at, but not brought out in all its brutal ugliness, and the lifelong trauma that can haunt a person. J.D.'s was the work of someone who was confronting his past and his demons but simultaneously keeping them at bay. "The artists who hold our attention have something eating away at them, and they never quite define it, but it's always there," observed Bruce Springsteen.

Motorland was the monthly magazine published by the AAA (American Automobile Association), depicting scenes of a Norman Rockwell-esque, idealized suburban life—picket fences, lawns, kids playing happily in the street. With *We close our doors in motorland*, J.D. is telling us that the magazine's images of a utopian suburbia were *out there*, but not behind the closed doors of his abuser; not the world he came to know. As they *take the neighbors for a drive*, his abuser pretends everything is normal and fine. *I catch it right in the eye* is the first inkling of something wrong, until we get to *But don't complain, when the pain breaks your knees*. J.D. is recalling an image—on his knees, being forced into sexual acts as a minor. *Don't complain*, is the admonishment: what we are doing is normal. But tell no one. The lyrics express the mix of confusing signals so often given to children of sexual abuse who are still trying to figure out what normal *is*, what the world is like, what promises it holds. ("Gather In Your Promises" was another of J.D.'s songs.)

The gut-wrenching verse is the second one. There was alcohol involved; a clergyman (the "dad" in the lyric, not his own father). Older children he knew had also been subjected to the abuse, and he is exhorted to take it like a man (*go on kid, just like your brother and sister did*). Bruises are mentioned—they sound like emotional bruises, but we suspect more than that; we find out later in the last verse, the back of the hand was used to keep the abused in line. *What we hid* is the unspeakable, hidden from the neighbors, and from younger kids until they were old enough to be subjected to it themselves. In the bridge, he implores us, his listeners: *Why did they hurt us? What was so hard?* And all this talk of "unity"? The idealized vision of a loving church community in which the members unite to protect one another. *The very thought of unity* is the vanished promise of a dysfunctional, unfit, and broken extended family. The very thought of it sent them *running scared*.

When we think of PTSD, post-traumatic stress disorder, we tend to think of soldiers home from the atrocities of war. J.D. was one of countless children who were sexually abused by someone they knew and had trusted. His way of coping with it was by writing about it, obliquely, so that the casual listener would never know. Bathed in beautiful melodies, you can listen to the songs for decades, or a lifetime, and never know. How many of us really study lyrics? It's like the song that Ella sang (and messed up) in Berlin, a song that people jauntily and gamely sing along with, not ever knowing that it is about a serial killer. "Mack the Knife" is so bouncy and fun! J.D. confronted his demons by writing about them but not writing about them. Such is the power of art.

I know all this because J.D. asked me to help him write the music for this and a few other songs. He'd come to me with almost completed lyrics, and we'd sit around with our guitars, trying out different ideas, both of us singing whatever notes came to mind. He did most of the work; I think he just liked to have someone there, a kindred spirit, to bounce ideas off of. It never occurred to me to ask him what the songs were *about*. These were primarily *his* words; I was just providing musical accompaniment and occasional editing of the lyrics. I remember changing the word "streetlamps" to "streetlight" in the bridge because the long ī sings better than the short ă, and because a singular streetlight sounded more like a hopeful beacon than a group of streetlamps. When we wrote it, he sang *baby birds that burned*, but in the studio, on the spur of the moment, we changed it to *baby birds* who *burned*, a touch that humanized them.

I produced the songs for him, playing bass and guitar. Friends added violin, drums, keyboards, and background vocals. Without a record deal, the cassette tape was a calling card for J.D. to get booked into coffee shops and small clubs. I'd often join him in these shows. And we continued to write. There was one song I had written a few years earlier but was never happy with the lyrics. I fiddled with it and finally, at one of our writing sessions, I asked him if he'd like to rewrite the lyrics. He came up with "Just Like Someone Who Loves You," later renamed "Wings Cut from Fabric." I had no idea what the song was about when J.D. wrote it, even after performing it dozens of times together. I only learned about its meaning, and the meaning of "Unity," shortly before he died in 2017.

J.D. was raised with religion and a belief in God. For many years he worked as a high school English teacher at a parochial school. "Wings Cut from Fabric" begins *Take the sins of your father, even farther away, than he was from home.* The abusing clergyman was not away from a house, he was away from *home*, the safe place that every child needs to return to. *Make the face of your mother, eyes uncovered, run the other way.* His mother knew. She closed her eyes to the abuse. Why couldn't she open her eyes, pick up the kids, and run away from this horrible life? She was undoubtedly a victim of abuse, too.

Music kept J.D.'s spirit alive, kept *him* alive, helped him cope. He confided to me before he died of cancer that every day was a struggle between the memories of the past he was trying to forget and the hope for a future he tried in vain to create. He always wanted to marry, and did once, only very briefly. Relationships with women were fraught, awkward, painful.

Songwriting has also helped me to better understand my emotions about the loss of those close to me, and the breakup of relationships. As with most writers, I don't write *after* I've understood something, I write to help myself better understand it, and to better explore my emotions. I was very close with my paternal grandfather Joe, a scientist, radiologist, and music lover, who took a loving interest in educating me. By the time I was six, he had already taught me Greek and Roman history, biology, and chess, and instilled in me curiosity and a love of Big Band music. He also taught me how to shut off the plumbing for the winter in the summer house, and how to clean the swimming pool. He taught me to swim, for safety's sake buckling me into a child-sized life jacket. I remember so clearly the first time I wore it in the pool—the sheer joy of the buoyancy, the effortlessness of being able to float along before I knew how to stay up by myself. Grandpa Joe died abruptly during a botched surgery when I was ten. At the time, my mother and father were building their careers, and the loss of his attention and companionship left a big hole in my life. He had been 56 when I was born. When I was that same age, I was going through an old photo album and saw a picture of myself in the pool, wearing that life jacket, and a verse suddenly popped into my head. Over the coming months I wrote a song that I eventually called "Headed for the Fall":

> *When I was a child I never learned to swim*
> *But I jumped right in that water if I was buckled in*
> *You know that life life jacket held me oh so high*
> *And I'd ride and glide and go so far as it held me by*
> *Headed for the fall*
> *Down down down 'til I start to lose it all*
> *Headed for the fall*

I played the song for Joni Mitchell shortly after writing it. She delightedly pointed out what I hadn't seen, peeling back the layers of the onion. The life jacket is a metaphor for my Grandpa. It was *he* who held me up, spiritually, artistically, and intellectually. When he died, I felt untethered, I sank, she ventured, and she recited the poem "Dying Speech of an Old Philosopher" by Walter Savage Landor (and she approved when I worked in a phrase from that poem, "it sinks and I'm ready to depart," into a new song of my own). The "fall" can mean different things to different listeners—a waterfall, a falling off course, a biblical fall from grace and innocence, a fall into deep depression. As Nick Cave observed, "The thing about writing a song is that it tells you something about yourself that you didn't know."

Joni insisted I record the song. But with what kind of an arrangement—solo guitar and voice, band? One of the important lessons Joni taught me: if I can find the authenticity in the song and what it means to me, it can be therapy for me, as her songs have been for her. That authenticity requires not just lyrical honesty but musical honesty as well. Grandpa Joe loved big bands. I loved rock. The right musical treatment would marry those sensibilities. My chords were already taken from jazz harmonies, not pop: the basic riff was B minor 7, F-sharp augmented 7, G-sharp augmented 7 (B-7, F#+7, G#+7). And so the choice was obvious, and Joni and I blurted out the answer at the same time: Steely Dan. I brought in a jazz drummer, keyboardist, and bass player who, like me, were Steely Dan fans. I dusted off my tenor saxophone and added some horn lines, and played vibraphone, a mainstay both of Steely Dan and the Lionel Hampton records Grandpa Joe had in his collection. Whereas before, thinking of my grandfather made me sad, now when I hear the song I feel uplifted. I also feel gratified, a sense of accomplishment. Even if no one hears the song but me, it pleases me, like a doodle on a pad of paper or a sandcastle, it reminds me that we are, all of us, creative beings. And the act of creating something—anything—is a potent elixir.

Chapter 9

Mental Health

"AND AS I PLAYED, I UNDERSTOOD THAT THERE WAS A profound change occurring in Nathaniel's eyes. It was as if he was in the grip of some invisible pharmaceutical, a chemical reaction, for which my playing the music was its catalyst. And Nathaniel's manic rage was transformed into understanding, a quiet curiosity and grace. And in a miracle, he lifted his own violin and he started playing, by ear, certain snippets of violin concertos which he then asked me to complete—Mendelssohn, Tchaikovsky, Sibelius. And we started talking about music, from Bach to Beethoven and Brahms, Bruckner, all the B's, from Bartók, all the way up to Esa-Pekka Salonen."

Robert Vijay Gupta, a violinist with the Los Angeles Philharmonic, is speaking about Nathaniel Ayers. Ayers was a brilliant violinist whose career was cut short by debilitating schizophrenia. Unhoused, he lived on the streets of Los Angeles in Skid Row (so-named because the people who live there tend to be downtrodden and impoverished, "on the skids"), a desolate area southeast of downtown that is encircled by three major freeways, the 110, 10, and 101. *Los Angeles Times* journalist Steve Lopez discovered Nathaniel in 2005 playing beautiful music on the streets, playing like no one was watching; playing like a master. Lopez's subsequent book, *The Soloist*, was turned into a movie starring Jamie Foxx and Robert Downey Jr.

Gupta describes the transformation that Ayers underwent, from a paranoid, deeply disturbed man to an artist—but only while playing or talking about music. It pushed the schizophrenia away, if only for a while. "For Nathaniel, music is sanity. Because music allows him to take his thoughts and delusions and shape them through his imagination and his creativity, into reality. And that is an escape from his tormented state."

The diagnostic criteria for schizophrenia include delusions, hallucinations, disorganized speech, jumbled or confused thinking and speaking, and grossly disorganized or catatonic behavior. Affected individuals display blunted affect, emotional withdrawal, poor rapport, passive/apathetic social withdrawal, and lack of spontaneity and flow in conversation. Their auditory hallucinations can be very vivid and pronounced, causing them to talk to someone who is not actually there. The standard treatment is pharmaceutical (drugs such as chlorpromazine, aripiprazole, clozapine, or risperidone); many individuals who are treated with pharmaceuticals go off their medication either because once they start feeling better they (falsely) believe they are cured, or because they don't like the side effects.

We still don't know what causes schizophrenia. It is both heritable and maladaptive, and so the question is why selective pressures have not eliminated the hereditary factor. Maybe it's because physical and cognitive diversity are requisite for adaptive behavior, and a cornerstone of evolutionary theory is descent with modification—so we should expect *some* incidence of schizophrenia based on that. But descent with modification cannot account for the relatively high prevalence, around 1% of the population. A leading theory is that the same genes or gene combinations that confer desirable traits, such as divergent thinking, creativity, or loquaciousness also increase the risk for schizophrenia. If you consider this from a population genetics standpoint, the benefits of having *those* qualities in a large number of people may outweigh the disadvantages of some people developing the disease (genetics only "cares" about the good of the many, not the good of the few). This trade-off is known as *balancing natural selection*.

Current theories are that schizophrenia, and some psychoses, arise from overproduction of dopamine, particularly as it affects D2 receptors in the mesolimbic pathway. Current treatments typically involve D2 inhibitors (receptor antagonists) such as molindone, sulpiride, and promazine. (I sometimes wonder if the people who name these drugs are themselves experiencing hallucinations.) We know that increased levels of dopamine can improve motivation, focus, and the maintenance of attention, and it can help people with Parkinson's disease to walk. So why might increased levels of dopamine lead to schizophrenia?

It turns out that dopamine does different things in different parts of the brain. In the prefrontal cortex it helps us maintain focus and avoid distraction. In the limbic system, dopamine increases motivation, the drive toward pleasurable activity. In the basal ganglia, dopamine is released from neurons originating in the substantia nigra and acts on various substructures (e.g., striatum, nucleus accumbens) to regulate motor functions and other behaviors. Adding to the complexity, every time dopamine levels in the brain change, it is likely that the levels of some of the remaining 99 neurochemicals will change in response.

As if it all weren't complicated enough, we have identified five different dopamine receptors:

Dopamine Receptor	Family	Main Function	Distribution in the Brain
D1	D1-like	Involved in neural signaling, particularly in reward-related behaviors, learning, and memory.	Widely distributed in the brain, particularly in the striatum, nucleus accumbens, and prefrontal cortex.
D2	D2-like	Regulates neurotransmission and modulates motor activity. Antipsychotics target D2 receptors to alleviate symptoms of conditions like schizophrenia.	Found in high levels in the striatum, nucleus accumbens, and olfactory tubercle. Also found in the hypothalamus and pituitary gland.
D3	D2-like	Thought to have a role in mood, reward, and addiction. Some antipsychotics also target D3 receptors.	Primarily located in the islands of Calleja and nucleus accumbens. Less densely populated in other brain areas.
D4	D2-like	Implicated in novelty-seeking behaviors and attention deficit hyperactivity disorder (ADHD).	Widely distributed but in low density. Found in the basal ganglia, frontal cortex, midbrain, amygdala, hippocampus, and hypothalamus.
D5	D1-like	Thought to be involved in learning and memory, as well as motor control.	Mainly found in the hippocampus, hypothalamus, and cerebral cortex.

The theory is that individuals with schizophrenia somehow have an abnormally high density of D2 receptors in their mesolimbic system—and that this makes them overly responsive to dopamine, leading to hyperactivity of certain circuits. Overactivity of the mesolimbic system could contribute to increasing the salience of irrelevant stimuli (such as internal voices); promote the formation of incorrect associations between unrelated ideas (a voice talking to you and an action plan you might make); and overall disorganized thinking resulting from hyperactivity in these circuits.

Music therapies have been used to treat schizophrenia for decades. The evidence, including that from meta-analyses, tells us that *any* form of music therapy will likely have some beneficial effect, and that any delivery method—individual or group music therapy, passive listening, or active participation—yields equivalent results. The strongest evidence shows that music can improve mood and increase motivation and, to a lesser extent, that it can reduce the duration and intensity of hallucinations. A recent Cochrane Review determined the evidence from 18 studies to be of "moderate to low quality," and recommended further research with more careful experimental designs.

Nonmusical auditory interventions have also been found to help suppress the effect of auditory hallucinations. In one case, a 60-year-old retired security guard heard the voice of an unknown male immediately behind him telling him to hang himself, while he was in an otherwise lucid state. (The man had indeed made five near-fatal attempts to kill himself over the previous 40 years.) To combat the voices, on his own inspiration, the man bought a portable cassette recorder and recorded his own voice shouting back at the hallucinated voice, which he played through earphones when necessary. That didn't help. Next, in his normal voice, he recorded his memories of pleasant times with members of his family, work, and holidays. During the 35 days he was observed by clinicians, the frequency of his hallucinatory bouts did not change, but their duration and severity did. He continued to use the therapy until he died of heart disease.

Many musicians find relief from mental health challenges through making their own music. Neil Young is afflicted with earworms—those snippets of songs that run circles around in your brain and drive you nearly to the point of madness. Unlike the earworms most of us have, of songs we've heard before, his are of songs no one has written yet. He writes them and records them to help get them out of his head.

Quincy Jones expressed how music helped in his early struggles with depression. "Music was the one thing I could control. It was the one world that offered me freedom. When I played music, my nightmares ended. My family problems disappeared. I didn't have to search for answers. The answers lay no further than the bell of my trumpet and my scrawled, penciled scores. Music made me full, strong, popular, self-reliant, and cool."

Bruce Springsteen has spoken openly about his battles with mental health, and the steadying, centering ability of music. "I think that's why you get into it. You're in pursuit of a certain sort of peace that's very, very, very difficult to come by. I realized that the only time I felt complete and peaceful was while I was playing or shortly afterwards. It was the first way that I medicated myself, so I always went back to it."

In May of 2023, Guns N' Roses bassist Duff McKagan released "This Is the Song" to mark Mental Health Awareness Month: *Tried Lexapro, and what else I don't know . . . This is the song that's gonna save my life.* In an open letter to fans, McKagan shared that he has suffered from panic disorder since he was 16, and during the coronavirus pandemic it "morphed and twisted and brought along some darkness that seems to appear out of absolutely nowhere." He wrote the song *during* a panic attack. "I have thankfully found my acoustic guitar as a refuge. If I just hold on to that guitar, play chords and hum melodies, I can start to climb my way out of that hole."

Tchaikovsky's diaries are a revelation. Entry after entry echoes these themes:

"May 21, 1884. Moscow. Feel a kind of depression."

"May 3, 1887. Walked with Alesha about the garden. The moon in all its glitter. Indefinite feeling of loneliness and depression."

Tchaikovsky was depressed most of his life. He felt rejected by his mother and was ashamed of his sexual feelings toward men. He expressed in one of his letters, "I hate myself. Only work saves me." Yet, he is responsible for some of the most beautiful and romantic music ever written: *Swan Lake. The Nutcracker. Music for the Young, Op. 39, No. 3: Mamma.*

Are the deep roots of music in the brain associated with neurodiversity, and, in particular, to what we today would call psychiatric disorders? We realize that a lot of brilliant musicians and composers have been depressive, suicidal, antisocial, and chaotic (think Keith Moon trashing hotel rooms). Is there a correlation? The putative connection between creativity and mental disorders has been noted for thousands of years. Aristotle said, "No great mind has ever existed without a touch of madness." Perhaps in the way that an aircraft landing is, at its core, a "controlled crash," genius is nothing more than "controlled madness." For what we consider genius requires, by definition and common understanding, a break with convention—an ability to see things that no one else saw before. A number of recent epidemiological studies have confirmed an overlap between general psychiatric disorders and creativity, and genetic studies provide strong evidence for a link between schizophrenia, bipolar disorder, and higher creativity activities, including music. Neuroscientist Laura Wesseldijk of the Karolinska Institute in Stockholm found that individuals with major depression and bipolar disorder were more likely to play music, practice music more often, and reach higher levels of musical achievement.

With physical ailments, we can more readily seek treatments such as medicine, and embrace treatments such as physical therapy. By definition, mental ailments involve some amount of disordered thinking, and that can make it difficult to seek treatment in the first place, or to continue with it once it's started. A huge problem for people who take medication for depression is that they feel it has dulled their senses, or made them anhedonic, or that they are no longer themselves, imposing a kind of self-identity disorder. Another problem is compliance (as with schizophrenia): once people have been taking an antidepressant long enough for them to start feeling better, they assume they are cured and then stop taking it—only to slip back into the depression. And once depressed, many people lack sufficient motivation or energy to pursue or re-pursue medication. The advantage of music in these cases is that it is an everyday activity that we pursue anyway, that lacks stigmatization, and that is easy to continue once started.

Eating disorders often arise through feelings of helplessness and lack of control over one's life; one of the few things we can control is how much we eat. This poses a special barrier to treatment because seeking treatment requires handing over control to someone else, the precise thing that patient is avoiding. Arts-based approaches, including music therapy, have gained prominence in recent decades. Although there exist over 1200 studies of music and eating disorders, all but two are case studies, or lack key experimental design features, such as a proper control group. Of course, this does not mean that music therapy is ineffective; rather, it means that we don't have enough evidence to draw any firm conclusions.

Case studies suggest that combining cognitive behavioral therapy and music listening for eating disorders yields greater improvements in psychophysiological factors (psychopathology, anxiety, and depressive symptoms), and in physical manifestations (weight, body mass index, and abdominal subcutaneous fat thicknesses) than conventional therapies alone. Well-being and psychosocial development of people with eating disorders can be increased using group music therapy with song-focused approaches (singing, discussion, and song writing), individual song writing, and guided imagery with music (GIM, see appendix). The *iso principle*—selecting music to match a client's mood (by having them choose from a playlist) and then changing the music to alter the client's mood along a continuum from depression to hopefulness—may help patients to cope when living with disordered eating, depression, and anxiety.

Another area requiring further research is the use of music for treating substance misuse. The paradox here is apparent: music may be effective for treating substance abuse, and yet musicians' own substance abuse tendencies are well documented. Most studies thus far have examined only the results of a single session of music therapy, rather than across the course of treatment; many others have lacked randomized control designs. How does this happen? Too often experiments are designed and conducted by people who are well intentioned but lack specific training in the science of experimental design. Based on anecdotes, case studies, and animal models, however, music therapy can reduce substance use, improve psychosocial well-being, and can lead to neuroplastic changes that disrupt the memory for patterns of addictive behaviors. Music therapy seems to reduce craving—possibly through modulation of dopamine, serotonin, GABA, and other neurochemicals. Both music and drugs alter our emotions, memories, and states of consciousness; while this may contribute to abuse in the first place, it may also facilitate the therapy for addiction.

I have touched on music's ability to relieve symptoms of depression and anxiety in other conditions (e.g., motor disorders, dementia, trauma, and stroke). But what role might music play when these symptoms are the primary complaint, not secondary to some other disorder or disease? Music therapy is well established as an effective stand-alone intervention and adjunctive treatment for depression in adolescents, adults, the elderly, and mothers experiencing postpartum depression. Specific music therapy techniques for depression include musical negative mood induction procedures, drumming and percussion exercises, group music therapy, musical improvisation, five-element music therapy (a Chinese therapy using tones to balance energy in the body and mind), and cognitive behavioral therapy combined with music therapy. Some techniques rely on trained facilitators, while others incorporate passive exposure to certain kinds of music. Some are individualized, others involve group participation. It is unclear which techniques and types of music are most effective at alleviating depressive symptoms in which individuals. Discovering the optimal application of music therapy remains a compelling challenge for future research.

Anxiety is a common emotional state that many of us experience when we're stressed or worried about things happening in our lives. As Bruce Springsteen puts it, "Dread is an emotion that all of us have become very familiar with." (And he hopes his record *Letter to You* is "a little bit of an antidote to that.") It is a normal part of human existence to feel anxious. But when that anxiety becomes excessive and interferes with our daily functioning, it can lead to long-term health risks driven by cortisol toxicity.

Traditionally, cognitive behavioral therapy and pharmacological treatments have been used to manage anxiety, and they can be effective for many individuals. However, medications often come with unwanted side effects like gastrointestinal discomfort, insomnia, and headaches. And while cognitive behavioral therapy helps individuals identify and change their negative thought patterns and behaviors associated with anxiety, not everyone responds to it, and relapse rates can be high, especially in children and adolescents. Adherence to therapy and medication regimens can be a challenge. Listening to or creating music can have a calming effect on the mind and body, helping to alleviate tension and promote relaxation. Music therapy offers a non-pharmacological and potentially accessible intervention for those who may not respond to traditional treatments or who prefer alternative approaches. It is encouraging to see therapists exploring the potential of music therapy and its positive impact on anxiety management.

The neural and biological mechanisms underlying music's ability to reduce depression and anxiety are complex and not fully understood. One explanation revolves around the emotional and psychological regulation elicited by music. As an individual listens to enjoyable music, the brain's reward pathways are activated and neurotransmitters like dopamine are released. This may enhance mood and alleviate symptoms. Another factor lies in music's stress-reducing effects. Engaging with music activates the parasympathetic nervous system, promoting relaxation and lowering heart rate, blood pressure, and cortisol levels—hormones related to stress. Music can also serve as a distraction and engage cognitive processes. Whether through singing, playing an instrument, or actively listening, music diverts attention away from negative thoughts and rumination. This redirection may disrupt cycles of depressive or anxious thinking, offering a respite from symptoms. Neuroplasticity, the brain's capacity to adapt and reorganize itself, also plays a role. Exposure to music has been found to enhance neuroplasticity, leading to changes in neural connectivity. Specifically, regions involved in emotion regulation, memory, and executive functions may undergo modification. These changes contribute to improved emotional well-being and a reduction in symptoms. Music's social and communal aspects can also impact mental health. Engaging in musical activities with others fosters social connections and a sense of belonging. These social factors are integral to mental well-being.

Prisoners on death row experience unimaginable levels of anxiety, fear, and despondency. Bryan Stevenson, in *Just Mercy*, writes of visiting Henry, an innocent man stuck in the labyrinth of the U.S. judicial system. Henry had been on death row for over two years and still didn't have a lawyer to take his case. The facts stand for themselves, regardless of the broader debate about the death penalty's constitutionality or morality. Since 1973, 196 people who had been wrongly convicted were sentenced to death and have been exonerated. A 2014 study published in the *Proceedings of the National Academy of Sciences* found that at least one out of every 25 people on death row is likely innocent.

When Bryan's meeting with Henry was over, a guard came and roughly shackled Henry's hands and ankles; the guard was angry that

Bryan had stayed longer than agreed, and he took it out on Henry by putting the cuffs on so tight that Henry grimaced in pain. Then came the chains. "With each click of the chains being tightened around his waist," Bryan recalls, "I could see him wince." The guard then—with no provocation—roughly manhandled Henry toward the door. At the threshold, Henry closed his eyes, and tilted his head back. "I was confused by what he was doing, but then he opened his mouth and I understood. He began to sing." In a beautiful, sonorous baritone voice that caused the guard to freeze in his tracks, Henry sang an old church hymn: *Lord lift me up, and let me stand by faith on Heaven's tableland. A higher plane than I have found, Lord, plant my feet on Higher Ground.*

Bryan writes that he experienced the song as a precious gift. In a dark moment, being led from a respite of unshackled freedom, Henry was now being forced back to the death row cell block. And in that moment, Henry's first response was to turn toward beauty. In that moment, Bryan writes, "he gave me an astonishing measure of his humanity . . . [and] altered something in my understanding of human potential, redemption, and hopefulness."

Chapter 10

Memory Loss, Dementia, Alzheimer's Disease, and Stroke

O NE AFTERNOON AROUND 3:30, GEORGE WAS LYING beneath the kitchen sink, his back pressed against the floor, his hand firmly grasping a plumber's pipe wrench, his gaze transfixed upward—a sight both arresting and enigmatic—when his daughter walked in.

"Are you okay, Dad?"

"Yes. Just fine."

His daughter crouched down to be at eye level with him. "What are you doing?"

After a pause, he said, "I don't know. I was looking at this thing in my hand" (the pipe wrench) "and I couldn't remember what I was doing with it."

Thinking he might be dehydrated, she gave him a cold Gatorade from the refrigerator. He remembered he was trying to fix the sink.

George experienced the initial stirrings of Alzheimer's disease when he was 69. The early symptoms began subtly, manifesting as bouts of forgetfulness and an uncanny propensity for misplacing objects. He'd venture into a room only to find his purpose eluding him, an experience familiar to many of us at any age. Yet a peculiar pattern emerged from George's lapses, primarily unfolding in the late afternoon and early evening—a phenomenon recognized as *sundowning*, a behavioral marker often associated with Alzheimer's disease.

Initially, these developments brought only a minor disruption to George's daily life. He continued to revel in the laughter of his grandchildren, to delight in his daily walks, and applied a deft and skillful touch to fixing leaky faucets and squeaky hinges.

No one gave the wrench incident much thought, chalking it up to dehydration. And anyway, all of us have momentary memory lapses. When we're younger, we attribute them to distractions or an overwhelming schedule. It's only when we're older that we start thinking, "Oh no! Alzheimer's!" Only about 11% of adults will get Alzheimer's, and women are twice as likely to get it as men, as are older Black Americans of both sexes. The probabilities increase with age: about one in three people over 85 years old have it. Age remains the number one risk factor for Alzheimer's. George was still relatively young, with the odds in his favor.

And then the pattern broke. When he was 72, he wandered off. His family eventually found him walking toward the old aerospace plant where he had worked. A neurologist diagnosed Alzheimer's disease. The disease progressed swiftly and significantly. At age 78, George became noncommunicative and unable to walk; he just sat. He entered assisted living, which, his neurologist recalls, turned out to be a blessing. He was finally able to socialize, albeit nonverbally. Music was often playing in the facility and he was singing along, and singing with other residents. Alzheimer's had stolen his speaking voice; he could only utter a scratchy yes or no. But he could sing when the music played as if he were 30 years old again.

Today 55 million people live with dementia worldwide. You may have heard that "children are our future." They're not—old people are. Lots of them. By 2035, there will be more people over age 65 than under age 18 for the first time in U.S. history. In Japan, more diapers are purchased for people over 65 than for people under 5, and the rest of the world is not far behind. With the aging population will come a greater number of dementia cases. Age remains the primary risk factor for dementia, followed by how many times you've been hit in the head (repeated head trauma raises the risk by a factor of four).

Keith Jarrett is one of the most important jazz pianists and composers of the last fifty years. Most of his compositions are improvisations, in fact, spur-of-the-moment pieces of great scope and complexity. His album *The Köln Concerts* is among the best-selling jazz albums of all time. It takes one set of skills to sit down and write complex, beautiful music when you can take your time doing it. What Jarrett does requires an entirely different talent, improvising highly structured music with a well-developed coherence and emotional story arc. The amount of creativity that Jarrett possesses—the amount of *music* in him—is astonishing. But in 2018, Jarrett suffered two strokes in succession, leaving the left side of his body partly paralyzed. Before the strokes, music blew through Jarrett with a powerful and inspired force, lifting him up and us with him. Following the strokes, his right hand could still play, but the great pianist found the well of inspiration was dry. Where did all that music go?

Maybe it's a bit like writer's block—even prolific writers report days or long stretches when nothing comes out, and then the words come out quickly, seamlessly, flowing like water. But Jarrett's case is different from writer's block because after several years no dam broke, no river of notes gushed out. No, maybe Jarrett's case is more like a dancer trying to imagine dance; a dancer may be able to work out the choreography as a set of instructions and images, but actual dance wholly depends on movement. Reading a review of a ballet or a music show is a poor substitute for the experience, almost pointless. (As the comedian Martin Mull famously quipped, "writing about music is like dancing about architecture.") A few composers—Bach and Mozart come to mind—work everything out in their capacious musical brains. Others rely on the touch and feel of the instrument, or some combination of both, in a feedback loop, during which a thought comes out through the fingers, goes back into the ears, and triggers another musical idea. That was Jarrett at his finest, and without that feedback loop, from muscles to ear to brain and muscles again, he can find no music.

The idea that music and dance are intimately and intricately related to movement is part of the theory of embodied cognition. Embodied cognition states that we come to learn about, know, and understand the world through interacting with it. Miles Davis played a note. As he heard it reverberate throughout the hall, he listened to the other musicians. Miles gave us permission to perform music through collaborative empathy. His knowledge and plans were influenced by what just happened, and that was expressed through his lungs, his fingers, his tongue, his lips. Jarrett's music was born of a similar state of flow, in which his brain, fingers, and ears become one. Speaking about those classic performances at the time, Jarrett said, "I don't have an idea of what I'm going to play, any time before a concert . . . If I have a musical idea, I say no to it." The *doing* of musicking is essential to the creating, and just thinking about it is not the same as experiencing it.

In 2020, he spoke to the *New York Times*' Nate Chinen. "When I hear two-handed piano music, it's very frustrating, in a physical way . . . I was trying to pretend that I was Bach with one hand . . . But that was just toying with something." Chinen observes that "When Jarrett tried to play some familiar bebop tunes in his home studio in 2020, he discovered he had forgotten them."

But what a difference two additional years of rehab makes. *Downbeat* visited Jarrett at his home studio in early 2023. "Jarrett jams with his one functional hand . . . [his] uncanny touch, capable of eliciting extra groove juice and resonance from the piano, became evident as he raced through one of his favorite tunes, Oliver Nelson's 'Butch and Butch,' followed by an unabated surge of rococo ideas on 'It's Alright with Me.' A grand finale to this impromptu set, which also included 'Sioux City Sue New,' included Jarrett's wife, Akiko, punctuating bass notes on the kind of rambunctious blues with which her husband would reward audiences around the world after challenging them with protracted bouts of invention."

Yet even with this rediscovered facility, Jarrett plays only a couple of times a month, and he derives little pleasure from it. "Jumping off a cliff takes two hands and two feet. Now I'm using half the piano, half my ability. I don't think I would write anything." The strokes didn't affect music centers per se, but Jarrett's unique brand of improvisatory writing and performing—bound together in a single act of inspiration—was rendered impossible by his inability to actually *do* the music.

Each year, around 12.2 million people worldwide experience a stroke, leading to roughly 6.5 million deaths and rendering 5 million individuals disabled. Strokes can be either ischemic—caused by a blood clot—or hemorrhagic—due to a ruptured blood vessel leaking blood into the brain. The repercussions can range from negligible to life-altering, influencing not only the lives of patients and their families, but also society in general. Strokes can lead to specific functional impairments such as paralysis (as in Keith Jarrett's case), aphasia (loss of language ability), or impulsiveness. After a major stroke, the odyssey of recovery unfolds like a meandering, sprawling, unpredictable play, in which some characters find themselves reclaiming the roles they once played, and others confronting the unyielding truth of a life forever marked by disability. The path and the results are as varied as the complexity of individual differences.

Another masterful improviser who, like Jarrett, played with Miles Davis is the saxophonist Sonny Rollins. In 2013, at age 83, Sonny was diagnosed with pulmonary fibrosis, and he has been unable to play the saxophone since. With the loss of the physical ability to play, he lost his desire to listen as well.

> I stopped listening to music some time ago. Over 10 years—well when I say *listening*, of course I hear music, you can't help but hear music. But I don't purchase music . . . I've listened to so much music in my life so that I don't really listen to much music.

These three different cases illustrate the variety of ways that aging interacts with music. George and Keith Jarrett each experienced a loss of brain function, with very different outcomes. Paradoxically, as the harmonies of George's mind became increasingly dissonant, he found himself irresistibly drawn closer to music, which became a social lifeline for him. Keith Jarrett's paralysis—and possibly neural damage we don't know about—rendered him less able to draw on his prodigious musicality. Sonny Rollins's brain was flourishing, but the mechanical loss of playing ability caused his interest in music to quietly fade.

Dementia affects multiple systems of the brain's intricate machinery, its many faces hidden behind a single umbrella term. It is not a singular affliction, but a gathering of cognitive declines. The origins of dementia are varied, with Alzheimer's disease the most common culprit—although not everyone with Alzheimer's ends up with dementia. Strokes can create a similar kaleidoscope of cognitive splintering, and can also cause physical hardships such as hindered movement, muted speech, and the enigmatic experience of hemispatial neglect—ignoring one side of the body.

But dementia is a more profound state than simple confusion. Dementia is not losing your car keys or even losing your car—it's when you look at your keys or your car and can't remember what they're for; it's not forgetting your child's name, but talking with someone you don't realize *is* your child.

I recently witnessed a tender conversation between a man with dementia and his daughter. He didn't recognize that she was his daughter, and he appeared to think she was just a stranger visiting him. She asked him questions about his life, and he enjoyed recalling his memories. He described his two daughters as being "beautiful girls . . . wonderful . . . intelligent, quick and sharp" and "everything a father would dream for in a daughter. They're just the best." All the while not realizing to whom he was talking. His daughter silently cried.

We live under a false societal narrative that people with dementia are so far gone that we cannot provide meaning to them and they cannot provide value to us. We are afraid of engaging them, and uncomfortable being around such disorientation. To indulge this fear would be to potentially miss out on some of the most meaningful moments of a life. Although some patients undergo negative personality changes, becoming irascible and irritable, many others become even sweeter and purer versions of themselves.

Whether it's from Alzheimer's disease, non–Alzheimer's dementia, or cognitive manifestations of a stroke, we may find cognitive decline, memory loss, and impairments in visual-spatial abilities, executive functions, and communication. These losses can inevitably lead to the troubling behavioral and psychological symptoms of dementia including depression, anxiety, apathy, agitation, problems with emotional control, sleep disorders, and challenges to living independently. Among Alzheimer's disease, mild cognitive impairment, or dementia patients, music has been found to be effective at treating disruptive behavior, anxiety, and depression, and is linked to improvements in quality of life and cognitive function.

Part of the tragedy of these conditions is that the patient often lacks insight into the cause, provoking further disorientation. In this way, afflictions of the human mind stand alone, a recursive cognitive blind spot that we don't even know we have. If you break your leg, or are congested due to allergies, you *know* why you are experiencing symptoms. That may not make them any more pleasant, and you may still feel anxious, but you are cognizant of what is happening to your body and why. Thought disorders, especially memory loss, add to biological injury the further insult of helplessness and confusion: why am I feeling this way? *What is going on?* Memory networks in the hippocampus are tightly coupled with emotional networks in the adjacent amygdala. When neural degradation of these circuits occurs, it creates a chaotic circus. As our minds attempt to navigate this hall of distorting mirrors, we fail to regulate emotions in response to forgetfulness and disorientation. This further amplifies our unease.

The disorientation brought on by these various symptoms understandably causes agitation and anxiety. Indeed, chronic agitation is one of the biggest challenges for patient care in long-term care homes. By themselves, agitation and anxiety constitute outsized risk factors for further cognitive decline and shorten life expectancy. Front line treatments for agitation rely on drug interventions that have potent sedative effects, further increasing disorientation while dulling the patient's emotional response to it. (One of these, haloperidol, increases mortality.) We've known for decades that music is just as potent as drug treatments for relieving anxiety, but getting it into clinics and care facilities has been a bumpy road. Fortunately, now, that road is being widened, graded, and paved.

My former PhD student Adiel Mallik is now working in the laboratory of my friend and colleague, Frank Russo, at Toronto Metropolitan University. Recently they published an important paper called "Developing a music-based digital therapeutic to help manage the neuropsychiatric symptoms of dementia." Much of what they say in the paper applies broadly to all forms of anxiety and agitation, pointing an arrow toward musical medicine for relaxation.

Extending my earlier work on mapping the musical brain, they've developed a model of the "relaxation network" of the brain, showing how neurodegeneration can result from the failure to relieve stress and anxiety, underscoring the importance of non-pharmacological musical interventions that promote relaxation. Russo explains, "The premise is that agitation in dementia is an outward manifestation of anxiety and that the same music-based approach we've used to mitigate anxiety in young adults may be able to reduce agitation in this population."

When the network and hubs in the figure are degraded, that can lead to the symptoms of anxiety and agitation. You may recognize the names of some of these regional hubs. There's our friend the putamen, involved in motor movement, and regions of the cingulate cortex involved in daydreaming—the Default Mode Network. And what better way to relax than to daydream? But if that hub is damaged, we become like walking insomniacs, unable to enter the relaxing daydream state. In the special case of

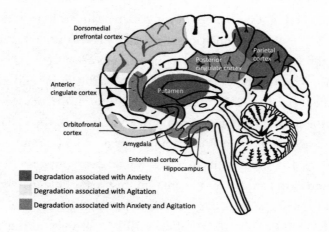

Russo's and Mallik's model of the brain's "relaxation network."

dementia-induced agitation, the precuneus is involved as a disruption in self-awareness, and it is subsumed (though not illustrated) in Russo and Mallik's model.

A major finding in music neuroscience over the past decade is that self-chosen music (chosen by the listener) is more effective at achieving a desired state than music chosen by others. But there may be some wiggle room here depending on the desired state. For some behaviors, like waking up in the morning, getting through an exercise workout, or even focus and meditation, the difference in effect may not be nearly as large as it is for relaxation. For the very idea of relaxing requires that we let go of any stress and anxiety; it's hard to do that if we feel that we have no control over the situation, that music has been foisted upon us.

Effective music relaxation for stress relief requires personalized music, not "off-the-shelf" relaxation music. Neurologically this is because familiar or self-chosen music reduces cortisol levels, and promotes activation of the brain's reward center—dopaminergic, serotonergic, and opioidergic pathways. Several types of opioid compounds exist in the brain, each with its own receptors. Mu-opioids (μ-opioids) bind to specific receptors in the brain associated with pain relief, euphoria, and addiction. In turn, activation of the musical reward system improves functional connectivity throughout the brain, linking areas of cognition, perception, and movement activity that otherwise are subject to degeneration with aging. (Many of the most commonly prescribed opioid analgesics, like oxycodone and morphine, are μ-opioid agonists, meaning that they activate these opioid receptors.)

But how to find the right music? Russo, Mallik, and colleagues write:

> Because personalization is so important to the effectiveness of music, it stands to reason that a limiting factor in scalability of any effective music program will be the time and effort required to personalize music for a given individual. This may be especially challenging when the caregiver has limited experience with the person living with dementia and/or the individual has limited communication abilities. A licensed music therapist would be able to cultivate some level of personalization through careful interaction and observation with an individual. However, there are barriers to accessing music therapists, which limits the benefits that may be obtained from music engagement.

The future of music in health care extends from hospital to home, from illness to neurorehabilitation, mindfulness practices, and wellness. The big need now is to confront this scalability problem. While music therapies and interventions are most successful when the patient or listener uses chosen music, not all chosen music is effective. There aren't enough music therapists to help every individual, and so we need some way to automate the process of selection, or at least pre-selection. AI will help here—not in writing music, but in selecting music that meets both an individual's tastes and desired therapeutic and wellness goals. Several start-up companies are doing just that. By extracting key features from music and matching them to an individual's preferences and needs, we can usher in a new age of personalized music medicine. Just as using an individual's DNA can guide treatment decisions and prescribe drugs that are an optimal match to an individual's genetic makeup, AI may one day extract the

DNA of music to identify precisely what music will help meet an individual's therapeutic needs.

In the sphere of aging-related losses, it is vital to acknowledge the multifaceted impact these can have, encompassing both psychological and physical realms. While the focus of current neurologic rehabilitation methods primarily centers on motor therapy for limbs and the restoration of brain function, the human brain's resilience lies in the intricate interplay between its hemispheres. Successful post-stroke rehabilitation stimulates the connectivity between hemispheres, culminating in the holistic reintegration of the entire brain. When damage to one of the hemispheres is too severe, we seek to stimulate secondary and alternate pathways in the uncompromised hemisphere. Within this context, music therapy holds a special role. By fostering neuroplasticity and nurturing neural networks that span the breadth of the cerebral landscape, music therapy emerges as an invaluable tool. Many patients harbor an aversion toward physical therapy, often demonstrating noncompliance when confronted with the prospect of rigorous exercises within the confines of their own homes. The introduction of music-guided motor therapies ushers in a more joyous and engaging alternative—one that has the potential to instigate remarkable transformations. Rather than supplanting traditional limb motor exercise training, music seamlessly complements and amplifies its efficacy. Additionally, music therapy improves dysphagia (difficulty swallowing), a common symptom among stroke patients.

In 2011, Arizona congresswoman Gabby Giffords was shot in the head. She sustained severe brain injuries, and although her language comprehension remained intact, she was unable to speak even single words. Representative Giffords underwent a course of treatment with therapist Maegan Morrow at TIRR Memorial Hermann Rehabilitation Hospital in Houston. The treatment included Melodic Intonation Therapy (MIT), which makes use of patients' unimpaired ability to sing words and phrases during daily life with musical melodies.

Patients are taught to sing everyday requests such as "I would like a glass of water" or "I need to go to the bathroom." Because musical circuits in the brain are evolutionarily (phylogenetically) older than speech areas, they are more robust in the face of injury or neural decay. As with stuttering, music therapy stimulates the damaged brain's language function area, and promotes neuroplastic changes within the language network that can restore functions that might otherwise remain entirely lost. Words had abandoned Giffords, but singing had not, and within ten months, although she could not *say* the word "light," she could sing "This little light of mine, I'm gonna let it shine." Four years later, she referred to herself as the new Gabby Giffords, "better . . . stronger . . . tougher." The same techniques used for Giffords are also effective for restoring language after strokes, tumors, and Alzheimer's disease; they are especially effective when the damage is to the left hemisphere, allowing the right hemisphere's music centers to drive the recovery process.

Some of the most captivating stories and videos surrounding cognitive decline and music show the profound paradox that despite severe memory impairments, patients can remember music. Even in the advanced stages of the disease, patients sing along to familiar tunes. Musicians with Alzheimer's and other forms of dementia can continue to play instruments. The guitarist Glen Campbell toured after a diagnosis of Alzheimer's. He did not know what city he was in, and often couldn't remember that he had just played a song, and so would play it twice or even three times in a row. But he had built up so much knowledge, and so much neural and cognitive reserve, that the music continued to flow out of him. Even with half of his brain offline, he was still among the best guitarists on the planet. Tony Bennett experienced a similar problem, a lack of awareness of where he was or what was going on. But once a song started, with the playing of the first chord, something took hold of him and he was off to the races as though nothing had happened to his consciousness or memory.

Journalist John Calapinto recalled visiting Tony Bennett in 2021, when Bennett was 94. Bennett's wife Susan led Calapinto into the living room where Bennett was paging through a coffee-table book open on his lap.

> His expression had a masklike impassivity that changed only slightly to dim awareness when Susan, a slim, fine-featured 54-year-old, placed a hand on his shoulder, leaned over and said: "This is John, Tone. He's come to talk to us about the new album." She spoke into his ear, a little loudly perhaps, in a prompting, emphatic register, as if trying to reach her husband through a barrier that had fallen between him and the rest of the world. Indeed she was. He looked expressionlessly into my eyes before returning wordlessly to his book.

Bennett appeared to be unaware of his Alzheimer's. Susan explained it to him, but he didn't grasp it. He felt physically fine; the mental impairments were apparent to those around him, but he seemed spared the knowledge of them. Bennett was able to sing for 90 minutes straight. He remembered lyrics for hundreds if not thousands of songs. How was this possible? Lyric memory is a special thing—it's the reason that we use songs to teach children the alphabet, and how to count, and so many other kinds of things. Lyrics, when embedded in songs, stick.

Traveling bards over the centuries typically lacked literacy, and yet sang long epic poems from memory; in many parts of the world, they still do. Some of these epics are thousands of lines long and take hours to sing (the *Iliad* and the *Odyssey*, for example, total 27,000 lines). On the surface, this appears to be an extraordinary demonstration of memory. Yet in preliterate and illiterate cultures, the notion of verbatim memory is different than what we hold in the experimental psychology laboratory—the very idea of a verbatim (literal) remembering is a product of literacy itself and was not the goal of those who tell epic stories through song. These apparently unusual feats of lyrical memory don't at all fit the stereotype of rote memory studied in modern laboratories. And errors in musical lyric recall (as in Ella's rendition of "Mack the Knife") tend to preserve the rhythm, the local and gist meaning, and the poetic sound pattern of the correct lyrics. In a study of everyday recall of Beatles tunes, subjects made these perfectly reasonable substitutions:

And you know you *should be glad* → And you know you *can't be sad*

If you say you love me *too* → If you say you love me *true*

To help with good Rocky's *revival* → To help with good Rocky's *survival*

Have you ever encountered a novelist who can recite *Moby Dick* verbatim, or poets who can recite all the works of John Donne? (Probably not.) Yet among musicians who play "casuals"—weddings, bar mitzvahs, parties—as well as those who perform in hotel lounges and piano bars across the country, it is the norm rather than the exception to be able to play thousands of songs from memory. The memory of such musicians seems exceptional, standing alongside that of professional chess players who can remember thousands of different games, or comedians who can remember thousands of jokes. But musicians who play popular songs from memory are likely not playing them with true fidelity to a recording or live performance they heard. Rather, they've extracted the gist—higher-order features such as style, tonality, and tempo—and they improvise notes that are reasonable and consistent. They may make chord substitutions that remain syntactically (harmonically) correct for the ones they heard in the original version, not because they can, but because ignoring small details rarely detracts from the recognizability of a well-known song. Their errors tend to be structure- and gist-preserving.

Consider any given performance by a cover band in a bar on a Friday night. In a 35-minute set, the musicians might play 10,000 notes, but they don't need to "remember" nearly that many. First, music involves redundancy; once musicians have learned the chorus of a song, they don't need to memorize each repeat of it, just how many repeats there are. Second, musicians tend to remember scales as chunks, rather than individual notes. If an expert musician sees on a sheet of music that they have to make a rapid stepwise passage down two octaves, they'll look at the first note and the last note and use their knowledge of scales to get from the top note to the bottom—they don't need to read, or even think about, each individual note. Similarly, they encode chord sequences as chunks rather than individual chords, akin to textual or numeric chunking, allowing for a relatively sparse schematic representation of each song. A musician knows that a blues song will have a dominant I chord, IV chord, and V chord (in the key of *A*: A7, D7, E7)[*]—they don't need to remember any more than the sequence and with experience, they can *feel* when it's time to change the chord. Performance expectations for classical musicians, particularly soloists, are different than those for popular music performers, and classical musicians are indeed expected to play the music as written. But this still does not require that they necessarily memorize each individual note—they employ the same strategies mentioned above.

I asked Shelly Berg—who always errs on the side of modesty—how many songs he thought he could play from memory. He answered without hesitation, "2,500." And how many more could he remember well enough to fake his way through them so that most people wouldn't know the difference? "Another 2,500," he said. Composer and arranger Chris Walden recalls having to turn in a list of 100 memorized jazz standards as part of his musical training at the Conservatory of Cologne. World-renowned harpist and violinist Carlos Reyes, overhearing the conversation, added that "I find I'm more likely to be able to remember a song in a band situation than if I'm playing solo. The drummer might start a groove and then I realize, 'Oh yeah, I know that song.'" Shelly Berg added, "If the bass player knows the song, all I have to do is listen to him and then I know the

---

[*] And they know what order they'll occur in. For a 12-bar blues, A7 (4 measures), D7 (2 measures), A7 (2 measures), E7 (1 measure), D7 (1 measure), A7 (2 measures; or 1 measure each of E7, then A7). This is so ingrained that an expert musician can play around with the form, adding chord substitutions (as jazz pianists typically do) without getting lost in the progression.

song." Although these are anecdotal, they are the kinds of observations out of which scientific experiments grow.

The guitarist Pat Martino presents a completely different case. After brain surgery at age 36 to remove a large tumor, he awoke with no memory. He couldn't remember his own parents, or even that he had ever played the guitar. His parents showed him photographs and played him his own recordings, hoping they would spark latent memories. "I didn't remember any of that, of what I was told were the things I did." He felt like they were from someone else's life. Over the next seven years, Martino learned to play guitar again, by listening to old recordings of himself, as he recovered at his parents' home in Philadelphia.

"As I continued to work out things on the instrument," he wrote in his autobiography, "flashes of memory and muscle memory would gradually come flooding back to me—shapes on the fingerboard, different stairways to different rooms in the house." The entire process, although profoundly frustrating at times, ultimately brought him to a childlike mental state, introducing a new playfulness to his performances, much like Keith Jarrett had experienced in his prime.

In March of 2015, singer-songwriter Joni Mitchell had an aneurysm that burst in her brain. She fainted in her kitchen and wasn't discovered until 12 hours later. By the time she got to the hospital there had been a lot of internal bleeding, and she remained in the ICU for many weeks. Multiple reports indicated that her condition was grave and that it was unclear that she would ever make a full recovery. After three months, she was transferred to a skilled nursing facility. She couldn't talk or walk. She was later released to her own home with 24-hour care.

One of the nurses saw my name and phone number on a scrap of paper in Joni's kitchen and called me. The nurse said that whenever Joni heard music coming from their cell phones, she seemed to come alive. They wanted advice about what to play for her.

I described where the record and CD collection was kept in Joni's living room, and told them what to start with. Eleven years earlier, in late 2004, Joni had been asked to release a CD as part of the

Starbucks' Record Label "Artist's Choice" series—a kind of desert island discs collection. Joni had asked me to come over and listen to what she had selected, to help her winnow down the list as well as discuss the rationale for some of her choices. She was weighing the pros and cons of organizing the collection thematically, or chronologically, or more intuitively based on what flowed into what.

The CD could have up to 18 cuts and she had chosen artists ranging from Debussy to Duke Ellington, Edith Piaf to Etta James. She'd made a few surprising choices, such as the ebullient "Saturday Night Fish Fry" by Louis Jordan. Her encyclopedic knowledge of music was extraordinary. We had an hours-long conversation about which Steely Dan cut to use (she chose "Third World Man"); about whether she could include not just one but two songs by Deep Forest (she used two); and whether it would seem cheeky or self-indulgent to include one of her own songs (I thought it was a good idea, especially if she chose a song that wasn't well known). We joked about an old *Saturday Night Live* skit. Steve Martin plays a NASA official reporting on the first contact with other lifeforms in outer space, from a civilization that intercepted the *Voyager* probe containing a variety of music from the Earth. The four-word interstellar message? "Send more Chuck Berry." And so, "Johnny B. Goode" made the list. Once she had decided, she sequenced them and invited me back to hear the tentative running order. I watched the pure joy in her eyes as she became utterly transfixed by the songs, a combination of awe, intense focus, and relaxation.

I told the nurses where the CD was, in the corner of the bookcase at the far end of the living room, and I said they should start by playing it once a day. It was also important, I added, to ask her if she wanted to hear it, so that she would know that she was in control of when and how often to listen. The nurses called me later that afternoon and said it was the first time they'd seen her smile since coming home.

1	Clair de Lune Philippe Entremont	2	Subtle Lament Duke Ellington
3	Solitude Billie Holiday	4	It Never Entered My Mind Miles Davis
5	Jeep's Blues Duke Ellington	6	Harlem in Havanah Joni Mitchell
7	Saturday Night Fish Fry Louis Jordan	8	Johnny B. Goode Chuck Berry
9	Third World Man Steely Dan	10	Night Bird Deep Forest
11	The First Twilight Deep Forest	12	Le Trois Cloche Edith Piaf
13	At Last Etta James	14	Lonely Avenue Ray Charles
15	Trouble Man Marvin Gaye	16	Sweetheart Like You Bob Dylan
17	Stories of the Street Leonard Cohen	18	You Get What You Give New Radicals

Song selection for Joni Mitchell's Artist's Choice CD.

Joni began physical therapy and speech therapy. The length of time that she went without treatment presented a pessimistic picture of whether she'd ever get back to anywhere near normal. After two weeks the nurses said she was working devotedly on her physical therapy and speech therapy and they wanted to play something else. I suggested Herbie Hancock's tribute to her, "River: The Joni Letters," and any other Herbie Hancock and Wayne Shorter they could find. She loves Herbie and Wayne. I suggested they add *Harvest* by Neil Young, because we had listened to that album together and I knew she had a special affection for it. I also suggested a song that was famously written specifically for her, "Our House" by Graham Nash ("I'll light the fire, while you place the flowers in the vase, that you bought, today").

A few weeks later the nurses called and asked for more. I suggested they try early Duke Ellington, and Joni's own collaboration with Charles Mingus, because she had been so deeply moved when the great jazz bassist had sought her out to collaborate with him when he was dying from ALS.

I visited Joni every few weeks. She was learning to walk and talk again, and could understand everything that was said to her. A year later, as I always do, I brought her flowers. This time she walked over to a cabinet by herself to get a vase for them. She moved some vases out of the way to find a particular one in the back, a glass vase with a single handle and flowers painted on it. "That's a beautiful vase, where did you get it?" I asked. "I bought it when I was living in Laurel Canyon with Graham." Oh. *That* vase. For someone of my generation, that vase holds a special significance, a symbol of young love between two artists living in fabled California.

Four years after her stroke, she made her first public appearance, attending the Grammys. In 2023, she was awarded the Library of Congress Gershwin Prize, and sang "Summertime" at the awards ceremony.

Recovery from brain trauma is not an exact science. I can't say for sure what caused the transformation in Joni over the following weeks and months. Maybe it was simply being away from the sterility and aural assault of a hospital room. Maybe it was being back in the familiarity of her own home, being able to spend time in her garden, watch her koi fish, and eat home-cooked meals. Maybe it was simply the passage of time or her unmatched will power. Or maybe it was the music. My guess is that it was a combination of all of these, and that music was the catalyst, if not the very instrument, of Joni's remarkable recovery.

Chapter 11

Pain

I UNDERWENT A COMPLICATED SURGERY SEVEN YEARS ago, after severing the tendons and nerves in my right hand. In the old days, an anesthesiologist would knock you out completely—deep sedation—but this is hard on the brain, and can lead to respiratory and cardiac complications, even death (this is how the comedian Joan Rivers died). When possible, the far safer technique is "conscious sedation," during which the patient is awake and able to respond to verbal questions. For my surgery I was given a small amount of the drug propofol. I'm told I was awake and talking during the entire five-hour procedure, and the anesthesiologist adjusted the dose of the propofol based on my verbal reports of how much pain I was feeling.

I have no memory of any of this. I asked my surgeon what we spoke about. Propofol—like the sodium thiopental anesthetic used a generation ago—lives a double life as a truth serum. With your conscious self-awareness put on a distant shelf too far away to reach, anything you say is likely unfiltered by social conventions, politeness, or the compelling desire to please your interlocutors. What deep, dark thoughts did I reveal, things I would never want anyone to know? None, he said. I spent much of the time bragging about my wife, Heather. "She is much more of a scientist than I'll ever be," I reportedly told him, and the entire surgical team. "She was the first to master near-infrared spectroscopy for studying neuroplastic changes in infant brains while they learn language." That's quite a mouthful for someone who is not conscious by any conventional definition. My surgeon, Greg Buncke, is also a bass player and so he had a lot of questions about what it must be like to perform with his personal heroes, Victor Wooten and Steve Bailey. And, apparently, he wanted the unvarnished truth without any filters. "They are two of the most musical, creative, loving and *collaborative* people I've ever known—and I can't believe I get to play with them," he tells me I said. Apparently, I went on and on about a video I had just seen in which a right-handed guitar player hands his guitar to Paul McCartney, a left-hander, who picked the guitar up, flipped it around, and played it upside-down, flawlessly; and the time that B.B. King changed a guitar string in the middle of a song without missing a beat. Such are the contents of my subconscious, it seems.

Long story short, the surgery was a complete success. How is it possible I was awake enough to carry on a conversation, yet felt no pain and had no memory of anything? Propofol acts on the very structure that is involved in the maintenance and switching of attention—the anterior cingulate. It also binds to and activates GABA receptors, responsible for inhibition, and acts on hypoxia-inducible factor 1-alpha (HIF-1α), which in turn helps to relieve pain from hypoxia, or lack of oxygen to tissue—which is just what happens during surgery. Finally, propofol acts on the hormone vasopressin (differing from oxytocin by only two amino acids), which is produced in the hypothalamus and has a significant effect on pain perception. The anterior cingulate, GABA receptors, HIF-1α, and vasopressin, therefore, are at least part of the story of propofol and pain relief.

Pain accounts for 80% of visits to the doctor. Pain researcher Jeffrey Mogil notes, "And yet, we still treat it, mostly, with the same substances we've been using for thousands of years: something made from the bark of a tree (aspirin, and its synthetic equivalents) and something made from poppies (opiates, and their synthetic equivalents, opioids)." With all the advances in medical technology, it is surprising that we haven't much advanced beyond these. People in pain show abnormal activation patterns of brain regions involved in sensory regulation, especially the anterior cingulate, the structure that modulates what we pay attention to and what we ignore. What if we could find a way to accomplish pain relief without using high doses of medication? What if we could figure out a way to flip a kind of switch in the anterior cingulate so that we no longer are attending to pain? Well, music, like propofol, also flips the switch of the anterior cingulate. And music inactivates that HIF-1α transcription factor that plays a critical role in maintaining stable blood oxygenation levels. Finally, music listening and vasopressin levels are correlated, with higher levels of vasopressin associated with increased music listening. I am not claiming that this is the end of the story; it may only constitute a feeble beginning. Pain is not the product of a single brain area or small handful of neurochemicals—it is polyfactorial and polygenetic. From a therapeutic side, however, we can say that propofol and opiates are powerful pain relievers. Music can be, too, and was used long before we had propofol or easy access to opiates. Music listening following surgeries reduces post-operative pain as well as the amount of anesthesia required during recovery, even for spinal surgery.

A new study by Prasad Shirvalkar at University of California, San Francisco found that people experiencing particularly high levels of pain showed a spike in low-frequency signal activity in the orbitofrontal cortex, including Brodmann Area 47. Area 47 is among the regions most impacted by music, and is the area that Vinod Menon and I found responds to musical structure. This opens the door for music therapies targeted for pain relief, in which particular structural elements of music may release pain sensations.

Although acute (short-term) and chronic (long-term) pain respond to medications differently, music is equally effective for both. Dentists figured this out a long time ago for acute pain; most dentists now let you wear headphones and listen to music while your teeth are being drilled. It doesn't mean you won't need any anesthesia, but you can get by with less. While he was a PhD student in my lab, Adiel Mallik published a novel experiment in the study of the brain's emotional responses to music with postdoctoral fellow Mona Lisa Chanda. They showed that the brain releases its own endogenous (internal) opioids in response to music listening, suggesting that this is a major reason why we like music, why it makes us feel good, and why—like opiates—music is such a powerful anxiolytic and analgesic. The amount of endogenous opioids we produce from music listening does not reach pharmacological levels, so in most cases, music won't replace prescription drugs—but it can lessen the dosage and length of time they're needed, with the potential of dramatically undercutting the opioid crisis in the United States. Similarly, sound baths, dancing, and performing music are known balms for chronic pain.

Getting the "dose" of music just right is an area of active research, with dozens of laboratories around the world conducting controlled experiments to understand just what kind of music should be used, and under what circumstances. But this is where music is very unlike drugs. Drugs have a half-life, the amount of time it takes for half of the drug to leave your system. For pain-relieving drugs, this can vary widely, from half an hour to 36 hours or more. If you take too much of a drug, it can make you sleepy, or cause unwanted side effects, and you can't really untake what you've taken—you just need to wait it out. But music can be turned off. You can leave the theater. You can change the channel. You can put on earplugs or noise-canceling headphones. Broadly speaking, any musical intervention for therapeutic purposes, including pain relief, can be done in collaboration with a therapist or on your own. The effects of self-administered music are immediate, and we can change the dose, titrate, at will.

⌒

Carlos Reyes, like Bobby McFerrin, is a world-class musician and performer, and people report that they find his music to be especially healing. Carlos is one of the most natural, intuitive musicians I've ever met. He also possesses an incredibly deep knowledge of music theory, and of musical forms from many different countries and styles. Sometimes, as when he's playing with the Doobie

Brothers or Steve Miller at a large stadium, he's just trying to show everyone a fun time. At other times, such as when he volunteers in hospitals and long-term care facilities, he performs with healing intent. What does that mean operationally? It's hard to describe, but he plays more delicately, more *simply*, stripping the music down to bare essentials. He dials back the "flash" that's required to reach the back row from a large stage. How does Carlos think music healing works? "If we feel better, it's easier for us to heal. You can talk about endorphins, serotonin, immune system signaling and all that stuff, but that's the bottom line—good music can make you feel better."

"When you're a kid," he continued, "you can be bullied at school and when your mom holds you and caresses you, all these feelings dissipate and you feel better. The opposite is also true—you can have all these accolades and certificates, and you're feeling good, but your parents can say 'why didn't you do *this?*' and you feel miserable." (Sadly, Carlos knows this from direct experience.) "When you're a kid, your parents are your whole world," Carlos adds. "Now as adults, we can heal ourselves. For me, music has always been that caressing hug I needed. And as I heal myself, somehow, others heal as well. I'm not doing that intentionally—I'm trying to fill the sadness and pain inside me." And it is contagious.

Coming out of the COVID lockdown, I went to a jazz club in Oakland two nights in a row to hear two different musicians I love. I can say from personal experience that playing while people are eating or drinking is a very difficult job, because people are expecting to talk, dishes are clanking, and the musicians don't really have the full attention of their audience—that's not why most of the audience is there.

The first night was a jazz trio and it was lovely, just what I had hoped for, and my friends and I enjoyed it. We looked around the room to see that others did, too. The next night, Carlos played and it was an entirely different experience. The diners seemed at once more relaxed and more focused—not just on the music, but on one another. The wait staff seemed less stressed. The mood of the entire room was transformed. Two nights in a row, same venue, more or less the same crowd, and music. What was different? Carlos's touch.

I could talk for hours dissecting the articulation and physical gestures of what he's doing—the attack, the hold, the release, the variations of timing, intensity, timbre. But that only gets to technique, and lots of musicians have that. What they don't all have can only be described as deep musicality and a connection to something larger than themselves; the thousands of tiny microvariations of movement every minute that are performed in the service of emotion, not dexterity or virtuosity. Miles Davis had it. Victor Wooten has it. Rosanne Cash has it. Many, many artists you've never heard also have it. Without a doubt, the shamans and music healers around the world and throughout time had it, and many still do. If I had to say what that special X factor is, I'd say it's the sublime nuance, the intense *caring* Carlos Reyes and the others put into each note and into each *millisecond* of each note, and a caring for the emotional impact that all of this will have. Most musicians never reach this point, perhaps because of ego, or lack of empathy, or lack of respect for the spirit of music. Carlos says, "I went in there with a concept that this is a dining room and this is background music. My intention, even before I got on the stage, was to give that room what it needed in the moment. Everything I played that night was highly focused on that, on creating a spirit of love, a spirit of connection, and a spirit of grounding. It was not about me; it was about the moment. And I watched the room, followed them, and took my cues from them—as though they were on the stage with me, holding their own instruments, and we were playing together." And in effect they were. Their instruments were their ears and their open hearts.

Music like this is effective for many things. Its earliest, prehistoric uses for pain management are today understood in terms of underlying physiology, even though we cannot yet describe scientifically what the performer does to create this neural symphony of healing.

Patients with chronic pain show structural brain changes, including loss of gray-matter and white-matter volume in the anterior cingulate, among other regions, and in the dorsolateral prefrontal cortex—the region of the brain responsible for decision making, working memory, cognitive flexibility, planning, inhibition, and abstract reasoning: people in pain tend to be impulsive, and bad at making good decisions. Neurochemical systems are affected by pain, including increases in adrenaline and cortisol, coupled with decreases in dopamine production, opioid receptor binding, and modulation of the GABA and glutamate systems. If you've ever been in pain and felt you weren't thinking clearly, these are the reasons why. The emerging evidence is that listening to music can partly restore normal neurochemical balance. How long the treatment lasts is highly variable. In some studies, the analgesic effects of music persist for up to an hour or more after listening ends, and individuals with certain neurodevelopmental disorders such as Williams syndrome (WS) are able to cling to the effects of music interventions for many hours after the music has ended.

Untreated pain disrupts sleep patterns, which in turn can cause profound deficits in memory and mood. I've heard a lot of people say that they "can live with" the pain they are in, and I wonder if this is some effort to display toughness and hardiness. If they knew that living in pain could shorten their lives, would they still eschew the therapies and medications that might relieve their pain? Listening to music reduces self-reported pain, anxiety, and depression symptoms in a diverse range of chronic pain patients with no adverse or negative effects. The analgesic effects of music are greater for patient-chosen than researcher-chosen music because locus of control is an important factor in letting your guard down to allow treatment to reach you. Pleasant (consonant) and unpleasant (dissonant) music influences the descending pain modulation pathway, increasing or decreasing pain perception, respectively.

Why is music so effective? Distraction is one of the most effective ways to alleviate pain. The brain is bombarded with millions of input stimulus packets every hour, and we pay attention to only a small proportion of them. People who are in enriched environments—with lots of things to see, listen to, and do—experience less pain than those in simpler environments, and this sort of distraction diminishes pain signals in the insula and primary sensory cortex. Listening to soothing music can offer effective distraction while in pain. Even when the distracting activity is forced on an individual, it leads to a reduction in pain.

However, if music's ability to reduce pain were solely related to distraction, we'd expect the pain to return when the music ends. This is not what we see. For instance, listening to relaxing, pleasant, self-chosen music significantly reduces pain and increases functional mobility in fibromyalgia patients, not just during, but after music listening. I conducted an experiment in my lab ten years ago with Laura Mitchell and Theodoro Koulis in which we tried to determine how music relieved pain. There were really just four broad possibilities: (1) music distracts us from pain; (2) music elevates our mood, making pain more tolerable; (3) music activates, through some unknown mechanism, neurochemicals and circuits involved in pain perception; (4) music acts through the placebo effect.

For our experiment, we examined short-term pain, rather than chronic, simply because it is easier to study—we can control the amount of pain people are in (more or less) and the pain is not comorbid with other conditions. To begin with, we needed to show that music does, in fact, relieve pain (one doesn't take anything for granted in the laboratory). College students in our laboratory had to immerse their hands into a bucket of ice water for as long as they could, in what is called the *cold pressor* task. If you've never tried this, it's hard to believe how painful it becomes after a short time; most people can't last more than a few minutes. (On one of his trips to our lab, Bobby McFerrin said he wanted to try it. He lasted seventy seconds longer than our lab record.) In the experiment, participants were randomly assigned to listen to music, listen to a book on tape, or simply hold their hand in the ice water with no external stimulation. The logic of the experiment is that if you can hold your hand in the water longer while listening to music, music must be providing some kind of relief from the pain. This is exactly what we found.

We also found that the effect was a combination of distraction, mood elevation, and possibly other factors we weren't yet able to identify. There is a small amount of evidence that dancing also provides pain relief using similar mechanisms, and it makes sense—dancing is done to music, and it also involves movement, which is healthy in itself. We often have trouble moving through the pain, but once we get started—depending on the type and source of the pain—movement can be greatly beneficial. Whether it works better than music alone we don't know yet.

Our experiment went like this. To test whether the music distracted people from pain, we flashed a number on the wall using a projector at random intervals while people listened either to music they chose, or music that other people chose. Participants couldn't report the number when they listened to music they liked, suggesting they were distracted by the music. As a control, we embedded unusual tones in the playback, unbeknownst to the listeners, to see if they could identify them. If they couldn't, that indicated they were listening to the music so closely as to ignore an unrelated signal in their headphones. Using music the participants didn't like—and didn't find engaging—allowed them to effortlessly identify the unusual tones.

To test whether the music improved mood, we administered mood questionnaires. (It isn't possible to measure things like dopamine and serotonin in the lab because they don't cross the blood-brain barrier, and putting a probe inside an undergraduate's brain is frowned upon by the ethics boards.) Again, only music that people liked improved their mood; other music either turned out to be neutral or to make them angry.

Self-selected music compared to experimenter-selected music (to give participants the locus of control) activated a network of pain-related brain regions across the cortex, brain stem, and spinal cord. Music is acting on the nervous system itself, and quite directly. On the neurochemical side, we knew that music modulates endogenous opioids, dopamine, serotonin, prolactin, oxytocin, vasopressin, adrenaline, noradrenaline, and cortisol.

To test whether we were merely observing a placebo effect, we deceived a separate group of our participants by telling them that we were playing them music that had been scientifically chosen to reduce pain, and compared them to another group whom we told only that they were going to listen to music. We found a very small but significant effect of the instruction, indicating that there was some placebo effect at work when people believed they were getting a "special" kind of music. Additionally, we told some of our participants that the ice water had been treated with a special chemical that made it not seem so cold, and that would reduce pain. Again, there was a small placebo effect.

Placebo effects are nothing to sneeze at—they are powerful, and depend to a large degree on activating the same neurochemical systems that drugs do. As Mogil notes, in studies of both acute and chronic pain, opioids help about 40% of the time and placebo pills about 38% of the time! That is not a big difference. This is because of the power of suggestion—it's the same reason you may perk up after that first sip of coffee in the morning even though there is no way that the caffeine has hit your adenosine receptors yet. It's the same reason that hypnosis works for some behavioral change (but not all, and not everyone can be hypnotized). The suggestion that you are going to feel less pain activates top-down mechanisms in the brain that activate the same circuits and receptors that analgesics do.

When we completed our experiment, the data were inconclusive; we couldn't rule out or favor any of the four mechanisms. Just last year, a team of researchers at Aarhus University found that neither naltrexone (an opioid blocker) nor haloperidol (a dopamine blocker) attenuated the effect of music on pain, suggesting the effect does not rely solely on those two neurochemicals, but there are many more to test. They also echoed our conclusion that the analgesic effect of music is at least partially mediated by expectations of pain relief.

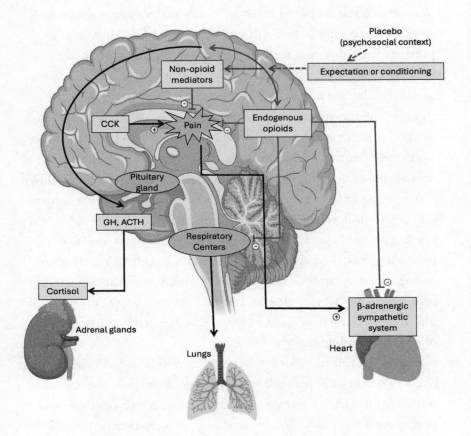

The Neurobiology of Placebo Treatment for Pain*

* How might a placebo function (nearly) as well as drugs? The placebo only works if we know we are taking something (it won't work if someone slips it in our coffee), because this sets up a context of expectations (called "psychosocial context" in the pain field). For example, the expectation that pain relief is imminent can initiate a chain of neurobiological processes in the brain. These include both the release of endogenous (internally manufactured) opioids, and non-opioid mediators of pain, driven by anticipatory processes or past memories. Stimulation of the pituitary gland may affect the release of ACTH (adrenocorticotrophic hormone) and growth hormone, which in turn affect the release of cortisol by the adrenal glands; cortisol has an anti-inflammatory effect, providing pain relief. In the case of a placebo effect for other ailments (e.g., high blood pressure, asthma, Parkinson's) the introduction of a placebo can shift our heart rate, with changes in β-adrenergic signaling, and can modify breathing rate. When placebos fail to work, it may be partly due to the effects of cholecystokinin (CCK), which can counterbalance the effects of the internal opioids, reducing the placebo's impact.

We weren't able to parcel out how much analgesia was attributable to each of the four factors—in retrospect, it's likely that those proportions were different for different people in the experiment, and we didn't have a large enough sample of people to make any claims.

Maybe the answer lies somewhere else. An emerging body of thought on the nature of pain proposes that rather than a "pain matrix" in the brain, there is a "saliency matrix." This theory emerges from the common observation that different sensory stimulations activate the same brain regions while simultaneously evoking wildly different mental experiences. The identical pressure on the underside of your foot is interpreted differently depending on whether you just stepped on a rock, or you're in a spa receiving a foot massage, indicating that the experience of pain is a cognitive construction, owing to interpretation and context. Pleasant music accompanying pain may serve to recontextualize the experience, just as music of differing emotional valences can profoundly affect your experience of a visually ambiguous movie scene.

Related to pain is inflammation and overall health. We are more likely to succumb to pain when we are unwell, because the resources we use for natural pain relief are being marshalled to fight off infection or heal an injury. Several studies have investigated the effects of music on immunoglobulin A (IgA), a principal immunoglobulin secreted externally in body fluids, including saliva and mucus of the bronchial, genitourinary, and digestive tracts. IgA is a first line of defense against bacterial and viral infections and a reliable marker of the functional status of the entire mucosal immune system.

Following music therapy, IgA levels have been shown to rise. Melatonin production rises as well. Increased melatonin production in turn leads to increased cytokines, small proteins that regulate immune responses, inflammation, and tissue repair; melatonin also regulates sleep quality, which is essential to overall health and well-being. Music therapy also leads to increases in the production of T cells (T lymphocytes), a type of white blood cell that plays a central role in immune system response (T cells originate in the bone marrow but mature in the thymus—the "T" is for thymus). Epinephrine and norepinephrine increase following music therapy, increasing arousal and attention. Finally, music listening and music playing have been shown to increase NK (natural killer) cell count and activity, thereby improving the immune system's ability to fight off infections and possibly even cancer.

Patients who listened to joyful music showed significant increases in arterial dilation, to a level obtained with aerobic activity or statin therapy. Though extensive clinical evidence supports music's analgesic effect, few attempts have been made to establish the optimal musical characteristics to achieve pain relief. Music theory could help close this gap in music intervention research. Describing and choosing music depending on tempo, musical mode, level of consonance, instrumentation, and many other formal characteristics could allow better replicability. Highly controlled interventions using single musical pieces may help us answer some of these questions: Is major-mode music better for pain relief? Is slow music better than fast? Is instrumental music better than music with lyrics? Answering these questions could allow the development of standard best practices for the use of music in pain management.

Chapter 12

Neurodevelopmental Disorders

O NE MORNING IN THE SPRING OF 1987, NOAM CHOMSKY was at MIT, sitting in his office in a nondescript, "temporary" wooden structure known simply as "Building 20." Chomsky's office was shabby, and the only view he had was of the side of another building. When I first visited him there, I was reminded of the stories about John Fogerty, the leader of Creedence Clearwater Revival whose music strongly evoked images of the deep south. Singing in a southern drawl and playing guitar like an old bluesman, Fogerty could have grown up in the Mississippi Delta, or on a farm in rural Arkansas. In fact, he grew up in El Cerrito, California, a working-class suburb of Oakland. Fogerty's music was a product of his prodigious imagination; he wrote his entire comeback album, the southern rock–laden *Centerfield*, from a two-bedroom apartment in Oakland. And it was in Building 20, a building with no views and far from the center of intellectual activity on the MIT campus, that Morris Halle and Noam Chomsky invented modern linguistics.

That morning, when the trees were still bare and snow dusted the sidewalks, Chomsky received a call from a distraught mother. Her child, she told him, had been diagnosed with a neurological disorder, with profound intellectual disability, and yet he had a large and unusual vocabulary. "I heard that you study language and I thought you might like to meet him," she said. Chomsky was intrigued, but he is a theoretical linguist, not an empirical linguist (he doesn't run experiments or collect data), so he reached out to Ursula Bellugi, head of the cognitive neuroscience laboratory at the Salk Institute in La Jolla, California, and an expert in language and brain development. It was Ursula who first demonstrated that sign language is not just a collection of gestures, but an actual language with grammatical rules. She showed that, even though sign language is perceived visually, it activates the same brain regions as spoken words.

Ursula met with the child, who it was later determined had the genetic disorder Williams syndrome (WS). Williams is associated with a deficit in elastin production, impaired cognitive function, and poor spatial, quantitative, and reasoning abilities, coupled with excellent face processing, relatively strong language abilities in adolescents and adults, and hypersociability. Ursula spent the next several years studying WS, to learn what Williams could teach us about neuroplasticity, modularity of mind, cognitive function, and language.

A fascinating picture began to emerge. Ursula noticed that most of the children she brought into her laboratory had a heightened musicality and engagement with music. One of the children, I'll refer to her by her first initial, P., was among the first children with WS that Ursula met in her lab. Like many with Williams syndrome, P. struggles with almost everything necessary to succeed in daily life, but she sings all day long and knows hundreds of songs by heart. She also has relatively preserved language—although she can't read, she speaks well and has a large, if somewhat odd, vocabulary. And it's not just P. Many of the Williams kids Ursula met shared a kind of Rain Man quality: in almost every aspect of their lives, they face profound challenges, and yet, amazingly, they have these savant-like abilities.

Because this is a genetic disorder, the fact that music appears protected in the face of so many other deficits suggests we might be able to learn something about the genetics of music by taking a close look at individuals with WS. In June 1996, Ursula asked me if I might collaborate with her to measure and assess musicality in this special population, with a long-term goal of linking genes with musical cognition and behavior—connecting genotype to phenotype. A trait, or phenotype, is simply anything you can observe and measure, like hair color, IQ, or personality. Or musicality.

We tend to think that science is neat, orderly, tidy, like Sheldon Cooper's bedroom in *The Big Bang Theory*. Instead, science is more like a college dorm room—messy, with clothes hanging off lamps and doors, newspapers scattered all over the place, and that thing you need hiding inside dust bunnies under the bed.

For much of its history, experimental psychology has viewed intelligence as somewhat monolithic, and intellectual disability as reflecting more or less uniform impairment across the various domains of cognitive functioning. Then in the 1990s, Howard Gardner popularized an alternative view of multiple intelligences. Of the millions of ways we differ from one another, it seems silly that our intelligence would be boiled down to a single number, the IQ. Even the SAT and GRE, two college admissions tests that are rife with psychometric (measurement) problems, separate out verbal scores from math scores. But what about street smarts? Social intelligence? Musicality? The study of individuals with WS brought Gardner's views into the vanguard of thought about intellectual disability. The relative sparing of some faculties in the face of profound deficits turned the whole notion of IQ on its head. A decades-long quest to find an "intelligence" gene was doomed from the start—of all the ways we humans differ from one another, there just aren't enough genes to go around to make a one-to-one match between genes and traits (phenotypes). Moreover, genes don't work that way—they interact with one another. Researchers began looking for clusters of genes that might contribute to intelligence, but this also has not been fruitful. Along the way, a parallel set of investigations sought to identify so-called music genes.

In recent years, the genetic techniques have matured, but there remains the problem of defining phenotypes precisely. What does it mean to be intelligent? What does it mean to be musical? If a phenotype is poorly specified, trying to match up the genes is hopeless—if you're measuring the wrong thing, the measurement means nothing. As the computer scientists say, GIGO: Garbage In, Garbage Out. So a crucial step is always defining the phenotype, accurately, completely. And as for intelligence, defining the phenotype for music has presented a challenge.

To study music as a phenotype in any group of people, we need to answer the question *What do we mean by "music"?* The problem here is that music is not one thing, it is a bunch of different things. We might say someone is musical if they have a good sense of melody—if they can sing melodies, or recognize them, or remember them. Then there's harmony, both in the sense of a parallel or contrapuntal melodic line (think Crosby, Stills & Nash, Pentatonix, or the Zac Brown Band) and in the sense of a chord progression—the *harmonic* progression of a musical piece. Some people have a highly nuanced sense of rhythm, whether as performers or dancers or listeners. Some musicians have a highly refined sense of timbre, of tonal color. David Bowie, for example, could evoke a great variety of timbral shadings of his voice; Miles Davis could coax many different sounds from his trumpet. And consider instrumentalists who may not have a wide range of timbres, but who are instantly identifiable by the unique timbre they make with their instruments. If you are a careful listener, you would never mistake Stan Getz for another tenor saxophonist, Peter Cetera for another singer, or John Bonham for another drummer. And being able to discern those differences, even if you can't perform them yourself, seems to be another hallmark of musicality.

So far, then, in our inventory, we have this:

Phenotypic Components of Music
- Melody
- Harmony
- Rhythm
- Timbre

We can add to these particular musical abilities:

- Playing an instrument
- Arranging/Orchestrating
- Composing
- Improvisation
- Lyric writing
- Reading music (particularly sight reading)
- Programming music, such as disc jockeys or music directors for film and advertising; being able to spot musical talent
- Choreography, dancing
- Musical memory
- Sensitivity, and ability to be emotionally moved by music

Although many great musicians are expert at melody, harmony, rhythm, and timbre, we acknowledge that some excel at one perhaps to the near exclusion of the others; in principle, these abilities are orthogonal. A fantastic melody writer may not be particularly good with rhythm. And then there are musicians who have a good sense of rhythm, but little sense of melody or harmony.

The musical abilities themselves are potentially distinct and separable. Two of my favorite pianists, Arthur Rubinstein and Daniel Barenboim, are not known for their composing, nor is the great arranger Nelson Riddle. And composers can't necessarily play very well—Irving Berlin was so limited in his piano ability that he played a specially made piano and primarily used the black keys.

Lyric writing is also a distinct talent. Lyrics are not simply words that are set to music—great lyricists, such as Hal David and Oscar Hammerstein II, divined rules for writing lyrics that sounded good in particular pitch ranges. Hammerstein, for example, avoided writing words with closed vowel sounds (like *sweet*) on high notes—the long ē sound closes the larynx, preventing the singer from using their full voice. And for a long, high note at the end of a line, you would not want to end the word with a hard consonant such as /k/

because it stops the flow of the word and hence the melody—you'd want to end with a long, open vowel sound or a liquid consonant, such as /m/ or /r/. In "Oh, What a Beautiful Morning," the highest note of the introduction is a long ī sound: *the corn is as high as an elephant's eye.* Just as we're learning how high the corn is, we reach the highest note we've heard so far. Although most of us don't explicitly notice such subtleties of songwriting, we are more easily transported by the story of a song when all these elements come together to mutually reinforce the message.

If we are going to search for the genes that give rise to musical behaviors, we want to be inclusive and cast a wide net so as to avoid missing anything important. As one example, the U.S. Department of Labor Statistics, under job requirements, indicates that a professional musician must be able to read music. This seems overly narrow, given that Paul McCartney and Stevie Wonder don't read music; for that matter, most of the musicians who have ever lived didn't read music (because music notation is a recent invention).

DJs, radio station music directors, and others have a talent for *programming* music, putting together particular songs in a particular order that can give rise to a total, immersive experience. A skilled disc jockey can create a playlist that helps you to see connections between songs you may have never seen before, and take you on an emotional journey through the selection of tempos, keys, tonalities, and timbre. And then there are A&R managers, the talent scouts of record companies, who are often not musicians themselves but have a preternatural ability for spotting and cultivating talent. The legendary founder of Sire Records, Seymour Stein, discovered a number of bands that no one else was interested in when he signed them: Madonna, Depeche Mode, Talking Heads, Ramones, The Pretenders. Seymour heard what the future of music was going to be. He signed these bands and allowed them to make records, often at an initial financial loss to his label, and waited for the world to catch up. Seymour could not play a note. One of the most venerated record producers of our era, Rick Rubin, also cannot play a note. But artists trust him to produce their records because they know that he will help them to achieve their artistic vision. Seymour and Rick's approach to music is thoroughly intuitive, unencumbered by prior knowledge of how things "are supposed to" be done.

Then there is dancing, a visual-kinesthetic art form tied to sound. Like drama, it is a cross-sensory art that requires coordinating two different perceptual domains (the visual and the aural) in the service of an aesthetic experience. This, too, requires musicality.

By the same token of casting a wide net, a capacity to be moved by music, if only as a listener, is, arguably, the very essence of musicality. Unfortunately, many musicians just play the notes with dry facility, devoid of emotion. Many listeners have an exquisite sense for when the music is conveying emotion, not mere technical mastery. I'd argue that a listener like that is intrinsically far more musical than someone who can move their fingers across an instrument but can utterly fail to stir anyone's feelings.

Many of the personality traits that are required to become an expert musician aren't about anything expressly musical at all! Becoming a great musician, just like becoming a great *anything*, requires good memory and attention. Becoming an expert requires the capacity to sit alone for many hours and practice by yourself—tenacity and self-discipline. It requires belief in oneself, that even though things may sound *awful* now, all this practice will amount to something someday: an ability to delay gratification. Learning an instrument typically also involves multiple failures, and you cannot be easily defeated by these if you want to succeed.

The emerging consensus, then, is that musical expertise is not one thing, but involves many components. These may be present to different degrees in different musicians. So, from a phenotype (trait) standpoint, we are confronting a *mess*. We have a number of different ways that this thing called "musicality" can manifest itself; there clearly can't be a single "music" gene given this complexity. To make matters worse, we don't have any way of measuring any of these components accurately. There are no valid, objective tests of melodic ability, or ability to be emotionally moved by music. Our evaluations of musicians remain subjective, and they are in effect social judgments—judgments about what members of a society at a particular time and place judge to be musical. One edition of *The Rolling Stone Encyclopedia of Rock Music* gave as much space to Adam and the Ants as to U2; as of now, the bands are no longer viewed as equally important, but that could change again. Our measurement tools are so bad that when my colleague Frank Heuser gave a standardized test of musical ability (the Gordon tests) to members of the Los Angeles Philharmonic and to incoming freshman engineering students at UCLA, he was unable to distinguish who was who on the basis of their scores. Some of the engineers scored in the top 10%, and some of the LA Phil players scored in the lowest 10%. The test was measuring *something*, but whatever it was, it wasn't correlated with what we usually think of as musicality.

Behavioral genetics advances when we have distinct, well-defined groups with clear genetic differences. Hemophilia is such a case—occurring in one in 5,000 males, it is characterized by an impaired ability to form blood clots, longer bleeding time after injury, easy bruising, and seepage of blood into body tissues or cavities. Hemophilia is known to be caused by defective genes—F8 and F9—on the X chromosome. Here we have a genetically homogeneous group—hemophiliacs—with a well-defined phenotype and genotype. Linking their genes to behavior is straightforward.

With music we have a poorly defined, diffuse phenotype. Even so, with Williams we have a distinct and well-defined group of individuals, more or less genetically homogeneous. The hope is that we don't have to concern ourselves too much with the specifics of what we mean by "musicality," at least for now, if we agree that those with Williams have it; then we can look to their genome to see what we can learn. Their uneven cognitive profile also allows us to investigate the extent to which different intellectual abilities are correlated or separable.

⌒

"There is a music camp in the Berkshires," Ursula explained to me over a scratchy long-distance call, "and they are holding a special one-week session for individuals with Williams. There will be about 75 kids there as well as their parents."

In July 1996, Ursula and I met for the first time at Logan Airport in Boston, rented a car, and drove up into the mountains. (I was 38 at the time, and Ursula was 65. She got a real kick out of telling people, with a twinkle in her eyes, that she was driving off into the mountains for a week with a young man she'd never met before.)

On the plane I read everything I could get my hands on about WS—which wasn't much. A single, unpublished brain scan study showed that individuals with WS had an enlarged planum temporale, a structure in the auditory cortex sometimes associated with absolute pitch. But absolute pitch—the ability to name or produce a note from memory—is not a particularly good marker of musicality; most major composers have lacked it, and a high percentage of mediocre composers have it. One theory is that absolute pitch interferes with one of the most important parts of being musical: recognizing that music is based on patterns, on the *relations* between notes, not their actual pitches.

Williams syndrome, which was first identified independently by physicians Guido Fanconi (1952) and John Cyprian Phipps Williams (1961), occurs in one in 20,000 live births (making it about twenty times rarer than Down Syndrome). It is a neurogenetic developmental disorder, meaning that it is caused by gene mutations that affect how the brain develops. These mutations appear to be transcription errors in the womb rather than inherited. Genes come in pairs, and what is uniquely interesting about the genetics of WS is that only *one* copy is missing, affecting about 20 genes on chromosome 7 (the missing genes can be from either the maternal or paternal contribution). Genes encode for proteins that do all of the work of building bodies and brains, cellular housekeeping, fighting off disease—the whole kit-and-kaboodle of living is run by those proteins. How fascinating that when just *one copy* of the instructions is missing, there are profound effects, in three domains: cognitive, social, and physical.

Individuals with WS typically have IQs in the range of 40–60. Most, then, never progress beyond the mental age of a nine-year-old. That means that most will never be able to live independently or hold a regular job. Those who do, find work as greeters, or domestic work, and typically part time—although there are exceptions.

One interesting deficit in individuals with Williams is time perception. Ask a typical nine-year-old to hold up their hand for one second, and they'll do a good job. Ask them to hold it up for one minute, and they'll come close (maybe by counting *one Mississippi, two Mississippi,* and so on). Ask them to hold their hand up for an *hour,* or a *day,* and they'll think you're joking. Ask a WS child to hold up their hand for any of these amounts of time and they'll gamely hold it up until you tell them to put it down. Ask them to hold their hand up "for a year" and they will happily begin. And yet, give them a count-off to a tune, or ask them to begin playing, and they keep perfectly good musical time, counting out (or intuiting) eighth notes, quarter notes, triplets, changes of meter, and the flexible time required of *rubato, accelerando,* and *ritardando.*

What makes WS unusual is the uneven cognitive profile—the combination of deficits in some abilities and an odd sparing of others with some quirky features. On the language side, individuals with WS are attracted to unusual words. On a verbal fluency task ("Name as many animals as you can in sixty seconds"), a typically developing child might offer animals such as "dog, cat, horse, lion," and so on. A child with Down syndrome might say "dog . . . dog . . . cat . . . lamp . . . dog," using repetition and intrusions of nouns from another category. In contrast, it is not unusual for children with WS to offer such exotic examples as "newt, saber-tooth tiger, ibex, duck-billed platypus." (This same task can be used to identify children with high IQs, because they tend to name the animals systematically and categorically, such as: dog, cat, parakeet, hamster [household pets], horse, sheep, goat, donkey, chicken, pig [barnyard animals], lion, tiger, giraffe, elephant [zoo animals], etc.)

Individuals with WS are highly social and loquacious. For many years WS was referred to as "cocktail party syndrome" because affected individuals are so adept at keeping a conversation going. They tend to use attention-getting devices when speaking (expressions like *all-of-a-sudden*, *lo-and-behold*, or *guess what happened next?*), often delivering these in a stage whisper to make a tale even more dramatic. And how cheerful they are! They smile a lot, laugh, and the thing they love most is meeting new people—even neuroscientists.

The physical manifestations of WS include a deficit in the production of elastin, a stretchy, rubber-band-like protein that helps build tissues requiring stretchiness, like blood vessels, skin, lungs, and the heart. The elastin deficit causes a heart defect, narrowing of the aorta, that shortens the lifespan of WS individuals. It also causes curvature of the spine, and a specific physiognomy, including "elfin" or "pixie-like" facial features: full lips, high and prominent cheekbones, a broad forehead, flat nasal bridge, wide mouth, upturned nose, a stellate pattern in the iris, and almond-shaped eyes. Their smiles often show malformed teeth and a large visible gum-to-tooth ratio. Scattered and informal references in the European medical and popular literature for the last several hundred years to "elfin syndrome" might have been descriptions of individuals with WS before its identification in the mid-1900s.

WS is also characterized by poorer than normal digit independence between the third and fourth fingers, and generally poor motor control and eye-hand coordination. Ursula and I witnessed camper after camper struggle to button a sweater, tie their shoes, hang clothes on a hanger, walk up and down stairs, or use a fork to get food from their plate to their mouths. And yet, they can play the clarinet or the piano—playing an instrument accesses those circuits that can't do things as important as clothing and feeding oneself, and somehow takes over those brain regions in the service of making music.

෴

Ursula and I began the two-and-a-half-hour drive across I-90 from Logan Airport to Belvoir Terrace, south of Pittsfield, near the Massachusetts–New York State border. As we went through the Callahan Tunnel and then through Back Bay, it was a homecoming for both of us, driving past familiar sites when we were students here—me at Berklee College of Music and MIT, she at Harvard (an old liberal arts school just up the road from MIT). Ursula described the Program Project Grant she had from the National Institutes of Health, to study all aspects of WS, linking genes to brain development and behavior. She had assembled a team of scientists with various areas of expertise, all necessary to fully understand the jagged cognitive profile that Williams presented. Julie Korenberg at UCLA (now at University of Utah) did behavioral genetics. Paul Wang, then at the Salk Institute (now at the Simons Foundation), is a pediatrician who helped diagnose and recruit people with WS for our studies. Allan Reiss is a psychiatrist at Stanford who conducted neuroimaging studies with several developmentally disabled groups, including individuals with Down syndrome, and fragile X syndrome (in studying neurodevelopmental disorders, it is important to have comparison groups of people with similar IQs, so that IQ can be held constant across studies). Al Galaburda at Harvard performed cyto-architectonic studies, looking at cell structure from Williams brains donated to science postmortem. Ursula, as project director, defined what the most important issues to study were, which ones would move the science forward.

Before 1996, research on WS tended to focus on language and spatial abilities. Various researchers had mentioned that people with Williams were musical, but no one had done any research on it. Ursula felt it was time, that a scientific understanding could not progress without looking at this island of miraculously spared ability.

But how to study it? Where to begin? Just a year earlier, Oliver Sacks had published his book *An Anthropologist on Mars*, a collection of observations told as only he could, about people with a surprising array of neurological disorders, all of whom demonstrated remarkable resilience and adaptation, even an ability to thrive, despite their conditions. The stories illustrate that what most of us consider "normal" is a complex and often arbitrary standard. The individuals in Sacks's essays might be seen as "abnormal" in some ways, but their experiences challenge our preconceptions about what it means to live a meaningful and rich life. We learn that just like the mythical "average" person, there is truly no person who is "normal." Ursula and I decided in the car that we would begin by observing, that we'd go with no preconceptions and just see what happened. We were delighted to discover that we were mutually guided by Galileo's maxim that the job of the scientist is to measure what is measurable, and to render measurable that which is not. We decided to be *anthropologists from Mars* (via Pisa and Padua), to figure out what needed to be measured and then figure out how to measure it.

We stopped in Chicopee, outside of Holyoke, to get a pastry and a coffee, then crossed the Connecticut River. The Massachusetts Turnpike was bordered by trees—pitch pine, eastern hemlock, yellowwood, and river birch. We began climbing in elevation toward the Berkshires. We've got a climb ahead of us, Ursula said, and she winked. Yes we did.

When we pulled our rental car into the driveway, we were met by Williams campers Crystal, Neil, Lisa, and Clark. What's your favorite song? asked Crystal. I don't know, Ursula said. What's yours? Why are you so short? Clark asked, Are you an adult? Oh yes, I'm an adult, Ursula said. I'm just short (she was 4'10"). Tchaikovsky's *1812 Overture*, Neil said.

Over the course of the next week, we found that few campers sat idly or sat alone. A frequent topic of conversation was "What's your favorite song?" and the answer was usually sung rather than named. There was lots of spontaneous music-making and drumming—on benches, plastic buckets, cheeks, with sticks and rocks and hands. There was a great deal of warmth and hugging and laughter. A couple of the kids perseverated. One would rock back and forth for hours with his knees pulled up to his chest. Gordon was fixated on steel drums. If someone named a favorite song, he would ask, "Does it have steel drums?" When a group of people were playing together, say a clarinet, piano, guitar, and drummer, Gordon would ask if anyone had any steel drums. I asked him if he knew how to play them and he said he did, something his parents later confirmed. Neil, it turned out, could say nothing but "Tchaikovsky's *1812 Overture*." He used it as a greeting, as a reply to "How are you?" and to comfort someone who had skinned their knee.

None of this fit with the conventional notion of monolithic intelligence. The disproportionately good language ability of the Williams campers suggests that language probably doesn't "piggyback" on general intelligence, and may represent an independent faculty. Maybe music also constitutes an independent faculty, something that can be preserved in spite of profound mental deficits in other domains.

In the evenings, after the campers had gone to bed, Ursula and I gathered with the parents in a meeting room to hear their stories, to learn more about what it was like living with a child with WS. Several parents felt guilty that somehow, something they had done in their lives had caused their child to be born with a neurogenetic disorder. The now discredited theory that certain mothering styles—"refrigerator mothers"—caused autism was also a concern. Had they not shown enough attention to their infants? Had they shown too much? Ursula assured them that they were not at fault, that there is no evidence that WS is caused by any of these factors; it just appears to be random.

The parents filled in a lot of the rich details about WS that are lacking in the scientific literature. We learned that, in addition to loving music, people with WS are endlessly amused by bird calls, animal noises, and all manner of novelty sounds—klaxon horns, slide whistles, buzzers, vibraslaps, didgeridoos; think of the recordings of

Spike Jones, or the ending of the "Theme from *The Addams Family*," with all of the buzzers and siren whistles. They are also drawn to the sounds of motors, fans, and other mechanical devices. One child is so fascinated by the sound of lawn mowers that he begged his parents for one and now mows all the lawns in the neighborhood. Another collects vacuum cleaners, and all he wants for Christmas is a new one he doesn't have—he now owns 18 of them. Once, in a hotel, he went running down the hall and into a room being vacuumed. Holding his hand over his heart, he said, "Oooooh, isn't that beautiful? That's a Hoover!" In a series of experiments we ran a few years later, we found that individuals with WS could identify different leaf blowers and models of automobiles just by the sound of their motors, even when we pitch-shifted them (transposing them electronically, to ensure that the children weren't merely identifying them through the absolute pitch they sounded), something that typically developing children could not do.

Parent after parent described their child's fascination with music, and their own surprise that when playing or listening to music, their child seemed to be like all the other kids. Their children would also wax philosophic about the power of music. "Music is soup for the soul," Crystal said.

Trying to rein in the child's sociability presented challenges; one of the most difficult things to teach Williams kids is *not* to talk to strangers. The parents of one 13-year-old said she routinely approaches men at bus stops or on the street and introduces herself, and then hugs them, saying, "I'm not supposed to talk to strangers. But now we're not strangers anymore."

Individuals with Williams don't know how to be any way other than their gregarious, trusting, ebullient selves. If they were asked to describe people *without* WS, they would probably describe neuro-typicals (what the rest of us are called) as a strange group of people suffering from a disorder: *People with neurotypical disorder are aloof and emotionally reserved; cold. They are socially inhibited and don't know how to enjoy music. They are always so serious and in a hurry, and don't have time to talk and make connections with other people. They don't know how to have fun. They might be able to unbutton their sweaters, but they don't know how to unbutton their emotions, their love, their heart.*

How to turn our observations into an experiment? Because WS is somewhat rare, it is difficult to find enough people with WS in one place to conduct an experiment in a university laboratory. Here we had fifty of them, all eager to talk to us, and to share their musical experiences. We decided to start with rhythm perception. One of the most enigmatic features of WS is the utter inability to tell clock time, in contrast to their ability to keep musical time—to conform to the conventions of musical note length, tempo, and rhythm. I remembered my lunch with Lee Ross just a month earlier, when we clapped and tapped out rhythms on a picnic table at the Faculty Club at Stanford.

I suggested to Ursula that we could start out with a simple, naturalistic task: I would clap rhythms to the campers and we'd record how well they could clap them back. The rhythms would start out simple and become increasingly complex, introducing triplets, mixed meter, syncopation, and so on. Throughout the next couple of days, we'd walk around the grounds, finding campers who were sitting by themselves, and asking if they'd like to play a musical game with us.

I sat across from each participant at a wooden picnic table. Although they were not instructed to do so, all of the participants tended to look me in the eye, rather than watching my hands. All of them clapped back immediately in perfect time, without missing a beat, as if their response formed part of some preordained rhythmic sequence. That is, when I was finished giving the example, the participants came in on the next beat without pausing. All of the subjects thus appeared to interpret the examples as forming part of a larger musical set or sequence; they seemed to assume that there was an implied time signature and tempo, and they responded to the "first measure" of music played by the experimenter in time for the downbeat (or in some cases the pickups) to the "second measure."

After 20 trials like this with one child, I moved to another, and I was careful not to let any children overhear what the others were doing.

When I got back home, I recruited two control groups: an equal number of individuals with Down syndrome, who were matched both for chronological age and mental age, and second, typically developing children who were matched on mental age. I then hired two independent coders, professional musicians, who knew nothing of the experiment or the hypothesis—drummer Jimmy Sage (from Lee Rocker's band) and guitarist Parthenon Huxley (from The Orchestra, an offshoot of the band ELO). I simply told them that I had conducted some tests of rhythmic perception, and that they were to code the responses as right or wrong. There was nothing on the tape or the coding sheet that indicated which kids were from which group, preventing any bias in their coding (I had assigned random numbers to each participant and kept the key to myself).

About a week into the coding, Jimmy and Parthenon called me, independently, to ask a question. They each noted that some of the reproductions appeared to be "wrong in interesting ways." On many of the trials, they said, the participant did not reproduce the presented rhythm perfectly, but the rhythm the participants did clap bore a clear musical relationship to the referent. Jimmy explained that "it sounds like a call-and-response; like the person is creating a musical completion to the rhythm you clapped."

I was panicked. Maybe I had designed a flawed experiment and this whole thing was ruined. I had only been out of graduate school for two months and had not yet started my postdoctoral position, and here I was, already blowing the first experiment I had designed without the help of my professors. Adding to my negative self-talk, I was scared of Ursula. Despite her short stature, she was a force of nature, and the sheer power of her intellect was intimidating. She led the cognitive neuroscience laboratory at the Salk Institute with a staff of a dozen people all scrambling around at top speed to keep up with her. The Foundation IPSEN Prize for Neuroplasticity sat prominently on a bookshelf behind her desk. She advised the Nobel

Selection Committee, and the MacArthur Fellows Program on the awarding of "Genius Grants."

I sheepishly phoned her and told her what Jimmy and Parthenon had said, fully expecting to hear that the trip from California to Boston had been in vain, that she would find someone else to collaborate with; that I would be fired from the project.

She listened patiently and finally said, "This is great news!"

Slow on the pickup, I said, "Umm . . . what's so great about it?"

"What you're telling me is that on some of the trials, the participants seemed to be making music out of their responses, rather than slavishly mimicking the experimenter. It sounds to me like you're describing something rather like an improvisatory or 'jam' session. Do you know which participants did that, the Williams or the controls?"

Gulp. "No."

"Well, I think you know what you need to do then."

I called back Jimmy and Parthenon and told them to go back and listen again to the responses they had previously coded as "incorrect" for the three groups of subjects and to make a special category for those that they felt constituted "elaborations or creative completions" versus those that appeared to be clearly wrong.

When I got the data back, the first thing I looked at was the percentage of correct responses to see if they differed between groups. By design, the trials increased in difficulty to avoid "ceiling effects": if the test is too easy, and everyone gets everything right, it is impossible to distinguish very good test takers from average ones. Overall, the WS and the typically developing controls performed about the same, one-third of the trials judged incorrect. There are very few tests in which those with WS perform as well as controls, and so this in itself was remarkable. The Down syndrome comparison group scored 95% incorrect.

But who was making all those "musical completions" that Jimmy and Parthenon spoke about? Was it a random assortment of participants, or were they all from one group? Among those answers that were coded as "creative completions," the WS individuals were

218 I Heard There Was a Secret Chord

three times more likely to have produced them. Overall, the controls produced creative completions 15% of the time, while the WS produced them a whopping 45% of the time.

Another thing that had intrigued Ursula and me was the attraction many WS felt toward sounds not typically considered pleasant, such as vacuum cleaners, leaf blowers, motors, and the telephone dial tone—what collectively we might call noise. No one had conducted an fMRI study with WS individuals because they tend to be both claustrophobic and fidgety, and the scanners are essentially tubes that surround you while you lie perfectly still.

With Allan Reiss at Stanford, we undertook an extensive program to recruit individuals with WS and make them feel comfortable in the scanner. Allan had a professionally produced video made, narrated by a young girl and with the camera angle at eye-level view with her. This meant that children watching the video were introduced to Stanford Hospital, where the scanning would take place. Further, they would see everything as if they were there: the camera had to tilt up to capture the door handles, for example, and as they walked the halls, they saw people's legs directly in front of them, and had to look up to see their faces. Little touches like this made the video inviting and fun. "I'm going into a brain scanner," the girl narrates. "It's like being in a sleeping bag. They put a blanket on me so I won't get cold." We sent the video out to families with WS children who had previously expressed interest in participating in scientific research. After watching the video, they were given the opportunity to visit the Stanford neuroimaging center, where they could climb into a simulator that looked and sounded like the actual fMRI machine, so that by the time they volunteered to be in the actual experiment, they would be used to it.

Another practical hurdle was that we wanted to play both music and noisy sounds to our WS participants, and brain scanners are themselves notoriously noisy, spitting out a range of strange buzzes, jackhammer-like poundings, and rumbles at 120 decibels or more—about the same as being 125 yards down the runway from a jet airplane taking off. Normally, participants are outfitted with industrial-grade noise-reducing headphones, but we wanted to play music to them. And because fMRI relies on an electromagnet powerful enough to rip an earring out of your earlobe (or wherever else you have it), music has to be delivered through a nonferromagnetic device—basically a pneumatic tube like airplanes used in the 1970s and '80s. (The high-quality fMRI-compatible headphones we use today hadn't been invented yet when I conducted this study.) We wanted the music and noise to be high fidelity enough that it made for a realistic experience.

I worked with Jay Kadis, the chief engineer at Stanford's Center for Computer Research in Music and Acoustics (CCRMA), to take measurements at both ends of the pneumatic tube, the frequencies of the input and the output, and plotted how the tube itself was altering the frequency response. We then installed a parametric equalizer that restored those frequencies that were attenuated by the tube, allowing for accurate sound in the scanner.

For the experiment, WS subjects and typically developing controls listened to high-quality recordings of Bach, Beethoven, Mozart, and Tchaikovsky (including the *1812 Overture!*) and a broad array of noises—car engines, leaf blowers, vacuum cleaners (including the beloved Hoover), water running, dial tones, and such. We analyzed brain activity by looking at the difference in overall brain activity between the music listening and noise listening conditions. That is, with music listening as the baseline, we subtracted out activity during the noise listening to see what we had left. This so-called subtraction paradigm is among the standard analysis techniques we use because the brain is busy all the time—random thoughts occur, your cheek itches, your stomach grumbles, and you wonder what's for lunch. None of this is brain activity that we care about, and so the way to deal with that is to choose conditions that are well matched except for the one variable of interest. For example, you might look at pictures in color and those same pictures in black and white, and we would subtract one condition from the other.

When we performed the subtraction analysis for music and noise, we found remarkable differences in brain activity between the control participants and the WS individuals. All five controls showed consistent and overlapping patterns of activity confined to the auditory regions of the brain, as we expected. But the WS showed substantially decreased activation in these regions—the noise and the music canceled each other out, meaning both were activated equally. Moreover, there was nothing consistent across our WS participants—we saw a highly variable pattern of brain activity throughout the brain, including all four lobes, and increased activation in the amygdala and cerebellum, pons, and brain stem. From this we could see the first evidence that Williams brains were organized differently than usual—it was as though their entire brains were responding to music, not just the so-called "music" areas. At a subregional level, this converged with the findings of Al Galaburda, who had found remarkable differences in the fine structure of the brain based on his autopsy studies. Cell measurements differed, with WS brains showing significantly smaller, more closely packed cells in some regions, along with abnormal cortical complexity and thickness. Together, these findings show how the deletion of such a small set of genes can have profound effects, and they nicely account for why the subjective experience of music is so different for individuals with WS.

In subsequent studies, we found that compared to people with autism spectrum disorders (ASD), Down syndrome (DS), or typically developing children (TD), people with WS are more likely to play a musical instrument, spend more time playing it, and spend more time spontaneously singing or playing original music. They first show an interest in music at age 3½, on average, compared to 5 for TD children. They spend an average of 11 hours a week listening to music, versus 7½ hours for TDs. They also have deeper emotional reactions to music: after hearing a sad or happy song, they are more likely to be induced into that mood, and to stay in that mood state longer.

All of this is fine and good, but the Heffalump in the room is the genetics. We began these investigations because of the well-defined genotype in WS, hoping that we might uncover something about the genetic basis of musicality. Usually, there is a gene we can identify that does something in all of us. Usually, when something goes wrong, the gene is still there and the protein is functional, but things go awry because of a mutation—as with the hemophilia gene I mentioned earlier. Now in WS, we don't have a mutation, and neither do we have a single deletion, but we have 20 of them. Linking the absence of these individual genes to symptoms, to the phenotype, is very difficult—it's the messy college dorm room all over again. And the story is even more complicated than that. Some individuals with all of the behavioral markers of WS lack this deletion, others who have the deletion do not have WS.

At the phenotypic level, it's been proposed that whatever it is that causes individuals with WS to be so social and empathetic also causes their deep musical engagement. This seems reasonable—music is the language of emotion, and it has a deep history of uses for social bonding, comfort, and the *social* expression of emotion.

The best hypothesis we have for how any of the WS genetic profile affects music isn't even a direct relationship between music and sociability, but rather, that one of these missing genes, *STX1A* (syntaxin), can affect other genes located somewhere else in the genome. Specifically, the syntaxin gene may be modulating the expression of the serotonin transporter gene, *SLC6A4* (related to elevated mood), as well as genes that affect expression of oxytocin (related to social bonding and affiliation). So it's not even a matter of which of these 20 genes that are deleted in WS is the "music gene" but which one of these affects genes elsewhere that have something to do with music. It's kind of like a live performance of the Miles Davis Quintet— Philly Joe Jones plays something different than he did before; he's not telling Miles what to play or how to play it, but he affects what Miles does indirectly. The genome is like an orchestra. Which gene is responsible for music? They all are. And one musician can affect another on the other side of the stage.

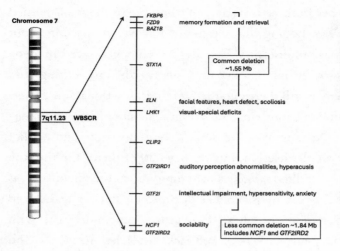

Deleted genes that characterize Williams syndrome.

In a related series of experiments, we set out to better understand music among people with ASD, and the ways in which it may help them.

For many composers and songwriters, music is aspirational, rather than reflecting their actual life. The man who wrote "imagine no possessions" (John Lennon, "Imagine") was a hoarder. The man who wrote "I find it very, very easy to be true" (Johnny Cash, "I Walk the Line") had a series of infidelities. These songwriters were writing about a world in which they were better people than they were at the moment, a world and state-of-being they were striving for. Such songwriting is a powerful form of therapy. It takes the whole concept of journaling, or keeping a diary, a step further because it creates a personal mantra or phrase that—stuck inside a catchy melody—can easily be kept in your head, and others who hold the same aspiration can keep it in theirs.

So it is with individuals with ASD. About half of the musicians I know are on the spectrum, and they've found that a music career is a way for them to channel their neurodiversity into an acceptable form: the compulsions, the repetitive actions (like playing scales over and over again!), the attention to minute details, the desire to be alone. Uncomfortable around people, they can perform for a couple of hours and then retreat into the solitary life they seem to prefer. I'm reminded of Belikov, the schoolteacher in Chekhov's "Man in a Case":

> His great claim to fame was going around in galoshes, carrying an umbrella even when it was terribly warm, and he invariably wore a thick, padded overcoat. . . . His face seemed to have its own cover as well, as he always kept it hidden inside his upturned collar. He wore dark glasses, a jersey, stuffed his ears with cottonwool and always had the top up when he rode in a cab. Briefly, this man had a compulsive, persistent longing for self-encapsulation, to create a protective cocoon to isolate himself from all external influences. The real world irritated and frightened him and kept him in a constant state of nerves.

Not a few of the performing musicians I know would fit this description, and some, so well known for being larger-than-life emotive performers, would surprise you. Although they themselves may face difficulty in reading and understanding the emotions of others, music and songwriting give them a way to explore and express emotion in a public and socially sanctioned way. About one, revered for his ability to express love and warmth in the most beautiful and sensitive music, it has been said that "every ounce of compassion, empathy, love and understanding he can muster goes into his music . . . and there is nothing left over for the people in his personal life." People who are dying to meet him, hoping that the wonderful glow his music brings to them will be extended by knowing him, discover that the persona he presents in the music is his aspiration for who he would like to be.

Many people on the spectrum tell us they find comfort in music. Given that ASD is defined in part by impaired emotional processing, could it represent an island of preserved emotional processing, a domain in which emotional understanding is spared?

The physician Leo Kanner in 1943 was among the first to describe autism. The word *autism* comes from the Greek work *autós*, meaning self, and was used to describe individuals who were largely indifferent to social contact and seemed to withdraw into themselves. Kanner's is the first description of autism and autistic-like behaviors regarding music. He cites a case study of a one-year-old child who "could hum and sing many tunes accurately." Of another little boy, Charles, he shared the mother's notes:

> His enjoyment and appreciation of music encouraged me to play records. When he was 1½ years old, he could discriminate between eighteen symphonies. He recognized the composer as soon as the first movement started.

My students Anjali Bhatara and Eve-Marie Quintin and I wondered whether individuals with ASD could recognize musical emotions, that is, emotional intent, irrespective of their ability to have those emotions evoked by the music. We played them dozens of songs that typically developing children and adolescents identify as happy, sad, frightening, or peaceful. The individuals with ASD performed as well as neurotypicals in recognizing these intended emotions, and they showed greater confidence in their responses. Eve-Marie noted that "although music is initially a social-emotional product created by the composer, a listener does not have to enter into a direct interpersonal interaction with the composer in order to appreciate the music." She concludes that emotion processing deficits in ASD are domain-specific, and are preserved in music.

Whether individuals with ASD actually *feel* those emotions is a different question.

Eve-Marie collected psychophysiological measures of emotion that are often more reliable and less ambiguous than asking someone if they "feel" an emotion, a problematic question for individuals who may have difficulty understanding what such a question means. She found that individuals with ASD showed reduced physiological activity in response to music-evoked emotion, suggesting that although they could *recognize* the emotion and possibly experience it to some degree, they weren't experiencing it in the same way as typically developing individuals. This is consistent with findings from a study that Anjali conducted, showing that individuals with ASD were relatively insensitive to variations in the emotional expressivity of a piece of music. That is, a continuum of performances that most people would consider to span the range from utterly robotic to wholly expressive were rated as equivalent by those with ASD.

The four emotions we studied are considered basic and universal. The extent to which music can express more complex emotions such as concerned, envious, or taken-advantage-of remains a topic of continued research and deserves further study.

One potential application of music therapy for individuals with ASD is to assist them in better understanding affective cues through language. Many, for example, have difficulty understanding that the pitch going up at the end of a sentence could indicate a question or the use of sarcasm. We know that they generally have no impairments in pitch perception, and Anjali has suggested that we can build on their pitch perception ability to train specific aspects of speech communication. This follows on findings by Ani Patel at Tufts that certain neural structures support both language and music, and strengthening one can reinforce the other.

ASD is believed to have a genetic basis, but this has been hard to pin down, with changes in over 1,000 different genes associated with various forms of ASD, although in at most only about 20% of the cases. Gene-environment interactions, that is, a combination of genetic vulnerability and environmental conditions, are probably responsible for most cases. Older biological parents are also a risk factor for reasons we don't fully understand.

⌢

So science is messy. The complexity of linking genes to traits such as musicality is not insurmountable, but it requires that we accept that many genetic factors can contribute to a single trait, and a single gene can influence multiple traits. Moreover, epigenetic factors, such as the environment, can modulate gene expression, switching genes on and off. Another crucial point is that individuals with ASD and WS are a heterogeneous, not a homogenous group—they differ from one another in as many ways as the rest of us differ from one another.

If it's so difficult to link genes with behaviors, why do we try? Because our understanding evolves not just through clear discoveries but through the questions that remain unanswered. From these come new techniques, new theories, and, ultimately, every once in a while, breakthroughs. I look at it this way: if a question can be easily answered with an experiment, it means that we probably already knew the answer. And that's not how science leaps forward.

Chapter 13

Learning
How to Fly[*]

P EOPLE WHO PLAY A MUSICAL INSTRUMENT HAVE A number of advantages over those who don't. Their brains are different in positive ways. Their social lives are different. And there is a personal satisfaction that is difficult to appreciate until you've experienced it yourself. Imagine for a moment being able to put your fingers on the piano keys in the same shapes and configuration as Chopin did, to *be* Chopin for a moment. Or imagine being able to play a B.B. King guitar solo note-for-note. These are not out of reach even for beginners. I chose these particular examples on purpose. Even if you've never played the piano before, you can learn to play the first two measures of Chopin's Prelude in E Minor (Opus 28, No. 4) if you apply yourself 15 minutes a day for a week or two.

April 2020, when the COVID-19 lockdown started, seemed like a good opportunity to take on a new project. I settled on getting reacquainted with the piano. I had started tinkering on the piano when I was four, but I never took formal lessons. Later, I used it as a "harmony engine" when I wanted to visualize notes while arranging pieces for bands and orchestras. I couldn't actually *play* anything. I was comforted when I remembered that the great composer Irving Berlin had such rudimentary piano skills, he could only play his own songs with great difficulty. There were Bach inventions, Mozart sonatas, and Bill Evans solos I wanted to learn. All of these would

[*] From the song "Still Learning How to Fly," by Rodney Crowell, 2003.

require the very difficult work of learning new fingerings and new ways to move my fingers, and realistically, that meant it would be a while before I could actually play even a few measures of a difficult piece. At the same time, I was already beginning to feel the shock of being socially isolated and emotionally cut off. Doing something so *technical* felt like it would put me inside my head even more when what I really needed was to get back in touch with my emotions and my spirit. As Guy Clark wrote, I was beginning to feel like "I couldn't find my heart with both hands."

I decided to spend my lockdown working on just three pieces that I already knew how to play, and to spend the time trying to find the emotion in them. I knew where to put my fingers, but not how to make the notes *sing*. I wanted to learn to explore the tiny little things—micro variations in timing and pressure, the proper use of the piano pedals—that turn a flat performance into a living, breathing one. I wanted to learn how to make *music* out of the notes.

It was intensely satisfying because I could just focus on expression, not on fingerings. I am still working on these three pieces. Classical pianist Mari Kodama reminded me that it is a lifetime project, that the pieces, and my interpretations of them, are living, breathing entities. She said that pedaling is a large part of the performance once you've mastered everything else—"it is the oxygen that makes the piece come alive." There is much to learn and, as I go, each new playthrough brings new insight to the pieces and, in turn, to me.

For example, the first eight measures of that Chopin fit neatly right under the fingers, without your having to stretch them into awkward Rachmaninov-like configurations. The piece is marked *largo*, slow, so you can take your time and not feel that you are being untrue to the composer's intention. It helps that I can read music, but even if I couldn't, the notes in the left hand make shapes that tell you, roughly, how far apart your fingers need to be. The musical staff is like one of those Cartesian coordinate systems you encountered in school geometry. The y-axis (up–down) tells you what notes to play at any given moment; notes that are stacked on top of one another sound simultaneously; the higher up the notes are toward the ceiling, the higher the notes are on the piano—to your right. In a piano score, like the one shown below, the left hand plays the set of notes on the lower staff, and the right hand plays the notes on the upper one (usually). The *x*-axis is time—how these notes change as the piece moves forward. Each black dot represents what one of your fingers needs to do. In the first measure, for example, you can see that two of your fingers need to be close together on the keyboard and one needs to be farther apart.

Prelude in E Minor, Chopin

This is the first thing your left hand does.
And it plays it 8 times in a row.

Prelude in E Minor, Chopin

Left-hand fingers

So for the first measure—about 4 seconds—your left hand just plays the same thing eight times. The right hand has a simple part that coordinates with the first left-hand downstroke and the seventh one. At first, using the two hands can seem awkward if you're a novice, but it comes together quickly (this is called bimanual coordination). The next measure requires you to move three of your fingers just a little bit: again, very easy to learn.

Learning to play the notes that B.B. King plays is even more straightforward. Here's a guitar neck, going from left to right as you hold the guitar.

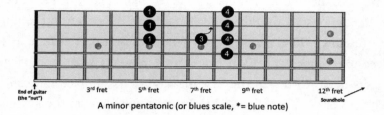

A minor pentatonic (or blues scale, *= blue note)

Blues scale in the key of A, for a guitar in standard tuning.
The top of the drawing is the skinniest string, closest to the
floor; the bottom of the drawing is the thickest string, closest
to your chin. The black dots show where to put your fingers,
numbered from 1 (index finger) to 4 (pinkie finger).

This is a pattern that you can move up and pinkie the neck to play in different keys. Here, it is called the A blues scale. If you want to get fancy you can bend the string where the arrow is after you pluck it, up toward your chin. If you get this right, the note will sustain through the bend and you'll have a sound that you immediately recognize from countless recordings. B.B. King almost never strays from these six notes, plus their octave equivalents (the notes an octave lower, as the bottom figure shows).

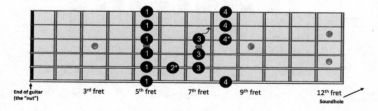

A minor pentatonic (or blues scale, *= blue note), two octaves

The blues scale with an additional octave.

Now, getting this to *sound* like B.B. King is nearly impossible. You don't have his hands, his fingers, his muscles. You don't have his brain or his experiences. But you can approach the sound if you listen carefully to what he is doing and watch some videos. Within two weeks you can sit in with a band and competently play a solo that will amaze you. You can slow down his recordings on YouTube and watch his hands or just listen to find the notes on the guitar. Or you can find a backing track on YouTube—the band without the soloist—and noodle. Play these six notes in any order, with any rhythm you feel like. As with Chopin's Prelude in E Minor, the notes will come fast. Turning them into *music* can take a lifetime— but a very rewarding one.

It really is never too late to start. A cognitive scientist I know, a music lover who had never played an instrument, started playing piano at age 50, for just 20 minutes a day. Within five years he was able to play a few pieces at parties. After ten years, he was quite good. He leveraged the power of the science of learning by playing ten minutes every night right before he went to bed, and then ten minutes in the morning right when he woke up. Because memories are consolidated during sleep, his brain was primed to incorporate the fingerings and motor movements he executed just before bedtime. In the morning, he reactivated the newly consolidated memories to instantiate them. Every morning he found he could play better than he could the previous day.

There is a special benefit for older adults who used to play when they were younger and re-engage in middle age or beyond. Our brain's plasticity and ability to remodel itself may slow down but it doesn't stop. Returning to music later in life profoundly engages those two things that hold a place of privilege in our memories: muscle (or movement) memory, and childhood memories. The convergence of these can bring skills back with striking swiftness, as Rosanne Cash learned when she was recovering from her brain surgery. Any task that requires you to interact with the physical world is both cognition-enhancing and neuroprotective, as Scott Grafton has shown: cognition is driven by physical engagement. The older adult musician gets to re-engage with their younger self, providing a rejuvenating look at the world and a way to create new threads of continuity in one's life story. And we come back to the instrument with a different sense of focus and purpose than we had as children—we are doing it out of love, rather than obligation; we are doing it for ourselves.

While young children can certainly learn music more quickly than adults, older brains have major cognitive advantages over younger ones, simply because they've experienced more—and through the sixties and seventies (and sometimes beyond) these advantages just increase. Older adults are better able to see the big picture, to make connections across things that aren't linearly related ("divergent thinking"); they tend to be more patient and are better at impulse control. They can approach new projects with a wholly different sense of purpose than young people.

A friend of mine who is about my age returned to making music in his sixties and has found it a rich and transformative pastime. "Even though I stopped practicing piano, both classical and jazz, in my early twenties, and resumed studying more than four decades later, I've found that my ear has improved in the meantime," he observes. "I've not only become more musical, but I hear the musicality of certain compositions in ways that I didn't before. As a result, I'm much more attuned to dynamics (and more intently focused on control to play more expressively, if the composition calls for it). It's easy for me to play a traditional blues, but while the music in Bach's first Prelude to his *Well-Tempered Clavier* is itself quite simple, the dynamics—which can shift from measure to measure—are not. I'm also more attuned to making my improvisations, to the extent that I can improvise, 'sing,' as opposed to simply executing a series of chops to show off." In short, he traded a childhood obsession with playing a piece "perfectly" for playing it *musically*.

Some late bloomers actually become better than just "quite good." Concert pianist Albert Frantz had loathed classical music as a teenager, and didn't start seriously playing the piano until he was 17, which might as well have been 50 in musician terms; most concert pianists you've ever heard of began no later than age six, many at two or three. It's not that Albert didn't try. A childhood teacher he describes as a "mean old bat" would slap his fingers with a ruler when he didn't play properly. He quit. Another teacher said to Albert's mother, "Take your money every week and throw it in the garbage. Albert will never be able to play the piano." At 17 he fell in love with classical music and began to learn to play by ear. "In retrospect," he says, "I think what drove me beyond the sheer love for the music, I didn't know what I was doing was supposedly 'impossible.' Because I didn't grow up in an environment that trains musical talent step by step, I had no references for what was impossible or not. . . . While merely thinking that we *can* doesn't guarantee success, thinking that we *can't* guarantees failure."

If classical piano isn't your jam, consider James Lewis Carter "T-Model" Ford, a blues guitarist who didn't start playing until he was 75, when his wife bought him a guitar as a birthday present. Buddy Guy discovered him and helped him get a recording contract at age 77.

Your first thought may be: but I don't want to make mistakes. All musicians make mistakes; skillful ones know how to move past them. The most important thing really is not the notes, but the feeling behind them. The celebrated pianist Arthur Rubinstein made many mistakes, even on his recordings—it just doesn't matter. Many musicians just play the notes, Victor Wooten has said; few of them actually play music. "It's better to think of yourself as a musician," he says, "than a bass player, guitar player, drummer, or whatever." Victor is saying that there is no music in the instrument; the music is in you. Remember Jascha Heifetz's words: "The violin doesn't make the music. The music is being made by the man who is holding the violin."

Or consider Bruce Springsteen. After his first number 1 album, *The River*, Bruce recorded his next album all by himself on a cheap multitrack cassette machine. There were lots of mistakes, but he left them in. The critic Joel Selvin, who loved it, wrote that it "is a stark, raw document, rough edges intact, and so intimately personal it is surprising he would even play the tape for other people at all, let alone put it out as an album." Springsteen told biographer Warren Zanes that when he went back to try to clean it up, it lost all its character and personality. The Georgia Satellites were signed to Elektra Records based on a poorly recorded demo tape. Elektra gave them tens of thousands of dollars to re-record it in higher quality, but the band could never recapture the spirit, the heart, and the sense of abandon of the original, and so they released the "flawed" version of "Keep Your Hands to Yourself." It reached number 2 on the *Billboard* charts. The grandaddy of granola rock, Neil Young, released one of my favorite tracks, "The Will to Love," a poor recording with a fire crackling in the background, and many little musical mishaps, all of which make the recording more charming, more human and honest.

If we take music lessons in school, or privately, teachers invariably steer students toward getting the notes right to such a degree that students are neither taught nor encouraged to play those notes expressively. I asked the Dean of Music at one of the big American conservatories, "Do you have classes where you teach students to play expressively?" She herself seemed surprised by the answer she gave. "No . . . there's just not enough time. We hope they pick it up somewhere along the way." I just looked at her. "Maybe they get it in their private lessons, but that's not formally part of the curriculum." If you're starting music later in life, you're probably doing it to have fun, not to play Carnegie Hall or Madison Square Garden. Have fun! Noodle! Find the emotion and expressivity in the music, and you will find it in yourself.

This is what separates creative and skilled musicians from those musicians that are simply technicians. No matter how virtuosic they are, they are mere technicians once they stop exploring, if the only movements they make are precisely repeated. If they are entirely predictable in every nuance, every gesture, it might as well be a robot performing.

The musical artist finds unpredictability and avoids repetition— embracing their surprise within the recognizable structure of the tune. Like life, they allow the music to constantly change. The *joy* in learning an instrument comes from a sense of play. That's why we say we *play* an instrument, just like children play. We don't "play" accounting, or law, or medicine, or engineering. We play music. Each time we *play* a song a little differently, imbued with mindful intentionality, we are reinventing ourselves, reconnecting with our emotions, and rediscovering art. And if we do all that, we connect with other people on a deeper level than we can with words and whispers.

⌒

Much has been made of the question of whether learning a musical instrument makes you better at other things, such as math. In the jargon of education theory, this is called cognitive transfer. I find the question offensive—it implies that music is somehow not worth doing in its own right, that it needs to be a stepping stone to other things. I've never heard a teacher say that people should learn math because it makes them better at music, although if they did, I imagine that a whole lot of people who find math intolerable might give it a shot. (Every time I meet someone who says that they are no good at math, my first thought is that they just didn't have a good math teacher.)

Setting aside my crabbiness about being asked, there is a large body of evidence showing that musical training correlates with other important skills. Young children who take music lessons have stronger preliteracy skills (e.g., phonemic awareness and rhyming) and perform better on tests of reading comprehension. Young adult musicians are better than nonmusicians at memory for words, verbal fluency, and maintaining attention. Musicianship has been associated with intelligence, visuospatial abilities, processing speed, executive control, attention and vigilance, and episodic and working memory. Children who receive intensive music training are better at impulse control, and those who regularly practice self-control skills do better academically and have healthier adult lives than those who do not.

I used the word "correlates" above deliberately because correlation does not equal causation. Unfortunately, researchers don't always heed this. Between January 2000 and June 2018, there were 114 peer-reviewed articles published testing for correlations between music training and a nonmusical ability, brain structure, or brain function. Two-thirds of these papers incorrectly inferred causation from correlational designs.

Why this matters is because those cognitive and academic advantages observed in correlational studies could be due to preexisting differences in children's intelligence, genetics, temperament, socioeconomic status, family life, and environment. Preexisting factors might, for example, make some students more able to afford instruments, or more motivated to practice, and as neurocognitive changes emerge from nascent musicianship, those students continue to hold an advantage in a virtuous cycle.

There *are* many cases in which learning one skill transfers to a related skill. For example, learning to play the piano and then applying that knowledge to learning to play the violin, or learning to drive a car with an automatic transmission and then learning to drive a manual transmission, are cases of *near transfer*. *Far transfer* is when there is no obvious connection between skills, such as finding that you're better at auto repair after you've learned to solve a Rubik's Cube puzzle, or that you get better at surgery after learning to juggle. Cases of far transfer are far less common than cases of near transfer, making it all the more striking when we run across them. The early evidence is that music lessons result in far transfer. A small number of experiments with randomized control groups show that playing an instrument leads to enhanced verbal ability many years later, and, more impressively, to lifelong benefits to brain health. In particular, musical instrument training is associated with statistically significant improvements in attentional focus and processing speed in aging adults.

The most plausible reason that music lessons and performance could provide all these benefits is because music engages nearly every area of the brain that we have mapped. It is a whole-brain activity—not simply left or right, not simply cortical or subcortical.

Playing an instrument entails generating suitable motor action plans, memorizing, and selecting refined movement patterns, as well as precisely initiating and stopping movements. These components must then be linked together into sequences. It demands memory, focus, emotional control, and goal setting, ranging from perfecting a single piano note to memorizing an entire symphony. For singers or wind instrument players, breath and diaphragm control are trained, together with lip, tongue, and jaw movements. Playing woodwinds and strings necessitates the coordination of both hands performing distinct tasks. Percussion players, in addition to this, must master different rhythms with each hand. Drummers must coordinate all four limbs independently. Even for non-drummers, foot tapping often accompanies hand movements, adding another layer of coordination to the mix.

The process of learning an instrument, then, exercises the very brain circuits we need to perform nearly every other task in life. And expert musicians? They will complete more training than surgeons and astronauts put together, typically *at least* 10,000 hours.

Neuroplasticity is the ability for the brain to be molded and formed through the changing of synaptic connections, and the growth of neurons, dendrites, and myelin sheaths. The most profound brain plasticity effects have been observed in professional musicians. And quality of life is markedly higher for people who learn an instrument. If that shatters your stereotype of the suffering artist, surely there are those that fit this description; but as a group, musicians are happier than nonmusicians.

The benefits accrue across the lifespan and take on even greater importance as we age. One inevitable decline that comes with aging is a general mental slowing. Reaction time and overall processing speed slow down with every decade after 40; it takes longer to hit the brakes on the car, to process and learn new information, or to retrieve memories. Older adult musicians, whether amateur or professional, show much smaller declines in processing speed, and their nonverbal auditory memory and verbal memory remain stronger. Musicianship also lowers the odds of developing mild cognitive impairment and dementia, or at least it can mask the symptoms by helping us build up cognitive reserve, as we learned from Glen Campbell and Tony Bennett.

Interestingly, amateur musicianship may confer more benefits than becoming a professional. This would seem to form a paradox: if music is so healthy for us, a kind of balm for the soul, senses, and brain, you'd think that being able to do it all the time would be better than doing it only some of the time. Yet statistics show that professional musicians experience greater incidence of alcohol and drug addiction, depression, poor health outcomes, and higher mortality.

One study analyzed death records of over 13,000 professional musicians who died in the United States between 1950 and 2014; their mortality rates were twice that of the general population. And professional musicians were more likely to die from violent deaths (such as suicide, homicide) and accidental deaths (such as drug over-doses and liver disease). This is particularly true among country, metal, rock, and jazz musicians. Apart from that, 71% of British musicians reported having panic attacks and/or high levels of anxiety, and 69% reported suffering from depression. That's three times higher than the general British public.

What accounts for these statistics? One possibility is that some unknown or hidden factor leads to distinct behaviors that then only appear to be related. It's not that being a professional musician is unhealthy, but that some underlying factor drives some people to music and also drives them toward unhealthy behaviors. As an example, maybe childhood trauma causes a desire to go into the expressive arts. That trauma also creates a vulnerability to depression, deficits in the brain's impulse control circuits, and a predilection for self-medication through drug abuse.

Another possibility—not mutually exclusive—is that we are lumping when we should be splitting. Much of what we do in science comes down to a question of lumping or splitting: "Do I lump together these separate phenomena because they are all of one kind, or do I split apart these things that appear identical because in fact they are quite different?" For decades we lumped a variety of unhealthy cellular division together under a single term, "cancer." We now understand that there are many different kinds of cancer, with different genetic signatures and treatments, and that curing them can be better understood by splitting them. Maybe amateur musicians have a different relationship to music than professionals and so we should be splitting these two groups.

And here, another split is key: the vast majority of professional musicians are just barely getting by. When I ask you to think of a "professional musician" you may think of Beyoncé or Drake, Taylor Swift or Eric Clapton. But they are the anomalies. There is an important distinction between these "rock stars" who have lots of money and support systems, and the workaday musicians who don't, living with the precarious and capricious ups and downs of the music business. A small number of artists at the very top make most of the money, leaving all the rest of them to divide up what's left. (Talk about lumping and splitting!)

George Musgrave of University of London distinguishes between the joys of creative practice versus the job of career building. The amateur gets all the benefits of music as a creative form of expression, an emotional outlet, the camaraderie of playing or singing with others, the rewards of discipline, practice, and mastery, and the opportunity to engage with some of the greatest works ever produced by some of the greatest minds in history. The professional receives those benefits, too, but has to worry about constructing and maintaining a career and a livelihood. Many of the characteristics of music as employment are destabilizing, and present challenges to one's well-being. The working conditions are tough—most professional musicians (like most actors) are never sure where the next job will come from, whether their next performance will be met with acceptance or derision, will open doors for them or close them. Their failures tend to be very public. More so than in many occupations, a performer's sense of self and self-worth becomes tied to their identity and status as a musician.

The very concept of success is particularly fraught for the popular musician. At first, it may simply be to make a living, then it may be to be able to play to larger audiences. At the very top, musicians worry about staying on top and not losing their relevance. After achieving success, they may worry about whether their peers feel they have "sold out" or whether they are respected not just as providers of entertainment, but as creators of something lasting, the creators of art.

Classical musicians face unique mental health challenges, often related to the high-pressure environment of their work. Performance anxiety, competitive auditions, the quest for technical perfection, and the physical demands of practice can lead to stress, anxiety, and depression. Consider the violinist, her bow a quivering diviner between silence and symphony, the pianist whose fingers must not only strike but caress, the singer whose breath and vocal folds must reproduce pitches, vibrato, and emotion with laser-like precision.

These technical and financial demands may find themselves in collision with the musicians' goal of creating meaningful art. As we've seen, the "right" song for you as a listener is the one that fits within your unique idiosyncratic taste in music. It builds a bridge between the mood you're in now and the one you aspire to. It needs to form a majestic balancing act between invoking the tangible and the possible, the way the world is and the way it could be. As Adam Gopnik has said, music that speaks to you embodies "some fight between the ideal and the real." Tip it too narrowly toward the ideal and it becomes a cartoon and lifeless; tip it too narrowly toward the actual, and it loses all the poetic sweep that causes it to inspire. Music can't *be* real, or *represent* the real—it has to *suggest* the real in a way that is artful, pregnant, pointing toward an eternal beauty, while remaining rooted in this world, this world of sounds, vibrations, notes, and neurons.

To achieve that requires the sort of surrender to the muse that few are able to sustain, but when an artist hits it, the audience can taste it. At the New Yorker Festival in 2016, David Remnick interviewed Bruce Springsteen, who confided that music was a way to medicate his anxieties, and a way to forget about those parts of himself he didn't like. "I'd had enough of myself at that time to want to lose myself. So I went on stage every night to do exactly that. Playing is orgiastic—it's a moment of both incredibe self-realization and self-erasure at the same time. You disappear and blend into all the other people that are out there and into the notes and the chords and the music that you've written, and you kind of rise up and vanish into it. And that was something that I was pursuing."

Remnick asked why Springsteen didn't lose himself in drugs. "I was too frightened. It took me so long to find a piece of myself that I could live with, that I was very frightened of losing that when it came to the substances. Plus, I'd lived around a *lot* of drug takers, I'd seen some of the really worst effects. . . . I had friends that killed themselves and friends that really kind of went and never came back. And so I was very frightened . . . it just wasn't for me."

The artist-musician plays a dual role as a creator and as a listener to their own creation. They strive to find this balance for us, to convey it to us. If you are that artist, and you yourself aren't well-rooted, this search will drive you mad. And in a furious and possibly futile attempt to keep madness at bay, you reach for the Jameson's, the Southern Comfort, Ambien, Adderall, Mary Jane, or Harry the Horse.

Acquiring and maintaining mastery tends be all-consuming—the laboriously intensive requirements of holding on to technical skills, the drudgery of practicing upward and downward arpeggios in every key, and the pressures of trying to stay on top, and to continually top yourself. As Gopnik notes, many great musicians such as Michael Jackson often have "hollow lives of enduring unhappiness." This is the opposite of the life Arthur Rubinstein chose. His philosophy could be boiled down to: *I make mistakes. I don't practice much. I live life.* He was a happy, gifted musician, an exception to the tortured, miserable one. Paul Simon had his artistic and personal ups and downs, but I find that his last five albums, recorded between the ages of 59 and 80, are the best of his career, and that he's "still creative after all these years."

I like to think that the music itself is a solace, a balm for the musician that does more good than the career does harm. There are illustrious instances where sadly that is not the case.

Here, it's important also to distinguish between musicians whose primary workspace is the studio, and those who are performers. Music has many health benefits, but the grueling life of a touring musician may be what causes these grisly statistics. The recognizable names of musicians who died young—Janis Joplin, Jimi Hendrix, Whitney Houston, Amy Winehouse—were all extraordinary performers. Maybe the music had healthful effects when they were in the comfort and sanctity of their own living rooms, singing and writing, but going out on the road presented a special hazard. This would have been especially true for Black musicians in the '50s, performing in the era of Jim Crow: You can't stay in the hotels that white musicians are staying at. The pay you're getting is a fraction of what white musicians are getting. Why wouldn't you, at the end of the day, or at four in the morning, do something to pick your spirits up?

It seems, then, that the risk factor for professional musicians is driven largely by the pressures of being on the road. In addition to everything else mentioned, the traveling and antisocial working hours make it hard to maintain close personal relationships, and we know that stable, intimate relationships confer not just psychological but physiological benefits that last a lifetime. To play music is to be vulnerable, to open oneself to the world. Yet in this vulnerability lies both the risk of despair and the potential for sublime transcendence. The life of the touring musician can be both fervently rich and startlingly barren. It is a precarious tightrope walk for most—to engage in an almost sacred communication with sound, a solitary path, where the instrument is not merely an extension of one's physical self, but of one's psychological self. Music is a balm for many, but for some, its healing properties may be overpowered by the pressures of trying to perform one's most innermost thoughts and feelings in front of a live audience.

I dropped out of college after my sophomore year to pursue a career in music. I joined a succession of bands—country, jazz, rock—each one collapsing under the weight of our own incompetence. The band I loved the most, and that seemed poised to really go somewhere, was The Mortals. Forty years later I still listen to our cassette tapes with rapture at the power of the writing by Alan Clement and Bob Miller, and the sheer musicality of the group. I was on bass, Alan on guitar, Bob on vocals and flute, and Alan's brother Bruce on drums. One of the features of our performances was that we would all switch positions, moving clockwise, and take over each other's instruments for one song, returning to our starting positions. (This has become a trendy thing for bluegrass bands to do.) Playing drums for that one song with The Mortals was the most fun I've ever had as a musician. Maybe this was because nobody expected me to be good—all I had to do was exceed the audience's (and the band's) low expectations and I could bring it all home.

I pursued the career for 13 years. Some months I made no money at all; other months I made enough to get through the next few months. About 95% of musicians live this way, holding down day jobs to pay the rent. Debora Iyall, from the popular San Francisco band Romeo Void, worked a job making sandwiches even when her group had a hit record. Musicians who played with Neil Young worked as drywallers when they weren't on the road. The music business is a crazy horse. Knowing that reality manifested in me as an increasing propensity to take everything very seriously. After all, my career was on the line. But I became so serious and single-minded that I wasn't having fun anymore.

After 13 years, I packed it up, sold most of my equipment, and went back to college, where I finished up my bachelor of science and then went for eight more years to get a PhD and postdoctoral training. One of my graduate school classmates was Dawn Rundman, whose husband, Jonathan, was a guitar player and songwriter. Jonathan and I got together a few times to play and had a blast. I continued to play, started to write again, and one day I realized I was having fun again. And that's when my music career started to take off. Steely Dan asked me to help them remaster their album catalog for CD release. Stevie Wonder asked me to co-produce an album with him (I had to ask my doctoral advisor for time off; he graciously granted it). When musicians I had met along the way in my first career started reading about my research in music and the brain, they thought it would be fun for us to play together. And that's how, as an amateur, I came to perform with professionals: Bobby McFerrin, Rosanne Cash, Rodney Crowell, Victor Wooten, Renée Fleming, and Sting. It's not that I am such an amazing player that they wanted to play with me (I'm not); it's that I'm clearly having a good time, trying to play the music, if not the notes. I make mistakes. So do they. No one seems to mind. The very top musicians and the amateurs have a lot of fun; it's those sad people in the middle, as I used to be, taking it all so seriously that they get in their own way and can't communicate the emotions that audiences come to hear. Being an amateur is something to embrace. Zen mind: beginner's mind. The legendary cellist Pablo Casals was asked, at age 90, why he continued to practice. "Because I think I'm making progress."

Playing music is fun, and perhaps we can leave it at that. But there are benefits, and a few ways to conceptualize them. Playing could make you a happier, more fulfilled person, providing joy, comfort, and a sense of accomplishment. But it seems these are not enough for the policy makers who want to eliminate music from school curricula—they want to know if musicianship will change the brain in ways that make students better at other academic subjects (math and reading are the usual targets), the issue of cognitive transfer. A third possibility is that musicianship changes the brain in ways that are beneficial to overall health, particularly if it is neuroprotective for aging. And, if any of these benefits accrue from learning to play an instrument or sing, how much training is needed before they show up, and how long do the benefits last? They can last a lifetime and begin after as little as 12 weeks of instruction.

We can calculate a person's brain age independently of their chronological age based on structural morphometry (measuring variations in brain anatomy, volume, and shape). When we apply this technique to musicians, they display younger brain age than nonmusicians: music-making brains are more youth-like. Eckart

Altenmüller is a professor at the Hanover University of Music, Drama and Media in Hanover, Germany. He has done some of the most innovative work in understanding the brain bases for musical behaviors. A trained flutist with a degree in medicine (and a specialty in neurology), Eckart is uniquely positioned to contribute to the field.

Recently, he's become interested in how to stave off cognitive and motor decline with aging. "Intact motor functions provide the basis for autonomous living," he notes. "In daily life, many activities require adequate upper limb function and especially hand and finger control, such as dressing, eating, and writing." Because aging has degenerative effects on hand function, Eckart wondered if piano playing could have neuroprotective effects. He recruited 156 older adults, average age 70, none of whom had ever had more than six months of musical practice in their lives. Notice that this is a true experimental design, not a correlational one. Half of his older adults were assigned, randomly, to a music listening group, the other half to an instructional piano playing group. All participants in both groups agreed not to attend any other music instruction during the experiment, which lasted one year. He found that compared to the music listening group, the piano lessons group showed improved fine motor control as measured by a standard motor control task. That is, piano practice improved performance in a non-piano domain. An interesting overlay to this is that, in general, older men are poorer in fine manual control than women of the same age, yet the performance of men in the piano playing group caught up to that of the women (with both showing improvement over the no-lessons group). And both groups showed strong improvements in mental processing speed.

Eckart found that the older adult piano players' brains formed a new and well-connected network that tied together primary motor cortex, thalamus (the brain's relay station for all the senses except smell, and a hub for learning and memory), and putamen (implicated in motor movements as well as emotion—we see higher putamen activity when people cooperate with others). He also found a refinement or streamlining of neural connections, resulting in more efficient information transfer. In other words, the brain was changing itself, remodeling and optimizing. The musicians' brains underwent structural modifications in response to the exercises and training, reallocating neural resources in a way that enhanced performance not just in piano playing, but in other domains as well. Why? Music lessons require learning motor sequences, including sequence order (explicit learning) and the optimization of those sequences (implicit learning). The explicit learning of sequences strengthens general working memory, and optimization promotes increases in overall processing speed. In this way, the piano lessons were able to spread their benefits around the brain, and around the everyday lives of the older adults who took them.

⌢·

I first met Krista Hyde when she was a graduate student at the University of Montreal, just across town from my laboratory at McGill. She was full of life and with an easy laugh, and everyone who met her took to her. Academia can be a very competitive, back-stabbingly brutal field. (Henry Kissinger once said that the reason university politics are so vicious is because the stakes are so low.) Krista managed to hold her head above the many petty rivalries in the fractured music neuroscience community, and that is quite a feat. Although we were at different universities, we saw one another often at parties and colloquia. In one of her first studies as an independent researcher (with her own laboratory at McGill) she gave keyboard lessons to children five to seven years old for 15 months, scanning their brains before and after. She found that the primary motor cortex of the children increased in size, that connections between that area and sensory areas (tactile and auditory) had become strengthened, and that fine motor control had improved even in tasks not involving the piano—indicating cross-domain skill transfer in young children.

In a separate study, she recruited children with autism spectrum disorders (ASD) who had a minimal history of music lessons. Half of them met with an accredited music therapist for 8 to 12 weeks. The music therapist guided them in using musical instruments to teach turn-taking, sensorimotor integration, social appropriateness, and general communication skills. A control group of autism spectrum children practiced the same skills using nonmusical play activities.

The communication skills focused in part on pragmatics, a term used in the field of linguistics for being able to understand that normal communication isn't always literal, that it uses sarcasm, metaphor, and indirect implications. For example, if two people are both sitting in a room and one is nearest the window, the other might say, "It's hot in here," as an indirect way of asking the other person to open the window. We don't just say, "Will you please open the window?" because such a direct command can sound overly confrontational with someone you don't know well. "It's hot in here" establishes an opportunity for the sharing of common experience through turn-taking—if the other person doesn't find it too hot, they have an opportunity to say so, and from there the two people negotiate. Or: if one person is droning on and on and on, the other may roll their eyes and say "that's interesting" with an exaggerated upward-then-downward pitch trajectory. Someone who is responding only to the literal meaning of such utterances has difficulty understanding what is really-and-truly meant, that it is *not* interesting, and this failure to use pragmatics is a marker of ASD (and of Mr. Spock from *Star Trek*).

Krista and her team found a significant improvement in the ability to use pragmatics for those in the ASD music group. Playing music with someone provides a structured form of turn-taking and sharing that is nonverbal, avoiding the difficulties many people with ASD have with language and sensory overload. An unexpected benefit was that family members reported a significant increase in their own quality of life, possibly because the children were able to maintain a less threatening, far less fraught form of social interaction through music than they could through speech. Adding to these

248 I Heard There Was a Secret Chord

behavioral measures, Krista found increased connectivity between primary auditory cortex and thalamus, and between auditory cortex and putamen.

⌒

Learning an instrument (including voice) passes through several stages. At the very beginning, everything may seem awkward and unnatural, as your muscles do things that they hadn't been asked to do before. Think back to when you first learned to ride a bicycle and how wobbly you were. The second stage involves rapid improvements of performance until a piece of music is known—the notes, the rhythms are all under your control. But then, somewhat frustrating for many, this is followed by a slow knowledge acquisition phase as the musician's brain reconfigures to create an optimal performance. During the initial two stages, the explicit motor movements are processed in the putamen and caudate nucleus; this is where and when the *sequences* are stored in the brain—which notes follow which. At the same time, sensory regions are listening to match up the sound you hear with what you intended. They work in tandem with the primary motor cortex and supplementary motor cortex to modify any unintended movements, while prefrontal regions plan the execution, and the hippocampus stores the new information. In parallel to this cortical activity, the cerebellum participates in error correction. Ultimately, if one can persevere through this slow knowledge acquisition stage, movements become both consolidated and optimized and finally automatic.

Even if you don't play music, learning about how to listen confers benefits. A team of Canadian scientists led by Sylvain Moreno found that just 15 hours of classroom instruction *about* music—cartoon characters teaching about rhythm, pitch, melody, voice, and other basic musical concepts—boosted preschoolers' overall vocabulary. Not only did they perform better on verbal tests than preschoolers who were taught about art by the same cartoon characters—concepts like shape and color—but neuroimaging results also showed corresponding brain changes.

Playing a musical instrument alone cannot claim the crown of ultimate brain health and anti-aging prowess. It cannot outweigh the significance of a nourishing diet, quality sleep, and regular physical activity. But it does confer a wide range of benefits, not least of which is the sense of joy it can bring, and the associated mood elevation.

About a year into the COVID lockdown, I was working on two piano pieces by Beethoven: the first movement of the *Moonlight* Sonata (Sonata No. 14 in C-sharp minor, Op. 27, No. 2) and the second movement of the *Pathétique* Sonata (Sonata No. 8 in C minor, Op. 13, the *Adagio cantabile*). I had become comfortable with the pieces and could play them from memory, which meant that I no longer had to concentrate on playing *the notes*; I could focus on coaxing the music out of those notes, the emotional meaning. Then something surprising happened. My fingers and movements merged into the emotional story of the music, becoming one with it. I was no longer just hearing Beethoven, or just playing his music, but it had become embodied. And with that, a great freedom came. I no longer had to think about the music, or the notes, or my fingers. Instead, I could think beyond the story that the composer was trying to tell, to what that story meant to me. And it meant something different to me each time I sat down to play it, each time I heard the music echoing off the walls of the room. It became a story with common themes, ever-evolving, and with no end to the surprises that unfolded. As so many others had, I was learning to fly.

Chapter 14

Music in Everyday Life

At home

The first encounter most of us have with music is a lullaby, sung to us by a parent, used to "lull" us to sleep. Lullabies are at least as old as recorded history and have common musical elements. They tend to be hummed, or sung in a soft timbre, almost a whisper (it is difficult to image Joe Cocker or Janis Joplin singing one, but I imagine they could access a softer voice if they wanted to). Lullabies tend to be in a slow tempo. The classic lullabies sung or hummed to Western children are in 3/4 or 6/8 time, giving them a lilting and gentle feel that facilitates sleep: Brahms's "Lullaby" ("Lullaby, and good night, in the sky stars are bright") and "Rock-a-bye, Baby." Lullabies tend also to use small, stepwise motion in the melody to avoid too many surprises—"Twinkle, Twinkle, Little Star" uses almost exclusively adjacent scale tones, except for the first interval, a consonant fifth. These soft melodic sounds increase bonding between parents and infants, and current speculation is that, like breast-feeding, both parent and infant experience increases in oxytocin, the social attachment hormone, and prolactin, a soothing and tranquilizing hormone.

Humming, in particular, has benefits not just for the hummee but for the hummer. Bhramari is a yoga breathing technique in which exhalations are longer than inhalations. This stimulates the para-sympathetic nervous system to reduce levels of stress, lower blood pressure and heart rate, and increase levels of nitric oxide to promote healing and widen blood vessels. Parents who hum to their infants experience reductions in their own stress, in turn reducing the likelihood that stress will be passed on to their babies. In one study with premature infants in a neonatal unit, live music was found to beneficially lower heart rate and improve sleep patterns.

We often hear from city-dwellers leaving the city for a quiet vacation, venturing to the countryside, forest, mountaintop, or desert, how much they enjoyed the silence. In reality, these environments are never entirely silent. The sounds of nature, such as wind, birds or insects chirping, leaves rustling, even a noise from far away, are present. In truth, humans evolved to find true silence alarming. Experiments with sensory deprivation tanks and anechoic chambers teach us that total silence produces a feeling of fear and discomfort. For social animals, vocalizations and movement are signs that everything is fine; as soon as a potentially dangerous situation presents itself—such as a predator—members of the herd or group freeze, and turn quiet. Other members notice and, in a spreading activation, become quiet as well. Intense silence is a danger signal.

Singing and humming to ourselves, then, serves to fill the silent void. We often sing when we find ourselves in intense situations. Mikheil Khergiani, a Georgian mountaineer, was known to sing when he faced a life-threatening situation. In *East of Eden*, Adam Trask is facing one of the most important decisions of his life. Steinbeck writes, "Suddenly he knew joy and sorrow felted into one fabric. Courage and fear were one thing too. He found that he had started to hum a droning little tune."

Filling that silent void is most often done by turning on the TV or background music. The uncomfortable sound of pure silence is why so many students want to study with music on, or at a café or other public space.

For thousands and thousands of years, we spent our leisure time with our family. When the sun went down, we'd gather around the fire for warmth. We might sing and play musical instruments. We'd talk. Changes in technology, architecture, the demands of work, and the addictive nature of being plugged in have caused us to retreat into a world of our own screens, walling us off from true human contact. If we do listen to music, we listen to it in isolation, in our own earbuds, in our own world.

I conducted a study in 2016 to learn more about the benefits of having music playing in the home on a regular basis. We sent surveys to 30,000 people from 9 different countries asking about their relationships and the types of activities they did with family members. Embedded in these were a number of lifestyle questions including how often they listened to music through loudspeakers where others in the home could hear it (as distinct from earbuds, headphones, computer speakers, or massive speakers in a "man cave"). When there was music playing out loud in the home, people reported having stronger relationships, more intimacy and quality time together, and they were happier.

To find out whether music was the cause of these feelings or merely correlated (after all, maybe people who are happier to begin with are more likely to play music) we conducted a controlled experiment. Thirty-one families from five countries agreed to listen to music for one week and then live without music for another, in counterbalanced order. All family members were given smart watches that tracked their heart rate, activity, movement, and calories burned, and we placed beacons around the home to track their movements. We installed wireless speakers that allowed each family member to listen to their own music, or to listen to what others were listening to.

During the music weeks, families spent 3¼ hours more time together. They sat or stood 12% closer to each other; the parents or partners had 67% more sex, and, in heterosexual relationships, the woman was more likely to initiate sex. In addition, having music playing accounted for participants cooking 33% more meals together, and 58% even reported that they thought the food they made while music was playing somehow tasted better. There were significant reductions in negative emotions such as distress, irritability, and the jitters during the music-on week (~13%).

This all makes music listening sound appealing, but one might ask if listening to some kinds of music might have negative effects. Heavy metal is often accused of causing aggressive behavior, for instance, because it features massively distorted instrumental sounds, vocals that are screamed, and a powerful onslaught of percussion. Many of these sounds are acoustically similar to animal calls that indicate a looming threat, communicating fear or aggression. The lyrics, though occasionally poetic and beautiful (Robert Plant's lyrics for Led Zeppelin are an example, as well as Buck Dharma's for Blue Öyster Cult), often focus on violent themes—death, suicide, alienation, and anger.

Those who are not fans of heavy metal may only hear it as dark, antisocial, and angry, causing them to wonder if listening to heavy metal can lead to suicidal ideation and psychiatric disorders. There is no evidence that listening to heavy metal leads to any of these negative behaviors. Those few studies showing detrimental effects of heavier styles are most often conducted with people who are not fans of those genres. Any negative effects could be better explained by being forced to listen to music you find unpleasant than by the music itself. In fact, heavy metal fans often feel empowered and joyful after engaging with it. It becomes a way for them to feel validated in feelings of otherness or disaffection.

In the workplace

One of the questions I am asked most often is some variation on "Is it a good idea to play music in the workplace?" I am not a one-size-fits-all kind of guy, and so the answer is—perhaps infuriatingly—*it depends*. It depends on what you mean by workplace, who it's being played for, how, and when.

To begin with, it depends on the music that you happen to gravitate toward, which is largely a mystery. Musical preference is no different from other tastes, but people seem to find it especially odd that we can't specify who will like what music. We don't find it strange that people have different taste in food. Or that people want to see different sorts of movies. As my grandfather used to say, "If everyone liked the same thing, they'd all want to get with your grandma." But we do find it odd that people have very particular tastes for music. Part of our blind spot here is self-selection: we tend to spend time with people who have similar tastes in music.

Next, not all workplaces are alike and not all jobs are alike. If you're a long-distance truck driver or a train engineer, you'll likely spend hours looking at a straight road, or set of tracks, ahead of you. Much of the job is repetitive and boring, and it doesn't require a whole lot of attention unless something unexpected occurs—then you need to be alert enough to take quick action to avoid an accident. The trick in all of this is not to fall asleep, and to stay alert, but not *so* alert that you exhaust yourself—if you are in a state of constant high vigilance, you may not be able to keep that up as long as you need to. Music can play an important safety role here by raising your arousal levels just enough to keep you awake.

In the old days, where people, actual humans, would stand on a production line and watch products come down, they'd have to pull out the defective ones. Or they'd be on an assembly line for a car, and all they had to do was attach a handle over and over again. This is largely done by robots now, but back then, Muzak® was developed to meet the challenge of keeping employees alert without distracting them.

The choice of music here is crucial—lullabies or spa music may lull you into sleep or inattention; EDM or trance music may draw you so far into the music that you're not attending to the outside world. You need Goldilocks music: not too hard, not too soft; stimulating and engaging, but not too stimulating and engaging. As neurosurgeon Katrina Firlick explains, neurosurgery operating rooms almost always have music playing. Like truck driving, much of neurosurgery is—counterintuitively, perhaps—quite boring and repetitive, drilling through bone and plugging up leaks like some sanguine plumber. In one study, self-chosen meditation music improved the quality of suturing—another repetitive task—in an experimental bypass procedure. The right music thus can be helpful, and the wrong kind can be dangerous.

Background music for repetitive work, like cleaning the house, is a good idea. But for anything that involves concentrated brain work, background music makes performance *worse*, placing a distracting obstacle in the brain's way. So the idea of music in a modern office place or research lab is a bad idea, and studies have borne this out: you're dividing your attention.

None of this means that music should play no role in the workplace. Painters and sculptors often do their best work with music on, as do cabinetmakers—and that is certainly brain work.

The ideal way to use music in a conventional office workplace is to listen to it between work cycles. Whether you use the Pomodoro method (working for, say, 40 minutes and then taking a 5–10-minute break) or some other work-break cycle, listening to music during the break can be beneficial. The science behind such breaks is that nearly all work that we do requires that we make a series of decisions, even if they are small ones. Do I read this email now or later? Do I re-read this sentence because I might not have understood it the first time? Do I write down what was just said in a meeting or is it either not important enough, or so memorable that this would be a waste of time? Do I need to take a bathroom break now or can I wait? The interaction of the voice in your head and the prefrontal circuits that are trying to keep you focused and on task is continuous and leads to decision fatigue. When this happens, your mind wanders or you feel like you need a nap. In our overcaffeinated society we tend to reach for another cup of coffee, but that does not solve the problem. What we need is to hit a reset button in our brains—to get ourselves into the default mode, to reboot out of the executive mode frenzy. Music is one of the most effective ways of doing this. So the optimal use of music is to step back from your work, close your eyes, and listen to music before going back to work.

Most of us have a strong intuition about what music we want to hear and what will help us. Whether curating a playlist for ourselves (what we used to call "mix tapes") or simply reaching for particular songs directly, most of us know whether that music is "working"—having the desired effect—and we can switch to something else if needed. Music is not the only way to do this: naps, meditation, walking in nature, immersing ourselves in art can all accomplish this with similar results. Music's advantage is that it is more accessible to most of us, and with more than 100 *million* songs in the catalogs of the major streaming services, there is more variety.

Another use of music in the workplace is as an icebreaker before meetings, or in the classroom before a lecture. I taught a 700-student introductory cognition course at McGill for 15 years. Students filing into the room tended to feel like they were cattle (some students even made moo-ing sounds as they squeezed through the narrow door-ways into the lecture hall to find seats). I played music as they were fil-ing in, typically not-yet-released albums I had obtained from friends in the recording business. This had three demonstrable effects. First, students made sure to show up on time, which was nothing to take for granted. In fact, some students came who otherwise would have skipped class. Second, instead of gabbing when they entered the class-room, which made it difficult for them to quiet down for the begin-ning of the lecture, they were attentive, and knew when I turned off the music that class would begin. Third, the music itself became part of the lectures in most cases, as a way to demonstrate fundamental concepts of cognition, such as memory, auditory perception, atten-tion, categorization, and language. Students didn't always like the music, but that didn't seem to matter. I remember that an advance copy of Taylor Swift's self-titled debut album was met with yawns and indifference (people are people and sometimes we change our minds). Students also heard advances of Amy Winehouse, Arcade Fire (a local Montreal band that came to class one day incognito),

Antony and the Johnsons, Sleater-Kinney, and some oldie favorites of mine like Creedence Clearwater Revival and Miles Davis.

After taking my class, Dale Boyle entered McGill's PhD program in Education and wrote his thesis about the use of music in a large classroom. He found that music in the classroom created opportunities for the professor to connect with students, which in turn created a more nurturing learning environment. The music established a more relaxed and friendlier atmosphere; it showed empathy toward students; facilitated student involvement; and helped to remove the social barriers between professor culture and student culture. Music also optimized learning and created salient moments in the classroom by motivating the course material with something students cared about, reducing exam tension, clarifying concepts dynamically, holding attention, and triggering memory.

Another campus music phenomenon is that there have been barbershop quartets and a cappella groups on college campuses for over a century, and they have always bordered on campy, straddled the line between cool and nerdy. Adding to the camp are the punny names—The Logarhythms (MIT), Treble in Paradise (American University), Shirley Tempos (Brandeis), The Accafellas (Michigan State), and The Pitchforks (Duke). To what do we owe the resurgence of college a cappella in recent years? Maybe it's the physical skill and vocal dexterity required. As journalist Madeleine Davies observed, "What's more impressive than human beings using their bodies to create amazing things? It's true for breakdancing, it's true for acrobats, and it's certainly true for a group of people coming together to recreate a multi-instrument song using *only their voices*." I asked Christopher Harrison, who himself is an a cappella singer (he is the leader of the a cappella groups Sonos and Arora and he has produced Pentatonix among others). He says, "The moment that a cappella is having on campuses is the perfect storm of shows like *Glee* being a huge hit, the swath of film musicals that have either been revamped or made in the last few years, and the *Pitch Perfect* films (1, 2, and 3) being a pop-culture spoof of college a cappella groups.

Pitch Perfect captured for me a very familiar fanatical fervor. Musical theater kids tend to hang out and break out in song all the time. It's just a very sweet, nerdy, lovely appreciation of vocals, which is very easy to do since you don't need any instruments. Those three things brewed into the resurgence we're seeing. Although in truth, a cappella never really left college campuses, but it does go up and down in waves. (It is always cool and never cool and it is usually accompanied by a conspiratorial wink.)"

For physical health and stamina

A recently published, randomized, controlled study found that older adults who took music lessons (1 hour a week) and played just 30 minutes a day for six months showed robust increases in gray matter, and significant improvements in auditory memory. Another study showed that lifelong musicianship improves the ability to distinguish speech from noise, such as in loud restaurants or other crowded spaces.

In activities of daily life, music is ergogenic: something that can enhance physical performance, stamina, or recovery. Ergo, music can help with physical fitness. Blondie drummer Clem Burke was monitored during live performances to investigate the physiological demands of drumming. It turns out he doesn't have a heart of glass, but that of a pro athlete. During each performance, he averaged between 140 and 150 (heart) beats per minute, peaking at around 190 bpm. He burned between 600 and 900 calories and lost about two liters of fluid in a 90-minute show (that's comparable to an athlete running 10,000 meters).

Back in 1911, statistician Leonard Ayres published a paper called "The Influence of Music on Speed in the Six Day Bicycle Race." For the New York City six-day race, he timed cyclists as a function of whether a band played or was silent, and found cyclists traveled 8.5% faster when music was playing. One factor is that music can improve our mood and motivation (dopamine again), increasing adherence to exercise workouts, whether it is walking to get in your 10,000 daily steps, running three miles, resistance training, competitive sports, or other forms of physical activity. It can help us to engage for longer periods of time, and to push ourselves harder than we would without music. Competitive runners can boost their cadence (time between steps) by listening to music at a tempo just slightly faster than their natural running cadence, because motor neurons synchronize to the beat of the music. Music can distract occasional exercisers from the unpleasant sensations associated with extreme physical effort and fatigue, as well as the boredom and lack of enjoyment many report as factors that prevent them from doing it.

Music has been shown to boost performance in a broad range of other physical activities. For example, music is used to enhance sexual intercourse by helping partners to synchronize their movement rhythms. Depending on the music and the couple, with training, music can help prolong male orgasm far beyond the normal range and to allow men to experience multiple orgasms. Guided imagery with music can improve desire, arousal, and orgasm in women suffering from sexual dysfunction. Music therapy also reduces anxiety, pain, and increases satisfaction in women undergoing assisted reproductive technologies; although more research is needed, music may even improve their chances of getting pregnant.

New scientific findings are coming down the pike all the time. Cancer patients are using music therapy to reduce chemotherapy-induced nausea, something that affects around 80% of patients. Anti-nausea medications are expensive, and carry a boatload of side effects. In a recent study, patients who listened to their favorite, self-chosen music (this part, again, is key) experienced a substantial reduction in nausea. Contrary to intuition, nausea doesn't begin in the stomach, although that's where you feel it—it is a brain phenomenon. That's why a non-pharmacological approach is effective.

For social meaning

Music is primarily a means of emotional communication. When we listen to music in a group, social meaning is constructed with the others around us. This social meaning can be intensely impactful; it can cause us to change our minds and even our lives—concerts to raise money for charities and political rallies with music are among the most powerful events many people experience. This social meaning also emerges when we listen with just one other person. I remember many, many hours just listening to records with my childhood friends, Tommy and John. The music connected us, and the growing friendship with one another allowed us to let down our guard to be moved by the music.

For older adults, whose social circles can shrink, music can help create new bonds. Many adult living facilities have pianos in a shared common space. Anyone is allowed to play them, and often a spontaneous singalong occurs, bringing the residents into literal harmony. It brings joy, stirs positive emotions, and sparks cherished memories that may have lain dormant for decades, and this contributes to residents' overall well-being. Participating in musical activities with others has become increasingly popular, especially group singing in communities or choirs. These activities offer numerous benefits for seniors. Older adults who join community choirs report a better quality of life, less depression, and greater satisfaction with their health. Singing together not only brings enjoyment and cognitive stimulation, but also improves physical and mental health and increases social interaction. Intergenerational choirs are popping up all over, an opportunity for older adults to meet young people (which keeps them young) and for young people to meet older people (which removes many negative biases that younger people hold toward older adults).

In one study, university undergraduates and older adults with dementia sang together. Starting with the first rehearsal, younger singers were "buddied up" with a person with dementia and that person's family members, and they sat next to each other during most rehearsals. At the beginning of each rehearsal, 15–20 minutes were given for socialization, a chance for the buddies to talk with each other. All choir members were treated as equals and called by their first names, regardless of age. Songs were selected that featured repetition, and few rhythmic or tempo changes, and included songs well-known by all, including "Stand by Me" and "Over the Rainbow."

Before the choir began, 65% of the words that young people used in spontaneously describing older adults were negative, such as *sadness*, *sick*, *helpless*, and *deterioration*. After three months in the choir, 75% of the words that young people used were positive, such as *unity*, *love*, and *caring*. One student said, "Months ago I was afraid of not knowing what to say or what to do [around a person with dementia]. But the issue was with *me*, not with the folks with memory loss. . . . Spending time with my new friends with Alzheimer's helped me to see that we are not hopeless when we start forgetting; we are hopeless when we give up and decide not to live." One individual with Alzheimer's said, "When I spend time with the students, I feel energized and accepted. It is fun being around young people."

Group singing constitutes a real-life activity in a setting that is organic and natural rather than clinical. But it is not for everyone— older adults with mobility problems or other issues that may make group singing stressful may benefit from group music listening. This may be one of the most promising avenues for future research. Listening to music together leads to synchronization of brain activity. No one has yet done the study, but based on the literature, a reasonable hypothesis is that group listening to music may foster intra-group cohesion, improve everyone's mood and feelings of well-being, and increase levels of oxytocin, serotonin, dopamine, and endogenous opioids. As Pink Floyd sang in 1967, "Music seems to help the pain . . . to cultivate the brain."

In some cases, social meaning can even emerge when we're listening alone, when we feel bonded or connected to an artist whose music we know so well. John Fogerty has been a constant companion to me since I was 12 years old, and today even a snippet of his music lifts me up. "He's my friend," I think, "and he can reliably raise my spirits when I'm feeling down or help commiserate with me when I don't want to be picked up." Of course, I know he isn't literally my friend, but the connection I've felt to him all these years, through shared musical experience, is as strong as connections I've felt to co-workers and a number of other people I called "friends" IRL (in real life). Of course, the connection is only one-way—the artist doesn't know me. But for art to succeed, the artist must know the human condition intimately enough to be able to convey experiences we all see ourselves in. The artist must give us the impression that they really *know* us.

Social meaning is also created when we play music alone as a proxy, or prelude, for trying to connect ourselves with others. And if we're lucky, social meaning emerges when we play with others because, in addition to what we're playing (and listening to what *they* are playing, one hopes), we are also coordinating actions, intentions, and pursuing a common goal.

What to listen to

The most robust finding, the one finding that has been the most widely and rigorously replicated, is that self-selected music is far more effective than music selected by someone else for nearly every application of music as medicine. One exception is the "relationship playlist" ("romantic mixtape") in which an intimate other prepares songs *just for you* that mean something to them, a collection of songs that may express the way they feel about you but don't know how to put into words, or are too shy to say. As Kate reminds us in the Tony Award–winning *Avenue Q*, "Sometimes when someone has a crush on you, they'll make you a mix tape to give you a clue." Imagine that you are young, inexperienced in the ways of the heart. You have a crush on someone but aren't sure if they like you back. Then, as Jim Croce sings (and Princeton's mix tape for Kate concludes), you "have to say I love you in a song." For the Boomer generation, such a song might have been "Be My Baby" by The Ronettes or "Cherish" by

The Association. "I Want to Know What Love Is" by Foreigner has brought many couples together. For certain millennials, it might be "Halo" by Beyoncé.

Another robust finding tells us that most people use music for self-medication and for mood regulation—and they tend to know what effects different selections of music will have on them. We use music to rev us up (like caffeine), or to calm us down, and as a social lubricant (like alcohol). We might engage with music on our own infrequently, encountering it only when others are playing it, or we might program our entire day with playlists for every activity.

But how to create the program? Some people who are extremely organized and systematic have created playlists for every conceivable event and know where to find them. Others use a more ad hoc approach, relying on memory, impulse, or friends. There is no "right way," other than what works for you. One factor is that order matters in a playlist. Whatever the purpose, you may find that some songs lead more naturally into your next choice than others. DJs, radio programmers, music directors, and musicians know this, either explicitly or intuitively. If you're trying to pump up the mood, you wouldn't usually move from a *fast tempo* song to a *slow tempo* one, or move *down* a semitone in key (rising *up* a semitone is a well-worn trick for boosting the energy of a song or segue). When planning music for relaxing, focusing, or meditating, you usually don't want the overall timbre and soundscape of the pieces to change abruptly. For dance music, it's nice to mix things up and put slow pieces and odd new rhythms in once in a while so the audience doesn't get exhausted, or too tired of a particular beat or tempo. For workout music, you probably want tonal variety but a steady tempo, coordinated to the speed of your movements (whether on a treadmill, elliptical, rowing machine, doing crunches, hitting a punching bag, or lifting weights).

Another consideration is balancing novelty with familiarity. How much do you or your intended listeners need to hear new music, versus finding comfort in well-known favorites? If you do like exploring new music, how exploratory are you—what constitutes "new"?

Some people want to hear new artists, but only if they sound like the old artists they've been listening to all along. Others have an insatiable appetite for the exotic, exploring music of other cultures and eras; many fall somewhere in between. Some fulfill their drive for newness by listening to new artists playing their old favorites. Classical music and jazz are particularly rich in this way. New recordings of well-known pieces are coming out all the time. You could spend months if not years listening to all the available recordings of favorite symphonies or concertos.

Ultimately, though, most of us will eventually habituate or grow tired of hearing the same music across a lifetime and will need unfamiliar music to create therapeutic effects. Indeed, searching for and finding new music we like is thrilling in and of itself. Harking back to our evolutionary history as hunter-gatherers, the act of search and discovery deeply activates our pleasure networks and exercises our motivational circuits.

Variety in music, as in many things, is key. As we have learned from Michael Pollan (and many others), a varied diet, drawing on proteins, fats, whole grains, green leafy vegetables, yellow vegetables, and so on, is more healthful than a limited one. As we have learned from exercise physiologists, varied exercise, with stretching, core workout, strength and resistance training, and cardiovascular/ aerobic training, is more healthful than a single fitness routine. People find comfort in what they know, but thrills, challenge, and excitement in what they don't. So it is with music.

There's an old story about President Calvin Coolidge visiting an experimental government farm with his wife Grace. Guides showed them around different parts of the farm. When Mrs. Coolidge was taken to the chicken yard, she noticed that a rooster was mating very frequently. She asked the guide how often that happened, and the guide replied, "Dozens of times a day, Mrs. Coolidge." Mrs. Coolidge said, "Go tell that to Mr. Coolidge." A farmhand was sent to find the president and give him the message. On hearing it, the president asked, "Same hen every time?" "No, sir, a different hen every time." The president said, "Tell that to Mrs. Coolidge."

The "Coolidge effect" in its original formulation referred to a shortened refractory period between consecutive erections in men when novel partners are available. This trait evolved through Darwinian sexual selection: males who responded this way to novel sexual partners could produce more children, thereby passing on the trait. (The existence of the trait stands apart from ethics, and should not be interpreted as an excuse for men to behave badly, nor as permission for promiscuity in a monogamous relationship; it is a biological phenomenon that is unrelated to cultural and societal norms.)

The Coolidge effect in its broadest form permeates our lives, shaping and guiding the human propensity for novelty-seeking. Infants go through a stage of trying to touch everything they can, putting things in their mouths, and exploring their environment; for the developing infant mind, the greater the novelty, the more attractive it is. Although this inclination can lead to dangerous behaviors, choking, and poisoning, in the broadest scheme it helps prepare infants for interacting with a complex world. Novelty-seeking reaches a peak in adolescence, when the combination of independent goal-setting and poorly developed prefrontal impulse control can lead to all kinds of outcomes, from the hilarious to the fatal. To varying degrees, all humans have a drive toward novelty-seeking, and it shows up in vastly different domains: early adoption of new technologies; exploring new recipes or restaurants; social organizations for networking and meeting new people; regularly getting a new car or home; keeping up with the latest fashions; travel; and experimenting with drugs (seeking novel states of consciousness or novel ways of altering emotions). And it applies to music.

Even if we don't realize we want new music, it is healthy for us to enjoy a varied musical diet. Finding new music used to be easier. When we were in school, our friends were sources of recommendations for new music, and we developed our social identity in part through the music we listened to. Radio stations played Top 40 songs, with new songs hitting the Top 40 every week. Album notes detailed all the musicians on an album, and in many cases, following that list of musicians could lead to discovery. It was Miles Davis whose "Kind of Blue" introduced me to Bill Evans, Cannonball Adderley, and John Coltrane, and impelled me to find their own records.

In the pre-digital era, the only way to hear music was to go to a live concert, listen to the radio, or buy a recording, and due to the expense of producing LPs and CDs, most recordings were funded by record labels. That meant that an artist had to make the cut. In the 1980s, Neil Young said, "There are *lots* of great musical artists and songs out there; you just don't know about them because they haven't passed through the turnstiles of the music industry." And that industry is unpredictable, capricious. When my job was to listen to demo tapes sent into Columbia Records, I was astonished at how good some of those were. Some of the best music I've ever heard is music you've never heard. Even a major label release didn't guarantee success. Two of my favorite albums of all time never sold well: *Sunny Nights* by Parthenon Huxley, and *Owsley* by Will Owsley, who was the touring guitarist for Shania Twain and Amy Grant. When I play them for people, they are dumbfounded at how good those records are.

Now those turnstiles Neil Young spoke of have all but vanished. Anyone with a laptop computer and a $100 microphone can make recordings as good as Rolling Stones records, and they can release them on their own. Every day, 100,000 are uploaded to YouTube, and more to other streaming services such as SoundCloud and Spotify. This is a golden age for music creators. Artists who might not otherwise have a following, those from the so-called long tail of the distribution, now can find a fan base who will buy everything they put out, including T-shirts, baseball caps, and special signed editions of their work. Although Columbia Records didn't sign Parthenon for a second album, he has continued to tour and record, and has built up an independent fan base (from that long tail I spoke of) who are dedicated enough to support him. Ironically, perhaps, he makes more money now by selling fewer copies of his recordings because instead of the label taking the lion's share of the income, it all goes to him.

Finding music you like in the current era presents a challenge. With so much to choose from, it's worse than looking for a needle in a haystack—it's more like looking for a needle in a stack of needles. As the streaming services incorporate more AI into the recommendation engines, as well as your feedback, they promise to be as good as a friend with whom you share musical taste. And friends, actual human friends, are still a great source. During the pandemic lockdown, people who reported limited access to the internet discovered *more* new music than people with fast connections, presumably because they were more purposeful in seeking it out and couldn't just turn on the internet radio and let it function as sonic wallpaper.

Going to concerts is also a great way to discover new music when there is an opening act. I unwittingly discovered some of my favorite artists as opening acts. I went to hear the Punch Brothers in 2011 at Jorgensen Hall in Storrs, Connecticut. Their opening act was Tom Brosseau. Picture this: an auditorium full of 2600 people all of whom came to hear a five-piece bluegrass band. Being an opening act is one of the hardest jobs in the music business. Nobody came there to see you, no one knows who you are, and everyone wishes the band they actually came to see would take the stage.

Tom walked out with just a guitar and came up to the microphone. Everyone was still chattering and shuffling about. When he began to sing, the room, as one unified mind, became absolutely silent. No one even coughed. He held us there in rapt attention through his 25-minute set. I was hooked. I bought some of his CDs at the merchandise table and now I listen to his music regularly, finding new nuances and beauty in songs I've heard hundreds of times. You never know who you'll discover in the opening slot, and some of them overtake the headliners in popularity and longevity. Jimi Hendrix opened for The Monkees; Queen for Mott the Hoople; Elton John for Derek and the Dominos; Taylor Swift for Rascal Flatts; Lady Gaga for the Pussycat Dolls.

The future of music search and health

The experiments that provide the strongest evidence for music as medicine employ music that is selected by a patient under the guidance of a licensed music therapist. But there simply aren't enough music therapists to go around, and this presents a scaling problem: how can therapeutic music selection be automated while still giving the patient the locus of control?

In the opening chapter, I alluded to the information that is stored about you in the cloud—your search history, geolocation, who you are with (if they also have a smart device), calendar, contacts list, and the kinds of things you view on social media. And the music services know a lot about your listening history—what you listen to, what you skipped, what time of day you listen, where you are when you're listening. Smart devices that read your biometrics know (or potentially know) your heart rate, heart rate variability, blood oxygenation level, respiration rate, skin conductance, body temperature, blood pressure—as well as how they fluctuate as a function of time of day and what activities you're engaged in. And they know those activities—whether you're running, walking, climbing steps, driving in a car, or sleeping; and when you are sleeping they know what sleep stage you're in and how long you've been asleep (!).

A consolidated, coherent ecosystem will take all this information and combine it in ways that are either to your personal benefit and convenience, or for the profits of someone else. The future will find a fully integrated system in which music selection and streaming are potentially invisible to you. Your smart watch will know to wake you in the right part of your sleep cycle so that you don't begin the day with sleep inertia, and it will play music that is specially designed to wake you up. Ideally, you'll remain in control of what gets played. When you first begin to use it, an AI-based system will select wake-up music that is informed by music science, such that the music will start out softly and simply, and gradually become more complex and louder so as to lull you awake, rather than startling you awake. Of the millions of songs that could potentially be used as wake-up music, AI will further build on your personal tastes and listening history to play either something that you have used to wake up in the past, or something you listened to before, or something that AI predicts you will like.

You would still be in control, with the ability to indicate to this ecosystem whether you like or dislike what it's playing you. But instead of the customary thumbs-up and thumbs-down icons to indicate "like" and "dislike," my preferred implementation would feature a third icon, maybe a clock, that you could press if you want to indicate to the system *I like that song, but just not right now*. A more elaborate version might even allow you to rate what you're hearing along one or more dimensions, using a scale from, say, 1 to 5, in order to capture more nuance than a simple binary yes/no. As you continue to rate the songs that AI plays you in this way, the system will learn more about you and your tastes, and continually iterate to improve its ability to select from you. You will end up with a truly personal set of songs that are optimized to help you wake up.

And that's just one part of your day. A day in the life of a connected person will select and refine the choices of music for when you're in the shower, when you're cooking, brushing your teeth, commuting. When your biometrics indicate that your attention is flagging at work, focus music will come on automatically. That will affect your biometrics, and if they don't reset to desired levels, the system will change the music until the desired levels are reached, and in that way it will continue to refine its algorithm for what you hear—based on your personal physiology. Similarly, when your biometrics indicate you are stressed, relaxing music will come on.

You'll have customized music for workouts, dinners, cleaning the house, walking the dog, practicing yoga, going to sleep, and having amazing tantric sex (almost surely that music will involve songs by Sting).

When you're sick or recovering from injury, music will be similarly selected—just for you, in the hospital and the clinic—and your caregivers will send you home with it to aid in your continued recovery. From hospital and clinic to home, music-as-medicine will be scientifically chosen.

The right music, whatever that turns out to be, can remind us of who we were, who we are, and who we want to be; it puts us in an intimate and private conversation with the angels of our better nature, as it opens up worlds of new possibility.

Sonic diplomacy

Experiments in my lab, conducted with Jeff Mogil, have shown that playing music together can increase empathy. There are several standardized ways of measuring empathy in a laboratory experiment; the one we used is based on the well-established phenomenon that people who feel empathy for one another dislike seeing one another in pain. Jeff and I set up that cold pressor test I mentioned earlier in a new study run by his then graduate student Loren Martin (Loren is now a professor at the University of Toronto). We asked one participant, the "experiencer," to hold their hand in a bucket of ice water (0–4 degrees C) for as long as they could while another, the "onlooker," watched. When the experiencer was a self-defined friend, understandably, the onlooker's ratings of distress, discomfort, and empathy were higher than when the onlooker was a stranger (on average the friends had known each other for three years or more and had spent 130 hours together across the previous month—what you might call close friends). But here's the kicker: we recruited strangers and asked them to play the video game *Rock Band* together before they were made experiencer and onlooker in the pain task. That simple intervention, of playing a musical game together for 20 minutes, brought on levels of empathy equivalent to being close friends for three years. We don't know how long that empathy lasted after the experiment, but the finding was striking.

My colleague Vinod Menon, our postdoctoral fellow Dan Abrams, and I were interested to know what happens in the brains of people who are listening to exactly the same music. We anticipated that there would be differences in activation, vast differences, because first, no two brains are alike, and second, not everyone reacts to a piece of music in the same way. Reactions to music are based on a lifetime of listening, individual preferences, and the momentary mood we're in as we listen. Indeed, your own brain activations to a piece of music are likely to differ from one listening to the next; you're not exactly the same person today that you were yesterday. What we found was surprising. The brain waves of people listening to the same piece of music actually synchronized, in spite of these differences. Synchronization refers to the phenomenon where neural responses in different individuals become aligned in time, showing similar patterns of activity in response to the same stimulus, in this case, music. Common structures spanning the brain, from the frontal lobes to the parietal, the limbic system, and the brain's control centers in the cingulate gyrus and insular cortex, had come into harmony with one another. Our study demonstrates, at a neurobiological level, the unifying power of music. What could this mean for healing and interpersonal, intergroup conflict?

Soon after our paper was published, I met a concert organist, Jonathan Dimmock. Dimmock had been touring the world promoting the idea that listening to music could help to resolve international conflicts. "What if government leaders could listen to music together before they sit down to negotiate treaties and trade agreements?" Jonathan asked. He founded an organization, The Resonance Project, with the mission of offering live music as a tool to benefit conflict resolution.

Stanford psychologist Lee Ross spent decades studying the process by which agreements are reached through negotiations and mediations. Impediments to reaching agreements, whether they're between corporations or governments, often stem from a lack of trust. Each side thinks the other is making unreasonable demands; that if the opponent is willing to make concessions, then those concessions must be much easier to make than those that are being asked of us; that even if an agreement is reached, the other side can't be trusted to uphold it; that the "facts" the opponent acknowledges are biased or made up. Jonathan Dimmock's idea was that entrenched positions, rhetoric, and language become walls that we hide behind, and that perhaps music can lower them and increase trust.

Some years earlier, in 2009, I had gone to the Esalen Institute near Big Sur, California, to attend Cris Williamson's excellent songwriting workshop. During our week there, there were several Israelis and several Palestinians taking workshops on a variety of topics unrelated to songwriting. One morning at breakfast, Cris proposed that we bring these two groups together and write a song during a couple of hours of free time that afternoon. Cris prompted them for thoughts, words, and lyric ideas, which she strung together into a song lyric. I played guitar with Cris and we found chords and melodies together. The finished song was the result of a combined effort of egolessness, collaboration, and an expression of yearning for peace.

The lyrics began by celebrating some of the things that Israelis and Palestinians both loved about their home—the taste of oranges fresh off the tree, the Joshua Pine trees, the connection to the ancient land of their ancestors. The very process of writing a song together taught us that the two groups of people had so much more in common with one another than they had realized, and that their similarities overwhelmed their differences. A couple from Los Angeles told us that LA has one of the largest expat populations of Israelis and Palestinians in the world. In a city that covers over 500 square miles, you might think these two groups had staked out neighborhoods as far apart from each other as possible, but that's not the case. They live in the same neighborhoods, practically on top of each other, and they live together peacefully.

The chorus that they wrote that day focused on the 708-kilometer wall that runs through the West Bank, separating Israeli Jews on one side and Palestinians on the other, isolating more than 25,000 Palestinians from the rest of the country. The Israelis in the room, some of whom were hardliners, came to see the wall as unnecessary and harmful to sustained peace because of the animosity it provoked. The refrain we wrote of "let's take down the wall," echoing President Reagan's speech to Mikhail Gorbachev years before, made for a powerful musical statement. We performed the song at dinner that night for all the attendees at Esalen. When I met Jonathan Dimmock, this was fresh in my mind because my singular experience at Esalen demonstrated so much of what he hopes for.

What exactly does the Resonance Project do? They provide small ensembles of exceptional musicians who perform carefully curated musical selections midway through negotiations or conferences, as the negotiators listen. "The effect of this simple catalyst can change the course of a meeting," Dimmock observes, "as each listener's brain aligns its wave patterns with those of others in the space, enabling a heightened desire to compromise." One fan of the program wrote in 2015 that music can "help us bridge our differences and show us we are heirs to a fundamental truth: that out of many, we are one." That fan? President Barack Obama.

Using music to ease suspicion and tension during negotiations is not a new idea. The Congress of Vienna brought together the heads of European states in 1814 to settle many of the issues that arose after a series of destabilizing events: the French Revolution (1789–1799), the dissolution of the Holy Roman Empire (1806), and the Napoleonic Wars (ongoing at the time of the Congress). Beethoven composed and performed several pieces for those assembled, including his Seventh and Eighth symphonies. The Seventh Symphony has long been one of my favorite pieces of music. It begins with a number of musical gestures that demand attention; shifts from loud to soft, and from fast rhythms to lilting ones. But it is the second movement that is etched in my memory. Its opening harmonic motion is delicate and nuanced, and before I even knew this background of being written for a peace negotiation, I found it would instantly evoke calm. The joy of being one of the clarinet players in an intermediate school orchestra playing that piece has lasted a lifetime.

Historians still disagree about the provisions agreed to during the Congress, but many consider that the outcome successfully prevented any further large and protracted European wars for the next hundred years. Beethoven at the Congress of Vienna was not a controlled experiment, so we don't know how the maestro's presence influenced the negotiators, but it is intriguing to imagine a room full of the heads of state being treated to new works by the greatest composer of the era as a backdrop for their peacemaking.

Music can and often has been used for the opposite purpose; rather than bringing people together, there is a long history of music being used to create conflict between groups, to incite interethnic violence, and to belittle out-group members. The Nazis used music to humiliate and break the spirit of prisoners; as the *Wall Street Journal* reports, "They were forced to sing cheery German folk songs as they marched to and from their work assignments, or during torturous physical activity. Camp orchestras were ordered to play during punishments and executions." Sean Michael Condon (University of Jyväskylä in Finland) reminds us that in Serbia, Slobodan Milošević used music to reinforce their isolationist self-image and stimulate enthusiasm for fighting, while simultaneously, the people of Croatia and Kosovo employed music to rally internal support and vilify the Serbs.

The arts—literature, theater, visual art, dance, and music—contain a power that we don't readily find in other modes of human expression. Spoken discourse is often literal and referential; it tends to be about specific things. The arts more typically rely on metaphor, and strive to convey emotional truth, if not literal truth. A work of art can introduce us to thoughts and ideas we may not have encountered before, expose us to the lives of people whose experiences are very different from our own. Through this uniquely human invention, art can cause us to feel empathy toward others. It can reduce prejudice. It can engage our compassion rather than our tendency to judge; it can arouse our interest and curiosity rather than suspicion. It is often said that you can't argue a person out of a position they didn't argue themselves into. In other words, if a person reaches an opinion based on emotion, all the appeals to facts and figures and logic will not cause them to alter their views. But the right piece of art can do that. By opening up a person's heart, the arts can cause them to see things differently and have a "change of heart." And once the heart changes, the mind follows.

For the heart and mind to change, we must allow ourselves to be vulnerable. We must make time and space in our lives for self-reflection and daydreaming, allowing music that inspires, challenges, and soothes us to help us reach a higher ground.

Music evokes dual emotional natures. It functions hedonically, creating pleasure, and it functions eudaemonically, for the pursuit of a higher state of being—personal growth, virtue, tolerance, acceptance, transcendence. We can find that higher ground listening to an honest voice. I doubt we will find it in AI-generated music. As Rosanne Cash writes, "We all need art and music like we need blood and oxygen. The more exploitative, numbing, and assaulting popular culture becomes, the more we need the truth of a beautifully phrased song, dredged from a real person's depth of experience, delivered in an honest voice."

Listening to music, one might wonder what is special about the person performing it, their instrument, the concert hall, the studio. Music does not live in the world of instruments, or scores, concert halls, or recordings. It lives inside composers and performers equally as in listeners. Scholar David Pettus says that the moment of creation of a work of art is not when the author puts down the pen, or the painter puts down the brush; it is not when the musician or actor performs the score or the script; it is born when that work of art is *perceived* by another human. Music then lives inside each of us who listen. It lives inside me. And it lives inside you.

Chapter 15

Fate Knocking on Your Door

Précis to a Theory of Musical Meaning

T HERE ARE A HANDFUL OF PIECES OF MUSIC THAT cause an instantaneous change of mood in me. It is not coincidental that these are pieces I know well. Although I have been deeply moved by music I was hearing for the first time—Mahler's Second Symphony, for example, or Chick Corea and Gary Burton's "Crystal Silence"—there was so much new and unfamiliar to me that those pieces had to do some heavy lifting to win me over. As Thomas Mann wrote, "To the layman's ears new music is wild, lawless cacophony, the dissolution of all restraint, the end of all things. It is rejected until the ear can catch up and becomes accustomed to the new."

Take "Crystal Silence." Chick's use of extended chords—eleventh and thirteenth chords—can sound like placing two unrelated chords on top of each other (if you play a C major seventh chord and then put a D minor seventh in the octave above, you'd get CMaj13). The pulsating sustain of Gary's vibraphone smears time so it can be hard to tell when the chords actually begin and end. Add to that, Chick was writing using a series of chords that move up by fifths rather than fourths, what in traditional harmony are called retrogressions (as opposed to progressions). Listening to Chick those first few times left me feeling untethered to the music; he gave me so many ideas, but they were beyond my ears. Still, I was intrigued enough to listen a few more times, and the music opened itself up to me, like a new friend gradually revealing their secrets to me.

Often I'll hear something—on the radio, at a friend's house—and it will catch just a splinter of my attention, but its foreign-ness puts me off. Not by intention I'll hear it a few more times, and it begins to slowly win me over. I can't describe the mechanism by which that happens; it just does. The very thing that puts me off of it draws me in and intrigues me: its novelty. It's Johnny Cash's voice on "I Walk the Line." In the The *Resurrection* Symphony, Mahler's Second, it's the scope of his diverse and eclectic musical language, combining traditional symphonic forms with folk melodies, funeral marches, and even moments that are reminiscent of klezmer music; not to mention the stamina required to grasp an 80-minute piece or the stamina required of the musicians and conductor who perform it. After the fact, after I've come to love these pieces, I can wax philosophic about the timbre of a singer's voice, a particularly pleasing chord progression or lyric, or maybe a hook, what in the music business we call "ear candy." But that is all post facto theorizing. What *really* grabbed me is, well, up for grabs. Maybe what first draws me in are the rhythms in the Buena Vista Social Club; the sweetness in Clare Muldaur's voice in Clare and the Reasons' "Pluto"; the dreamlike naïve realism of Tom Brosseau's "How to Grow a Woman from the Ground." But as I fall in love, it becomes so much more.

Most composers and songwriters were similarly hooked by an indefinable something that made them want to become writers themselves. They are forever in search of the perfect sound, the perfect musical gesture that will cause in them the same excitement they experienced as children when a particularly delightful song took hold, shook them, comforted them, and amused them all at the same time. And as listeners, we seek the same thing—that sense of wonderment, some combination of newness and sweet familiarity, challenge and reassurance—that led us to love music in general, and certain music in particular.

The music that does this for you is an aural, cognitive, and perceptual fingerprint unique to you. It is unlikely to do it in precisely the same way for anyone else. The music that does it for me I know to be idiosyncratic. Some friends nod in vigorous agreement to my choices, others roll their eyes. For me it is the opening notes of Chet Atkins and Lenny Breau playing "Sweet Georgia Brown," not because those opening notes are especially divine, but because I know that the solo section of harmonic invention that occurs 2 minutes and 36 seconds in will always send me into reverie. It's the sound of Stevie Wonder's synth bass (sounding a bit like someone rubbing a balloon) in "Boogie On Reggae Woman." It's John, Paul, and George trading guitar solos in "The End"; McCoy Tyner accidentally hitting the "wrong" notes during a solo in "My Favorite Things." It's Shawn Camp singing "Off to Join the World," or Clare and the Reasons singing "Pluto," or Jeff Silbar singing "Wind beneath My Wings," three of the most perfect songs I know.

You may not have heard of any of these; you may not like them. But if you like music at all, even the tiniest bit, you have your own list of favorite moments and favorite songs. The universality of music is unquestionable. It is everywhere, and always has been. Why do we like music so? And how does it have the power to shake long dormant memories, to change a mood, to change something as fundamental as our heart rate and blood pressure? If music matters this much to us, it must be that it holds or conveys some meaning—emotional, psychic, philosophic, intellectual, or otherwise.

If music conveys meaning, it is probably not literal meaning, any more than a lover touching your hand or gazing into your eyes has literal meaning. These gestures may convey "please stop doing what you're doing" or they may convey "I love you" depending on the context. So it is with music. The note F-sharp does not *mean* anything in and of itself; it depends on context. Music is constantly moving and its context evolving, changing shape. That F-sharp depends on what came before and after. In sentences, the humble period expresses the end of an idea. But in music, the end of a single musical note, or even the end of a sequence of notes bound into a phrase, is almost always linked to the beginning of a subsequent note. So in music (as in life), context is everything.

When we speak of context and language, and seek to study how words give rise to meaning, we can turn to linguistics. Linguistics has codified different elements of speech, beginning with the sound itself (the phoneme), and the sounds we have available to us in any given language (the alphabet). The way those sounds can lawfully be combined into words is morphology. Of course, some sounds *are* words, like *mmm* and *I* and *shhhh*. More often, these sounds, these phonemes, are combined with other sounds to build words like *mama* or *iconoclast* or *shush*. We might consider that musical notes are like phonemes, and the notes available to us at any given time are an alphabet, which we call musical scales. In Western music, our full "alphabet" is the 12 tones of the chromatic scale, plus their octave equivalents across the range of musical instruments, totaling around 88 notes, ranging from A0 to C8—conveniently, the notes on a piano keyboard.

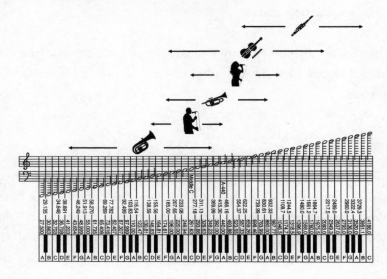

Most musical pieces we listen to use only a subset of these notes at one time—the seven notes of the major and minor diatonic scales, a five-note pentatonic scale, or its cousin, the six-note blues scale. You can play that five-note pentatonic scale easily on your own by playing only the black keys of the piano.

We can ask, then, whether there exist additional equivalencies between speech and music and if so, what those are. But first, let's expand what we know about linguistics. Beyond phonology (the sounds) and morphology (the words) there is syntax, the way that words go together, how arrangements of words convey meaning. In English, if I say

The dog chased the cat.

you know who did the chasing based on word order. Some languages convey the actor and the agent—the chaser and the chasee—through other means, such as suffixes attached to the words. There is some freedom in English to play around with word order. Although the following sentence is awkward, it is not illegal, nor uninterpretable (with some reflection):

The cat the dog chased.

(The Yoda sentence structure it is.) But we cannot use just any order of words whatsoever (linguists use * to indicate a malformed sentence):

*The the chased dog cat

When we play a musical phrase, does the order of the notes matter? You bet it does. The opening notes to Beethoven's Fifth Symphony, what Leonard Bernstein has described as the sound of "fate knocking on your door," are as simple as can be—three notes all the same, followed by a lower note with a hold on it—like that front door resonating. And Beethoven tells us you can hold that last note as long as you want.

I'm aware that many readers don't read musical notation and find it intimidating. If it's any help (in case you skipped Chapter 13!) consider that this is like one of those x, y coordinate systems that you used for graphing in seventh-grade math. The elliptical black dots are musical notes; when they're inked in they are short and when they're open they are longer. The x-axis (left to right) represents time—notes on the left come first, just like words that you read from left to right. The y-axis (up and down) represents the pitch of the notes—a dot that is higher on the page is higher (to the right) on a piano keyboard. Those horizontal lines running across a page of music are like grid lines you might see on graph paper, and they just help to tell you where the notes are in this musical staff coordinate system. The three notes that start this piece are the G above middle C on a piano. An equivalent way of notating the opening to Beethoven's Fifth is here; there are many online piano keyboards that will let you play just by clicking the mouse.

Beethoven's *Fifth Symphony* motif

All the other symbols on the musical staff are details, but one important one is the symbol that looks like an eye with an arched eyebrow over it: ⌒, the *fermata*. This is the composer's instruction that we should hold that note before moving on.

If Ludwig van had written the notes in the reverse order, it would mean something different—more like the knock of a door-to-door salesman than Fate herself. Having established the motif, Beethoven then repeats it, adding some more musical words. In the first repetition he starts a note lower—like the same four knocks on the door, in case you didn't hear them the first time, but with a different pitch, and again a hold.

Symphony No. 5, Beethoven

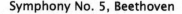

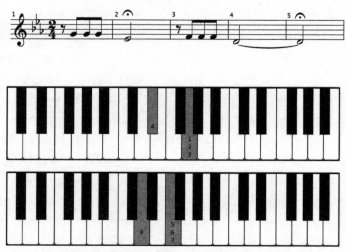

Beethoven's *Fifth Symphony* (measures 1–5)

What music has over speech is that, in addition to note order conveying meaning, so does rhythmic structure. Look at what Beethoven does next in measure six: he repeats the same pitch motif, but doubles up the sense of urgency by getting rid of that indeterminate hold and giving you the impression that someone really is trying to get into the house: Let me in, let me in, let me in!

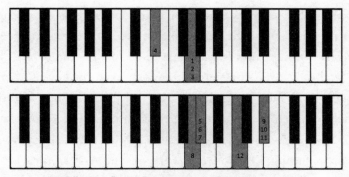

Beethoven's *Fifth Symphony* (measures 6–9)

Does the average listener know that those first four notes are a motif, the core idea on which the symphony is built? Perhaps not as such, but conductor Edwin Outwater and I undertook an experiment in four different performances of the Fifth Symphony, in Canadian concert halls with the Kitchener-Waterloo Symphony Orchestra. The audiences were mostly first-time concert-goers, and each was outfitted with an interactive response device (a clicker). The symphony orchestra played those first four notes and we explained that this was the *motif,* and told the audience that it would be repeated with variations (changing the pitch and tempo) during the first movement, and their job was to count how many times. Then the orchestra performed the first movement and we asked the audience to indicate, through the use of their clickers, how many times they heard the motif recur. More than half the audience was able to keep track, in this rapidly moving piece with a lot going on, that in the first eight minutes, the motif recurred 262 times (263 if you count an ambiguous statement in the French horns). And that even included a section of the piece where Beethoven *does* play the motif upside-down, as a musical interrogative.

Another feature of the symphony is that Beethoven takes us on a harmonic journey during which he changes the tonal center—modulates—several times. To fully enjoy the piece, it's necessary to remember that initial tonal center so that a return to it feels triumphant. It's sort of like hearing a very particular sentence at the beginning of a story, hearing many permutations of that sentence for 18 minutes, and then being able to recognize the sentence that started the whole thing (something that skilled comedians do with their "callback"). The first movement begins in the key of C minor, and then modulates to E-flat major, and goes through additional modulations, before finally ending in C minor. The second movement begins in A-flat major and passes through a cycle of eight modulations including C major, A-flat major, and A-flat minor, ending on A-flat major, establishing a new tonal center. After the second movement ended, we gave the audience a choice of several different chords to see if they could identify the chord that represents the tonal center that had started the piece 18 minutes earlier. An extraordinary 95% of them did. They had subconsciously held in their brains a chord that had been turned inside out, upside down, slid up and down a four-dimensional torus of harmonic manipulations, and recognized the core tonal center of the piece after all of that. If non-musicians and even non–classical music lovers can do all that, just imagine what you can do with music you're familiar with.

So if the opening gambit is fate knocking at the door, what does the rest of the piece mean? Let's go back to linguistics for a brief digression. Semantics is the formal study of what words mean. You might think this a rather puny job because you can always just run to a dictionary and *find out* what a word means. But perhaps it's not as simple as it seems.

An obvious case is the noun *bank*, which has two common meanings.

The boys threw rocks at the bank.

This creates a semantic ambiguity. We need some context to figure this out, to choose which semantic sense is intended. Part of clear expression in speech or writing is avoiding such ambiguity by using other words that help to disambiguate the sense we intend.

After their father lost his job as a loan officer, the boys threw rocks at the bank.

Walking in the stream up to their ankles, the boys threw rocks at the bank.

Poetry often trades in semantic ambiguity, as do musical lyrics, but common speech and expository writing try to avoid such lack of clarity. In contrast, music revels in it—ambiguous musical phrases are plentiful, and they add to our enjoyment by giving us free rein to interpret them however we like. In "I Remember It Well," from the musical *Gigi*, the music could be interpreted as playful, romantic, wistful, or even melancholic, all depending on your state of mind, your own experience, and memories. The two opening chords of Beethoven's First Symphony are enigmatic—we don't know what they portend until later. That's one of the things that makes the symphony so enduring.

Another kind of linguistic ambiguity occurs when a word's meaning depends on where it falls in a sentence, a so-called syntacto-semantic ambiguity. First, consider this unambiguous sentence:

The giraffes are amusing.

The word *amusing* is a synonym for *entertaining* or *humorous*. Here, the syntax tells us that the word *amusing* is an adjective (just as if I'd said *The giraffes are spotted*) and we are meant to understand that the giraffes are providing us with amusement. There is no ambiguity here, of course. *Semantics, schmemantics*, you might say.

Now consider this sentence:

Barry and Sally are amusing the giraffes.

The word *amusing* still means *entertaining*, but something has changed here—it's suddenly functioning as a verb. Barry and Sally are now the subject, actively engaging with the giraffes. The syntax tells us that whatever word goes in that spot in the sentence *must* be a transitive verb. "Barry and Sally are *blank*-ing the giraffes," they are doing something; this is the basic structure, and any number of verbs could go there: *feeding, serenading, taunting, prank-calling, salsa-dancing with.*

Are there some musical components that function as adjectives and some as verbs? We might say that expressive intonation is like an adjective—for example, when a trombone *slides* into a note instead of landing right on it. So is the expressive use of timbre when Karen Carpenter's voice takes on that slightly raspy, lost and winsome tone on the very end of the word "away" (in the song "Superstar": "long ago and oh, so far away"), she is coloring the music with the blues and violets of despair. When Danny Rio, the tenor saxophone player of The Champs in the song "Tequila," rolls his tongue on the reed to create a growly sound, that's when the crowd goes wild—he is adding sexual intensity to an otherwise mundane part, a musical adjective, as it were, to the melody. When Ravel repeats the same musical and rhythmic theme of *Bolero* over and over again, changing only the timbre, we feel a different emotional message based on the sound of the instruments. Children understand this very early, through the special timbres employed for musical stories. Consider the instruments that Prokofiev chose for *Peter and the Wolf*, meant to depict the various animals timbrally—the French horns for the wolf; the flute for the bird; Peter himself as a lush string section.

We were taught in school that verbs are the action in a sentence, the thing that moves. Rhythm is what moves in music—it is what prevents all the notes from happening at once. Some rhythms make us want to move and wiggle, and others lull us into sitting still or even falling asleep. When the rhythm is steady and slow, it functions to calm, the verb *to slow down*. If that same slow rhythm is lilting in 3/4 (waltz) time, it suggests a lullaby, the verb phrase *go to sleep*. If the rhythm is steady but faster, as in EDM, it might put us in a trance state. Dance rhythms typically have some irregularities to them— syncopation, where a note doesn't occur exactly where we expect it to, a hallmark of swing, reggae, and Latin rhythms. Classic '70s-era disco was characterized by a steady, unvarying kick-and-snare-drum pattern, but with a prominent bass line that got people to "kick out the jam." These are just a few of the ways that rhythm acts as a verb.

Similarly, chords can act as verbs, pushing things forward, as in a cadence that demands a resolution—the opening two chords of Beethoven's First Symphony, or the first three measures of the melody of "Maggie May" by Rod Stewart. Beethoven's First opens, unusually, with a C dominant-seventh chord (C7), which sounds like a chord just before the *ending* of a symphony. Beethoven is provoking us, startling us, with this break from convention. The C7 contains a tritone, E and B-flat (at one time called the devil's interval), and sounds unstable—it begs for resolution, as though we are poised above a chair about to sit down and gravity pulls at us to plant it. In this particular case, the chord points to—wants to resolve to—an F chord. What's unusual is that the symphony is *in* the key of C. Beethoven is turning convention upside down, suggesting that we are in a different key than we actually are. This move—C7 to F—is the fundamental chord change in the blues as we know it today, and blues musicians have added an additional break from convention by making the F an F7, which points (wants to resolve) to a B-flat that we never get to. The blues progression of C7–F7–G7 has become so popular, in part, I believe, because of the unconventional way it builds momentum . . . always pointing, always seeking resolution, which is of course, the emotional state of *having* the blues.

Rod Stewart sets the key for us in "Maggie Mae" through the use of a conventional intro—the chords D, Em, G—which is known as a I–ii–IV pattern (here in the key of D). But he cleverly begins his melody on the V chord, A, just as he sings "Wake up, Maggie." The V is the "wake up" chord, demanding movement to the I, which he doesn't reach until two measures later.[*]

---

[*] The best explanation I've seen of how chords are built is here, by Paul McCartney, who does not read music, explaining it to producer Rick Rubin, who does not play an instrument: https://www.youtube.com/watch?v=0-sXAqgP5KE.

Timbre is used in combination with pitch and rhythm in children's cartoons, particularly the Warner Brothers cartoons scored by Carl Stalling. Stalling uses the staccato sound of xylophone or pizzicato violins to depict Bugs Bunny tiptoeing, an ascending scale if Bugs is tiptoeing upstairs, a descending scale if he's tiptoeing downstairs. Stalling uses the legato, sostenuto sound of the vibraphone to indicate a dream state. And then there is the image of Fred Flintstone (in Hoyt Curtin's scores for Hanna-Barbera) preparing to dash off, his legs running in place to the sound of bongos. These mélanges of sonic adjectives and verbs are musical metaphors that require little explanation—as soon as we see the scene they are coupled to, they make complete sense. The context is married to the music. Yet there is nothing intrinsic in bongos that indicates running, or the absurd cartoon invention of *preparatory running*. Bongos in Latin music can recede into the texture of a song, or they can help to define the rhythmic dialect of samba, rhumba, bossa nova, or tango. That is, the semantic meaning of the instrument, in and of itself, works in tandem with some other extra-timbral context. In other words, one would be ill-advised to simply make an argument based on *semantics*, that these instrumental sounds mean something all on their own.

So far, we can see that borrowing ideas from linguistic syntax can make for pleasing analogies that help us to better understand music. Let's get back to Barry and Sally and amusing giraffes. Earlier, I said that

Barry and Sally are amusing the giraffes.

English, like all naturally spoken language, is generative—we can generate an *infinite* number of sentences. Truly! Without limit. Don't believe me? For example, to any sentence, we could always add something recursively, in a never-ending sequence.

The barista saw that Barry and Sally are amusing the giraffes.

The psychic told me the barista saw that Barry and Sally are amusing the giraffes.

The Instagram influencer posted that the psychic told me the barista saw that Barry and Sally are amusing the giraffes.

Ad infinitum.

Music is also generative and recursive. A mainstay of improvisatory solos is the repeated phrase that is then added onto and recontextualized. And jazz musicians since Louis Armstrong have been known to quote pieces of other songs within their solos.

Although we can often rely on syntax—word order—to clear up any semantic ambiguities in language, there are cases where it is of no help whatsoever.

*They are amusing giraffes.

What does this mean? "They" is ambiguous in terms of its referent—it could refer to Barry and Sally, or to the giraffes. This example trades further on a second ambiguity, that "amusing" can either be an adjective or a verb; and the syntax does not offer us a clue.

They [referring to the giraffes] are doing something that *I* find amusing. (These giraffes over here—they—are amusing, as opposed to those over there who are boring.)

or

They [old friends Barry and Sally] are engaged in some behavior that *the giraffes* find amusing. (Barry and Sally are amusing the giraffes, rather than, say, taunting them.)

The context may be made explicit by resorting to some supplementary language, as when we asserted that the boys were downtown or out by the stream. In everyday speech, we provide the context in ways that require some world knowledge, or an appeal to body language, and other paralinguistic cues. In music, as in poetry, we often allow the ambiguity to stand, in order to invite the listener to create their own interpretative experience, depending on how they feel in the moment. This openness to interpretation is part of the enduring appeal of music, and a large part of its power for healing. Whatever Handel's *Messiah* meant to Handel, or to his audience in 1741, it certainly means something different to us today. And it means something different to someone sitting in a church versus a concert hall or in the hospital pre-op room awaiting surgery. Speech is usually referential—whether it refers to something fanciful or factual, the words are intended to convey something specific, more or less, as they do with Barry, Sally, and the giraffes. Music is usually nonreferential, not intended to depict a specific, concrete object or idea, but rather to convey an emotion. And emotions are subjective, "facts" only to the perceiver of them.

I've posited several times that the purpose of music (and the arts) is to convey emotion. Here I want to clarify and distinguish that the purpose is to *convey* emotion, but not necessarily to *depict it*. Music does not typically refer to or depict a specific thing-in-the-world; it conveys or suggests or stirs up emotions, and you, the listener, get to attach those to specifics things-in-the-world if you so choose.

I'm aware also that many would disagree and claim that music doesn't necessarily have to have *anything* to do with emotion, whether by conveyance or depiction. It may do this, but what was in the minds of the early twentieth-century composers experimenting with the 12-tone system and serial music? There have always been artists trying to break rules and boundaries, often ripping themselves from the fabric of why that art had conventionally been made. In visual art, consider the Dadaists, or Damien Hirst. Perhaps they were trying to invoke the emotions associated with "shock" and "disgust" and nihilism; perhaps they were just trying to be contrarian and say "look how silly all this is." We do need that. Perhaps to a lesser extent this is what

William Burroughs was doing, and algorithmic writers such as Gilbert Sorrentino—experimenting, which of course is what artists do.

The formal study of *what we really mean* when a strict appeal to a dictionary doesn't help is called pragmatics. And this is where music and linguistics really get interesting. Many, from Leonard Meyer and Leonard Bernstein to Fred Lerdahl and Ray Jackendoff, have tried to map the elements of linguistic analysis to music, looking for the musical equivalents of phonemes, words, sentences, phrases, and stories, appealing to phonetics, morphology, syntax, and semantics. I'm not sure they were on the right track. They ignored the most obvious analogy—musical meaning is best understood at the level of pragmatics.

In language, pragmatics is not an objective set of rules or specifications, like syntax and semantics are—it depends on *you*, on *your* world knowledge, *your* memory and experiences, *your* present state of mind and environment. If that all sounds like hogwash—how a sentence can have universal meaning and yet mean something special to you, all at the same time—consider the many things we say to our partners, our parents, our best friends, that are like our own secret language. Famous TV catchphrases from yesteryear, like Steve Martin's "Well, excuuuuuuse me" or Seinfeld's "Not that there's anything wrong with that" or Steve Carrell's "That's what she said," all hold a pragmatic meaning beyond their dictionary meaning. And so does music.

The opening notes of Steely Dan's "Rikki Don't Lose That Number" remind a jazz fan of the opening to "Song for My Father" by Horace Silver. Steely Dan are signaling us something beyond the notes themselves—both that they are jazz heads (in case you didn't already know) and that they are resourceful composers who can find inspiration in the air around them. They're also pragmatically setting up a key without tonality (because they are using open fifths, ignoring the major- or minor-defining thirds) and then surprising us when the first chord of the verse enters as a flat-seven major, outside the rules of normal diatonic harmony.

The philosopher Paul Grice did some of the most amusing work on this problem of linguistic pragmatics—meanings that cannot be solved simply by resorting to syntax and semantics. These often involve an *implicature*:

Tina: Have you seen Dave?
Wade: There's a gray SUV parked in front of Gail's house.

On the surface of things, Wade's response would appear completely irrelevant, but Grice invites us to entertain situational-contextual knowledge. Suppose Wade knows that Dave is dating Gail, but has been sworn to secrecy. Both Tina and Wade have the shared knowledge that Dave drives a gray SUV. This exchange is a playful way of conveying information without explicitly betraying a confidence. We speak in such riddles all the time.

Suppose your friend is telling you a boring story. You might say, "That's boring," or you might use a Gricean implicature, saying "Thaaat's interesting," with an exaggerated melodic intonation, possibly accompanied by a roll of the eyes, to indicate the *opposite* of what you're saying. Any use of sarcasm requires an understanding with pragmatics. There are developmental stages for this, and young children before the ages of 6–9 typically do not understand sarcasm at all; their brains need to develop more flexible ways of information processing. Many people with developmental disorders never learn to understand sarcasm—individuals with autism spectrum disorders and Down syndrome can't always process the difference between literal meaning and intended meaning. This may partly account for the finding that some members of these groups exhibit impaired musical understanding.

Pragmatics goes beyond these fanciful examples of Dave's SUV, the office window, and sarcasm. Pragmatics plays a much larger role in day-to-day interactions than we realize. It includes little gestures, smiles, twinkles in the eye, and all manner of nonverbal cues. When my wife smiles and says, "You missed a button on your shirt this morning," and affectionately comes over to button it for me, I feel loved and nurtured. When my boss, who never thinks I am dressed presentably enough, says the exact same words through pursed lips and shaking his head, the meaning is entirely different.

A few pages back, I asked, "If the opening gambit [of Beethoven's Fifth] is fate knocking at the door, what does the rest of the piece mean?" I then took a digression into linguistic theory, and, along the way, described some of the modulations the piece makes—musical detours off the straight and narrow path, from the opening musical ideas to the concluding ones. For Bernstein, the journey itself and the detours that journey takes along the way *are* the experience. The meaning arises *because* of the detours. If you feel the same way at the end of a piece as you did at the beginning, I would say that piece has failed you. The journey becomes the meaning because the musical phrases, gestures, and structures are inherently ambiguous, and that ambiguity is the point. The more ambiguous the music, the more personal and customizable the experience can be.

Bernstein often emphasized the symphony's journey from darkness to light, particularly noting how the initial C minor key, associated with struggle and fate, ultimately resolves into a triumphant C major in the final movement. This transition symbolizes a narrative of overcoming adversity, a theme Bernstein found deeply embedded in the music's structure and thematic material. After the initial struggle with fate, one might suggest that the second movement offers a respite, a moment of peace or introspection amidst the tumult. From there, we travel through recapitulations of struggles, with growing intensity and complexity, and of resolutions and contemplative serenity.

There are moments of giddy lightheartedness in the third movement, perhaps even suggestions of triumph and of solace. The fourth movement that ends the symphony may not *depict* a specific emotion but it *conveys* great emotion, deriving from an overall sense of musical and narrative resolution. No one can say with any authority what it is *supposed* to mean—that is up to you. The meaning is a product of the journey, and if you're lucky, that journey will be intimately personal to you, bringing up thoughts and feelings and hopes and desires and dreams all your own, perhaps different ones each time you hear it. They might be feelings of victory, joy, playfulness, gaiety, or a combination of all of those. (I see all of these in Bernstein's facial expressions as he conducts the piece, and many more that are impossible to describe.) And the sheer durability of the piece is a testament to its complexity and delicious, multifaceted, interwoven ambiguities.

That fateful knocking on the door announces that you are about to set out on a journey that will be exhilarating, and on the journey you'll learn something—about yourself, about the world. Even if it's a trip you've taken before, there will be new things to discover. The final chord tells you the journey is over—until next time.

⌢

Earlier I mentioned syntax, that set of rules that concerns the order of notes in music and words in language. The syntactic rules of most natural languages are too complicated to codify; any rule book would be of no practical use. Instead, we have to use intuition. Within our broader American culture, there are grammatical variations—morphological and syntactical—that are no less important or structured than the textbook examples we typically see, and these interact with pragmatics.

As one of many examples, African American Vernacular English (AAVE) has its own rules that are just as logical and valid. For example:

(1) I went to their house (standard English)
(2) I went to they house (AAVE)

and other AAVE examples, such as

(3) Bobby sick
(4) Tony funny

(5) He do

(6) I be workin'

All are grammatically correct within their own milieu, and to pretend otherwise is simply ethnocentric elitism. In the melting pot of cultures that is the United States—and increasingly in Canada, Britain, Europe, and other places that are absorbing immigrants—subcultures maintain some of their own dialects as a way to preserve their culture and to bond with others who share the same history. Black Americans may speak Standard American English at work, especially if the workplace is predominantly white, and reserve AAVE for social and family occasions. For similar reasons, Hawaiian Pidgin English is widely spoken throughout the world, such as in Japan, China, Portugal, Philippines, and America. Many American Jews use Yiddish in the same way. It would hardly raise eyebrows in a social gathering if I called somebody a *putz* or a *mensch*—these words have entered the American lexicon. On the other hand, when Jewish people get together, as when Black Americans do, we may change the words we use or our intonation as a sign of shared culture, exclaiming: *mazel tov* (congratulations) or *it's a shonda* (a disgrace). Italians, Koreans, Persians, and others all use these cross-linguistic insertions in a pragmatic way, that is, to convey meaning beyond the strict dictionary definition of the words themselves.

AAVE has a parallel in music, the famous *blue* note, notes that are outside of the normal scale and, in fact, are *between* the notes on the piano. String and wind players, and singers, can hit this note because they aren't constrained by pre-tuned notes as the piano is. Violinists can put their finger in between positions, just as a trombonist can move the slide to an "in between" position. Trumpet players and woodwind players can relax their jaw (embouchure) to create the note; guitarists get to it by bending a string. The blue note is adjectival, modifying a musical noun, coloring it as it were, as when a singer slides into a note that is in tune from one that is out of tune (the famous *blue note*, in the key of C, e.g., is an E-flat). The "blue notes" in our contemporary, Western tradition are the flattened third (E-flat), fifth (G-flat), and seventh (B-flat).

Billie Holiday's "My Man" (in the key of C minor, same as Beethoven's Fifth) is an ode to a man she loves in spite of terrible mistreatment—she can't let go. "All my life is just despair," she sings, but when he takes her in his arms she just don't care. In the B section of the song, she cries "What can I do?" sliding down from A-flat to F and wobbling back up to G through the blue note, the G-flat. Her indeterminate pitch supports the semantics of the song by invoking a note that is outside of the key she is in—a note free of the constraints set up by the rules of harmony, a tonal metaphor for the freedom denied her in the relationship.

Michel de Montaigne, the French Renaissance philosopher, noted that the most universal quality of humankind is its diversity. Underneath the great diversity of musical styles that span the world's cultures, built up from a variety of scales and rhythms, there exist some pragmatic universals. That is, while the phonology, morphology, and syntax of musics may differ, their pragmatics may rest on a common desire to reach for emotional expression through collections of organized sound.

As a student, Leonard Bernstein noticed that the first four notes of Aaron Copland's *Piano Variations*, E, C, D-sharp, C-sharp, are the same four notes that, in another order, are the principal motif of Bach's C-sharp minor Fugue from the *Well-Tempered Clavier*, Book I; Stravinsky's *Octet*; Ravel's *Rapsodie espagnole*; and some Hindu music he'd recently enjoyed. "There must be," he mused, "some reason why these discrete structures of the same four notes should be at the very heart of such disparate musics as those of Bach, Copland, Stravinsky, Ravel, and the Uday Shan-kar Dance Company." There must be, he said, a worldwide, inborn musical grammar, some underlying structure that accounts for both the universality of music and the diversity of different musics, all stemming from a common origin, a neurobiological core. By abstracting the logical principles by which music is created, we may be able to discover "how we communicate in a

larger sense: through music, through the arts in general, and ultimately through all our societal behavior." A tall order! But, as many
have noted, music appears to be the one universal language we share
on this planet. Its power to move us may derive from the fact that, as
Bernstein claims, it is a metaphorical phenomenon, mysterious, and
a way of symbolizing affective existence. Music's universality may
derive from the fact that it is "born of science . . . made of mathematically measurable elements: frequencies, durations, decibels,
intervals."

Are there rhythmic universals? The 1–2, *bump*-bump or *lub*-dub,
of their mother's heartbeat is the first thing infants hear in the womb.
Walking, marching, and running convey a 1–2 rhythm, and even if
the clip-clop of our footsteps are equally loud, because the motor
activity invokes different halves of the body, and therefore different
halves of the brain, we perceive these as 1–2, as two separate events,
rather than 1–1. Limping gives rise to a 1–2–3 pattern where one steps
with one foot, quickly steps with the other, and then pauses briefly
before stepping again—much like a waltz. Cantering or loping has a
different rhythm, as we lift off one heel and glide until the other heel
sets down: step gliiiiide, step gliiiiide, giving rise to a 1–2–3 rhythm,
in which the 1 and 2 are tied together. Beyond locomotion, we find
the 1–2–3 rhythm in swinging, as the swing reaches its apex and
turns the corner of a parabolic motion before switching direction. As
we've seen, all cultures use the low–integer rhythmic ratios, 2:1, and
3:1, and many become more sophisticated and idiosyncratic, demonstrating Montaigne's point that diversity is the most universal human
quality (a sentiment echoed by the anthropologist Clifford Geertz).

Music has a pulse, a beat, and neurons in the auditory cortex, cerebellum, and other brain regions keep time with the beat.
This capacity is present even in newborns. We know this, in part,
because in 2009, Henkjan Honing and his research group designed
an experiment exposing infants to drum rhythms with occasional
beat omissions. They noticed the infants' brains exhibited a distinct
electrical response signaling an anticipation of the missing beat.

This discovery not only spotlighted the musical acumen of newborns but also laid the foundation for exploration of the biological underpinnings of our innate musical capacity.

Despite initial skepticism from some in the field, the study withstood scrutiny. In a follow-up study with not only newborns but also adults (both musicians and nonmusicians) and macaque monkeys, Henkjan employed an alternative paradigm and came to the same conclusion: beat perception is an innate mechanism rather than a learned phenomenon. Subsequent work with macaques revealed their sensitivity to rhythmic regularity but a lack of evidence for beat processing, suggesting the gradual evolution of this mechanism among primates, with humans exhibiting a unique, strongly wired connection between motor and auditory brain regions. Beyond a mere cultural phenomenon, then, music appears to possess deep biological roots, offering an evolutionary advantage to humans. Henkjan's research not only deepens our understanding of the biological foundations of musicality but also emphasizes the intricate nature of our capacity to engage with rhythmic elements.

As we've seen, the trick in music is to meet those rhythmic expectations, rewarding prediction centers of the brain, but also challenging them with surprises, such as syncopation—notes falling on unexpected beats. One universal rhythm plays on this—it's what musicians call the pickup notes, particularly three eighth notes leading into a song, as exemplified by "When the Saints Go Marching In." If you're counting 1-2-3-4, the song starts just halfway between the 3 and the 4. Musicians might count the song as 1 *and* 2 *and* 3 *and* 4 *and*. The *and*'s are the eighth-note upbeats, and we say that the song starts "on the *and* of 3."

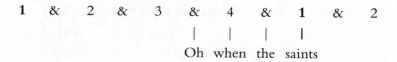

$$\begin{array}{ccccccccccc} \textbf{1} & \& & 2 & \& & 3 & \& & 4 & \& & \textbf{1} & \& & 2 \end{array}$$

Oh	when	the	saints

This same rhythm opens up Tchaikovsky's First Piano Concerto and "This Land Is Your Land." The Beatles play with this convention in "Can't Buy Me Love," using a variation on the rhythm, starting it a half beat earlier and then syncopating it.

When The Saints Go Marching In

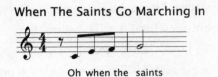

Oh when the saints

Piano Concerto In B minor, Op. 23, No. 1

This Land is Your Land

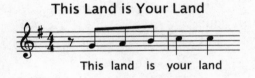

This land is your land

Can't Buy Me Love

Can't buy me   love

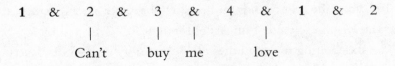

It shows up in music all over the planet, in music as diverse as that from India, Sub-Saharan Africa, the Middle East, Indonesia, and the indigenous musics of Canada and Australia. Why? A pragmatics interpretation is that because the note starts between beats, it is sur-prising. And try this experiment: if you were to sing *just* those three notes, from any of these songs, and then suddenly stop, what would you feel? You would likely feel some state of suspended animation, as though you'd been asked to freeze mid-step in the children's game of Simon Says; you'd likely feel off balance as though you need to plant your foot down. This musical, narrative momentum makes the rhythm compelling, which is why it shows up in so many wildly diverse contexts, and sometimes wearing clever disguises. Consider "Fire" by Bruce Springsteen (also recorded by The Pointer Sisters and others). The song starts on the "and" of 1 with those same three eighth-note pickups, followed by two quarter notes, giving the song the same kind of push forward; it would be all but inconceivable to stop the song after those five notes "I'm ri-ding in your . . ."

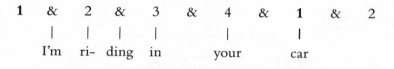

Fire

Great songwriters combine their use of rhythms, pitch, and lyrics to support the meaning they are trying to convey. I'm riding in your what? We need to know—the rhythmic, melodic, and lyrical momentum work together to propel the song and us along with it, so that when the music lands on that downbeat, we are expecting to hear the lyrics resolve that lingering thought.

The six opening melody notes of "Something" by George Harrison are all sung on a single pitch, a C.

Something

When Harrison steps off the single note melody in "Something," he moves the melody down half a step, from the tonic (C) to the leading tone (B)—the note that is like being stuck with your foot in the air between steps before you have a chance to plant it on the ground. A tonal move like this demands an immediate resolution. Harrison supports this lyrically: "Something in the way she moves . . . ." Cognitively, we want to know. What *is* this something? What *is* it about the way she moves? Harrison writes "attracts me like no other lover."

The near internal rhyme of "other" and "lover" in retrospect seems inevitable. And tonally, upon leaving the leading tone (B) he moves chromatically down another half-step to B-flat, a blue note, meandering up and down a bit until he ends on the note A, changing the chord underneath it to an F major. What this *means* is that he's reached a comfortable, partial resolution. We don't feel like our foot is stuck in the air, but we know that something more is to come—and that's just how he wants it.

The first six melody notes of "For No One" by Lennon and McCartney are also sung on a single pitch, a G:

For No One

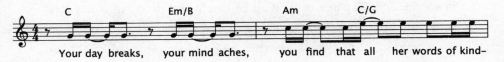

Your day breaks, your mind aches, you find that all her words of kind–

When McCartney changes his single pitch motif, he moves *up* from the dominant (G) to the tonic (C). Normally, this would feel like the full completion of an idea—the dominant-to-tonic gesture ends a great many musical pieces (including Beethoven symphonies) in what musicians call an "authentic cadence." But McCartney slyly changes the chord underneath that next note to an A minor, completely recontextualizing the melody, forcing the temporary illusion that we are now in a new key. He supports this lyrically. Keeping the same rhythm, McCartney writes "you find that . . ." At this point in the song, we've heard the same rhythm three times, each with a three-word lyric: "Your day breaks; your mind aches; you find that . . ." As with "Something," McCartney is teasing us into the next bit. You find that *what?* Here he changes the rhythm to straight eighth notes (with some eight-note rests), and he moves up in pitch to the third scale degree, the mediant, and sings "all her words of kindness linger on when she no longer needs you." "All her words of" brings us back to the "home chord," C major, but McCartney still has a couple of tricks. "Kindness linger" occurs on an F major chord, a stable resting place in the key, but "on when she no longer needs you" brings us to a B-flat major chord, totally outside the key, presaging that she is no longer in his life, before ending on the home chord "C major" again. But it no longer feels like home.

It's often been said that the purpose of a song is to take you on an emotional journey in which you start and end on the same chord. But if you feel the same way when you get to that chord at the end, you'll likely feel cheated. Here, we do end up on the C, but we've been through an emotional journey that has left us (or at least the songwriter) sad and vulnerable—an empty home, once full of joy, now devoid of warmth. And we're only at the end of the first verse! How does McCartney resolve all of this when the song is entirely over? We learn that she is really gone; this is no feel-good song: she has not come back. To convey this, he pulls an unprecedented harmonic move; rather than ending on the home chord, C major, he ends on G7, the most unstable chord in the key, the one that feels like our foot is still in the air . . . and there he leaves us hanging. Jimmy Webb uses the same trick in "Wichita Lineman," which is in the key of F major. Webb establishes the "home" key of F by playing it twice in the intro, and then, like the central character in his story, the song is eternally left hanging "still on the line" by never returning to the F major again. It moves to D, and then with B-flat and many suspended fourth chords, Webb engages in a game of harmonic misdirection—sleight of ear. The end of the song repeats "still on the line" as the chords cycle between IV and V, begging to go back home to the I chord, to no avail. The listener and the singer are left emotionally unsettled.

Certain sequences of notes appear to be fundamental and universal. What can we learn from them? Let's take another look at "Can't Buy Me Love." Lennon and McCartney begin with three notes, a simple and very common melodic fragment. It's no accident that these three notes, when sounded together, make a major chord, the building block of Western tonal harmony. What is it about these notes?

Can't Buy Me Love

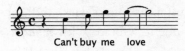

Can't buy me love

Can't Buy Me Love

If you'll forgive a digression into high school science: all sound begins with some kind of movement—something being hit (like a drum or piano string), or pushed (like a column of air through a flute), or rubbed (like the bow dragging across a violin string). This movement transfers energy from you to the object, causing the object to vibrate, to wiggle back and forth at some rate that is determined by its size, its rigidity, and its composition. This disturbs air molecules, which in turn send out pressure waves that may eventually reach your ears. Blow air through a harmonica, and tiny little metal strips oscillate (vibrate) back and forth, and the rate at which they vibrate is directly correlated with the pitch, the note you hear. Sometimes these vibrations are pleasing to us, sometimes they just sound like noise, as when a bunch of pots and pans fall to the floor or you scratch your fingernails across a chalkboard.

If you bang your fist on your desk or hit this book, it will send out an irregular pattern of enharmonic waves (a sound composed of multiple frequencies that are not harmonically related to each other). But play a pitched musical instrument and it resonates in a way that emits a series of waves in a periodic, cyclical fashion. If I press the low C key on a piano, the string will vibrate 65.4 times per second (65.4 hertz, or Hz, named after the German physicist Heinrich Hertz). You might think that this one note is all that you're hearing, but if I were to use a tone control in an audio editing program to get rid of everything above 66 Hz (low pass filter), that note would instantly sound dark, dull, and not very appealing. Musical instruments vibrate with *multiple* modes at the same time, and give it a rich quality. What we call "timbre" is simply the unique character of those multiple modes, causing us to hear a difference in tonal color between a trumpet, say, and a piano playing the exact same note. Why? Those different modes cause us to hear a great many notes at once (although we are not generally aware of this). This occurs because the string is vibrating not only across its entire length, but in fractional segments of itself. It's as though the string were divisible into an infinite number of fractional parts: ½, ¼, ⅓, ⅕, etcetera. The smaller segments vibrate more quickly and give the impression of higher notes, called *overtones* or *harmonics*. The same principle applies to columns of air through a tube (a wind instrument) or the resonance of a bell, xylophone bar, or other musical instrument.

If you can get yourself to a piano (an acoustic piano, with strings, not an electronic keyboard), find "middle C" (C4) in the middle of the piano. Now, find the C two octaves below middle C (C2) and press down on it, holding the key down, and listen. Let it up, and then press down on the C one octave above that (C3), gently enough that it doesn't make a sound. You've now taken the damper off the strings for that note, allowing them to resonate freely. Next, press down on C2 again. C3 is free to vibrate "sympathetically" with C2, and you will hear C3 very clearly along with C2. What you are hearing is an overtone, one of the many different components of what we are really hearing when we listen to that low note. This C3 is the first of an infinite series of overtones, most of which are too high for humans to hear. The frequency ratio between these various notes we call "C" is 2:1: C2 vibrates with a fundamental frequency of ~65.41 Hz. The C above it (C3) is two times that, or 130.82. The C above that vibrates 261.64 times a second (notice that each number is double the previous). Similarly, A2 vibrates at 110 Hz, A3 at 220, A4 at 440 Hz. This 2:1 ratio is thus a musical universal both for rhythm *and* for pitch. And women's voices are generally about an octave above men's, giving us another universal.

By the way, if that A at 440 Hz sounds familiar, it's because that's the note we use to tune to. We could in fact tune to any note, but for reasons of convention and convenience, we use A. In earlier times European musicians still tuned to "A," but defined it differently, lower (or sometimes higher) than 440 Hz, based on historical and regional conventions.

Returning to the notes resonating with C2, the next note in the overtone series is a G. Press down silently on G3 and strike C2 again and you will hear the G very clearly. The first overtone was an octave above the lowest note you hit—the tonic—and this second overtone is an octave plus a fifth above. The interval of a fifth forms a frequency ratio of 3:2. G3 has a frequency of ~196 Hz. Take C2 at 65.31, double it for C3=130.62 Hz, multiply that by 3/2 and you get G3 at ~196. This pitch ratio of 3:2—this interval of a fifth—is known in almost every scale in all of the world's musical systems, and it was the basis of harmony as far back as the Middle Ages. The two notes played simultaneously sound open and consonant.

The next tone to ring out with a special radiance is a fourth above that G, bringing us up to yet another C. The all-important interval of the fourth falls out "for free" when you move up from G to the next C. Once you've established the fifth, you can start taking fifths of fifths—from G we go up a fifth to D, from D to A, from A to E, and so on. If you keep doing this, you end up back where you started. Now, take all 12 tones of this cycle of fifths, put them in the same octave in scalar order, and you end up with the 12 notes of our chromatic scale, the building blocks of all Western harmony.

Our harmonic series so far consists of C2, C3, G3, and C4. The next tone above that is an E, E4, making an interval of a major third. Moving these three notes into the same octave, we get C–E–G, a major chord—the foundation for nearly all music in our culture, whether it's pop, blues, classical, jazz, or hip-hop. And, it's the opening notes of "Can't Buy Me Love." (That major chord may not be able to buy you love, but it sure can buy you record sales.)

Lower that E by a half step, the smallest legal move you can make in our Western tonal system, and you have E-flat. Play it with the C and the G and you have a minor chord. With these two chord forms, the major and the minor, you can play almost any song you've ever heard, just by changing the note you start on. Play that E-flat in the melody against a C major chord (with an E-natural in it) and you get the all-important blue note.

If you keep going up the overtone series by using specialized measurement devices (spectrum analyzers) or the piano keyboard trick I showed you, you will eventually find that E-flat, and all of the other notes of our major and minor scales. Part of what makes musical instruments sound different from one another when they are playing the same note—their timbre—is that different musical instruments emphasize some overtones in the series more than others.

The minor third is a special note in speech as well as in music. Consider how we use pitch in English. If I'm asking a question, the intonation at the end of my sentence will rise. If I were to ask, "You're going?" with only mild surprise, the rising interval might be as small as a minor third. If I'm very surprised, I'd use an exaggerated contour, up to a minor sixth. When Scooby-Doo, the cartoon dog, utters *rrrr?* (with a rising glissando), we know that he is reacting to something quizzical, a gesture we easily interpret in music. If I'm agreeing with you, or I'm simply affirming that I'm listening to you and following your story, I'd hold a steady pitch and say "yes, yes!" But if I want to *disagree* with you, by being *sarcastic*, I might say "yeah, yeah" using the falling minor third, for instance from G down to E. In every culture we've studied, children taunt one another with these two pitches, that same falling minor third: *nyah nyah*. A longer, elaborated version of that taunt starts with the first five notes of the harmonic series: nyah nyah nyah-nyah nyah on G–E, A, G, E*. Musically, this sounds as though it must resolve, and the note it wants to resolve *to* is the C below. We taunt with the overtone series, leaving out the implied resolution that, if it were a song, the note the song would finally end on. Musically, prosodically, pragmatically, we are expressing disbelief.

The notes in the overtones series are primordial, a gift of nature and physics, cultural universals. All cultures begin with the octave and the fifth, and divide up their scales into a discrete number of steps. While not every culture uses all the notes of our chromatic scale (derived from the overtone series), all use some, with the most famous being the pentatonic (five-note) scale, as introduced in Chapter 2. The differences in which of the overtone series notes are used accounts for the unique Japanese, Balinese, and African scales, which produce a great deal of beautiful music.

Japanese Pentatonic Scale: C, C#, F, G, A#

Balinese Pentatonic Scale: C, C#, D#, G, G#

---

* The A, which really sounds like the most taunting part of it, is the 13th note of the harmonic series; in many cultures 13 is considered an unlucky, jinxed number.

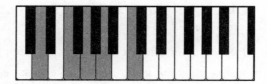

One of several African Pentatonic Scales:
D, F, G, A, C
(This is also the minor pentatonic in Western music.)

Recalling Chapter 2, Western culture divides the octave into 12 equal-stepped parts—the white and black notes on the piano—and we typically use only five to seven notes at a time. Middle Eastern and Eastern cultures may divide their octave into 24 or 48 tones, but generally, these are regarded as ornaments more often than as functional tones. Like our "flat seven" (B-flat) and the flattened fifth (G-flat) in the key of C, "blue notes" outside the diatonic scale and favored by American blues singers and guitarists, these are expressive intonations designed to bring tension, interest, or surprise to a performance. Singers from Billie Holiday to Frank Sinatra to Kurt Cobain famously sang intentionally flat or sharp on their way from one note to another to build up emotional tension and release.

But the overriding fact is that all musics we hear (and even those we don't), whether classical, hip-hop, country, jazz, EDM, music from the distant past or near future, have their origin in the overtone series. The universality reflects a common origin, and we see evidence of that in humankind's oldest musics. The 60,000-year-old bone flute discovered in the Divje Babe cave near Cerkno, Slovenia, plays a pentatonic scale that would be recognizable by anyone alive today. The pitches and rhythms of ancient Hebrew chants, as preserved by Ethiopian Jews who were cut off from outside influence from the time of Solomon (10th century BCE) to 1984 are strikingly similar to those sung by modern Jews.

Getting back to the three notes that open "Can't Buy Me Love," opening a song with the notes of a major chord (the major triad) is quite common.

Can't Buy Me Love

Can't Buy Me Love

We see the same notes, but in reverse order, here:

The Star-Spangled Banner

The Star-Spangled Banner

In "Honey Pie" (Lennon and McCartney), we see the same three scale tones, just in a different key and in a different order.

Honey Pie

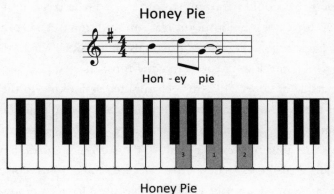

Honey Pie

These three notes of the major triad have held a role of privilege in Western music from the Renaissance period onward. However, that privilege does not exist in all music traditions around the world. In Indian classical music, the concept of the raga is fundamental. A raga is essentially a melodic rule set for creating a tune. The focus is more on melody and rhythm than harmony. Therefore, the concept of a major or minor triad doesn't naturally fit into this system. There is a set of "parent" scales known as *thaats* in Hindustani music and *melakartas* in Carnatic music, but these are not used in the same way that Western scales are used as a basis for constructing chords.

Chinese music, including the music found in traditional Chinese opera, also has a different focus. The pentatonic scale is a common underlying structure, and while it's possible to extract something similar to a major triad from a pentatonic scale, harmony and chords are not the focus. Traditional Chinese music relies heavily on melody and timbre. Different modes or scales may be used to evoke certain emotions or fit certain themes, but these don't align directly with the Western concept of major and minor triads.

Arabic music is based on a system of *maqams*, which are melodic modes. Like Indian ragas, maqams are rule sets for creating melodies. They include specific ascending and descending patterns, and each maqam can evoke certain moods or be appropriate for certain situations. The focus in Arabic music is also more on melody and rhythm than on harmony. While there are scales in Arabic music that contain notes that could be arranged into a triad, the triad itself is not a concept that holds a significant place in traditional Arabic music.

So what are the "substantive universals" in music? There is the harmonic series; the low-integer-ratio rhythms; the tempos that rarely reach outside of 20 to 240 beats per minute. By looking to these, we see that there are common elements that subserve music, just as there are common speech sounds—articulatory gestures—that subserve language. As long as we are bound to the surface features, however, we may miss the boat. As Leonard Bernstein mused in his Norton lectures at Harvard, "Language leads a double life; it has a communicative function *and* an aesthetic function. Music has an aesthetic function only. . . . [A] prose sentence may or may not be part of a work of art. But with music there is no such either-or; a phrase of music is a phrase of art. It may be good or bad art, lofty or pop art, or even commercial art, but it can never be prose in the sense of a weather report, or merely a statement about Jack or Jill or Harry or John."

It is true that we can think of a piece of music as a continuing series of anagrams, re-arrangements of the 12 notes of our chromatic scale, or the 5 notes of a pentatonic scale, or what have you. But that *jeu de notes*, as Stravinsky termed it, is not where the meaning lies. For the meaning, we have to turn to pragmatics, and a very personal pragmatics at that. For, while I can play around with Grice's clever examples of indirect speech ("my, it's hot in here"), those are things that the average person will interpret in the same way. This *jeu de mots*, as it were, can be entertaining to us, but that is where it ends. Music can be entertaining, but it is also inspiring, cathartic; it can amuse us and it can lull us into a dreamscape where reality and possibility merge. It functions, as all good art does, on multiple levels all at once. We can appreciate the ingenuity of Bach's fugues separate from having them push us into a daydream. There are moments of music that give millions of people chills. Research in my lab has shown that people agree where the most impactful, chill-inducing moments of music are, within fractions of a second. That is some universality. And yet, we all have different tastes, and even though we may find a particular passage of music impactful, it doesn't mean we like it; and if we like it, it doesn't mean we want to hear it right now. So where do we go from here?

Like drama, music offers us opportunities to engage with human emotional experience. First, there is the original composition (the play, the movie, the symphony, the folk song), and second, there is the interpretation of it (the actors, the musicians). There's a popular misconception that classical music is staid and fixed, and that jazz is free and unconstrained. But classical music has enormous room for interpretation—indeed, if every orchestra and soloist played the pieces exactly the same way, there would be no reason for each of them to make recordings. Yet the classical music catalogue is chock full of interpretative variations of all the major works.

How do we study the emotional aspects of music scientifically, rigorously? One way is to study subjective aspects: people's reactions to music, through brain scans, biometrics (changes in heart rate, respiration rate, or the smiley and frowny muscles on the face—the zygomatic and corrugator muscles). Another way is to look at objective aspects: variability in performances of the same musical piece. Both types of study have validity, both present their own experimental challenges. I've already reviewed what we know about subjective studies—the neural circuits involved in musical pleasure, relaxation, and pain relief—and the fact that different musics can lead to the same neurophysiological reactions, for reasons we still don't fully understand. By combining the knowledge gained from these studies with the knowledge from objective, quantitative studies, we can possibly converge on better answers.

Philippe Lalitte, of the Sorbonne, undertook a study of performance variation in Stravinsky's *The Rite of Spring.* "It is well known," Lalitte writes, that "Stravinsky . . . does not respect his own tempo indications and makes use of rubato to varying degrees." Lalitte studied a corpus of 96 different recordings of the piece, spanning the ninety years from 1929 to 2019, and demonstrated quantitatively the enormous latitude taken by performers. A logical next step might be to try to correlate these results with the popularity of each interpretation, but such an effort is doomed to failure, due to unequal availability of the recordings. (If there are any graduate students reading this and looking for a project, this would be a good one to test with listeners of various backgrounds and ages.)

One of the modern pioneers in the quantitative study of musical performance is Bruno Repp, a psychologist and pianist. The piano is especially amenable to studies of this sort because piano performance is for the most part limited to only two parameters that affect the sound of a note, two decisions that the pianist can make: *when* to press a key, and *how hard.* Setting aside pedaling, all the beautiful and intricate nuance we hear in a piano performance, astonishingly, comes from only these two degrees of freedom. Critical to this analysis is that master pianists seldom play scores exactly as written, exhibiting considerable timing and amplitude variability. Although a passage may be written all in eighth notes, which hold identical nominal values, in practice the notes may range widely in duration: some eighth notes are performed twice as long and some half as long as the average. Similarly, a passage marked *mf* for moderate loudness may be performed with some notes louder than others. This variability gives rise to highly individual differences.

Bruno Repp studied a single Chopin etude by 115 different pianists and found such a wide array of stylistic and interpretative differences that he was able to codify and quantify them, creating four major dimensions of performance style that bound together similar approaches, while simultaneously preserving their utter distinctiveness. And that was just from his examination of their *timing* profiles, ignoring any differences in loudness and accents. When he undertook a separate study of dynamic variations, he found five major dimensions, or clusters, each representing a nonstandard dynamic profile. He found, in general, the more master a performer is, the more liberty they take with timing and amplitude as a way to individuate their expression, and, possibly, the listener's aesthetic interpretation of the piece.

You might be thinking that more performance variability is better. Why doesn't everyone just run amok with changing note lengths and amplitudes? Repp found that there exist stylistic constraints—guardrails, if you will—and that most people stayed within these. With this as a launch point, my students Anjali Bhatara, Anna Tirovolas, Marie Duan, Bianca Levy, and I undertook a more precise experiment in mapping performance variability to emotional perception of listeners. We asked a concert pianist to perform four Chopin pieces (Nocturnes Op. 15, No. 1 and Op. 32, No. 1, both in major keys, and Op. 55, No. 1 and KK IVa, No. 16, both in minor keys) as would be done in a normal concert setting, playing our Yamaha Disklavier recording piano—an instrument that has highly delicate sensors underneath each key and pedal, and a series of highly precise motors corresponding to them that allows for as near perfect a reproduction as possible after the pianist has left— a twenty-first-century version of the player pianos of yore. We then hacked into the computer file and removed all of the variability contained in the performances—the expressive timing and expressive amplitude information—leaving us with a completely flat, robotic performance that sounded not unlike a young schoolchild practicing the piano metronomically. We then added back the expressivity a bit at a time, creating intermediate versions with, say, 25% amplitude variability, or 75% timing variability, as a ploy to see how real listeners responded to this continuum of versions, from wooden to expressive. We then *extrapolated* from the data we had to create *super-expressive* versions of those same pieces, with amplitude and timing variation extending to 125% and 150% of the original performance.

Our experiments showed, first, that both musician and non-musician listeners were attuned to subtle variations in musical performance. People liked the mechanical version the least, and their liking for the music steadily increased with increasing expressivity. People did not like the 150% version very much, regarding it as too shmaltzy, or overdone. But the 125% version surprised us: for the minor-key nocturnes, but not the major-key ones, listeners preferred the 125% version. Maybe our pianist held back a bit in the laboratory from what he would normally have done in a concert hall with a live audience. Maybe our participants were feeling particularly emotional that day—we don't know.

We also considered the possibility that people prefer performance variability for its own sake, simply because it seems more human. To address this, we took all the timing and amplitude variation found in the piece, and assigned lengthening and shortening, and louder and softer playing, to notes at random. If people just liked variability, it shouldn't matter where it comes from. In fact, people *hated* the random version, scoring it lower than even the purely mechanical version. Several subjects said that the random version sounded "like the pianist was drunk." In other words, the variability in timing and amplitude have to be *coherent*—they need to help display the music in an authentic way; they need to stay true to the *pragmatics* of what we think the musical story means.

I asked Mari Kodama about the sessions she had with her mentor Alfred Brendel. Surely, at that level, he is not going to be talking to her about fingerings, or technical matters—she is already a world-class master. What did he teach her? "I would play for him and he would stop me if he felt it was 'inauthentic,'" she said. By that, he meant, if there was something about the way she played a passage that felt like she hadn't fully thought it through, as though she played the notes but had not found the musical story in them that tied them to her own deep emotions. "He is not a teacher, not a pedagogue," she continued. "He cannot tell me 'do it this way.' He can only tell me when he thinks my performance is somehow not [emotionally] truthful, neither toward the message left by the composer nor to my own instinct. I was playing a Beethoven sonata for him once, and he stopped me and said I had not thought through the passage well enough. I couldn't play the piece for three months while I figured out how to repair that part. What he taught me was to have a good balance between the brain (analysis, calculations), heart (emotion), and fingers (technique) to try to reach the highest art form. It is my theory that this is necessary in order to touch people profoundly."

The authenticity is key—the artist being authentic to themselves helps the listener, *reminds* the listeners what it means to become authentic to *their* true selves. Joseph Polisi, the former president of Julliard, tells this story. In 1944, the impresario Billy Rose produced an elaborate Broadway review entitled *The Seven Lively Arts*. The show would include songs by Cole Porter, new choreography by George Balanchine, and a musical interlude by Igor Stravinsky, who eventually entitled the work *Scenes de Ballet*. When the Stravinsky work arrived, Rose was concerned to see that the instrumentation for the composition required a much larger orchestra than the one Rose wished to use in the pit for the run of the show, wanting a smaller one in order to lower the weekly payroll of the musicians (a typical collision between art and commerce). Rose sent the following telegram to Stravinsky: "Your ballet a colossal success. Would be even greater success if you agree to certain modifications in instrumentation." Stravinsky wired back: "Quite content with colossal success." Stravinsky got his orchestra. Who knows if it would have made a difference? But that's not the point—this was the authentic expression of Stravinsky's imagination, for better or for worse. And that is worth something.

Chapter 16

Music Medicine, Mystery, and Possibility

Here's a puzzle. Musical taste is highly subjective, ranging from Gregorian chant to jazz, rock, hip-hop, heavy metal, country, ad infinitum. How is it that different people can listen to Keith Jarrett, J. S. Bach, Selena Gomez, or Snoop Dogg and all have the same emotional experience? Well, first, do they? Yes—MRI scans show that neural activity increases significantly in the brain's pleasure centers, regardless of the genre, so long as the listener likes the music. In one study of younger adults I conducted, 17% of listeners reported having a "transcendent experience"— a feeling of being in touch with a higher power—while listening to religious music, and 15% while listening to hip-hop or rap, with the others finding that experience in alternative music, ambient, classical, metal, R&B, soul, EDM, and other genres.

So then, what is it that actually gives rise to these experiences?

Music is a uniquely powerful combination of elements that coalesce in a way that impacts a wide range of neurological, neurochemical, and bodily systems. The Therapeutic Music Capacities Model, developed by neuroscientist William Forde Thompson of Macquarie University in Sydney, provides a framework for understanding how music can improve psychological and physical outcomes, boost immune system function, and treat motor and behavioral functions in people with neurological disorders. Music's capacities, he says, lie in its potential to be engaging, emotional, social, physical, personal, coordinating, and persuasive. Each is like a strand of a rope—they all work together such that trying to isolate their individual contributions is not just difficult, but misguided. Much research over the past 100 years has attempted to decompose music into elements to study their individual effects. Are there particular frequencies that move us? Is it a certain combination of notes that we will always find pleasing, or a mathematically described rhythm that makes us want to hit the dance floor? To attempt to understand music's sway over us by looking at these components is like looking at the tiny dabs of paint that make up the *Mona Lisa* and trying to account for why we find the painting so engaging. It is not the parts, nor is it even the mere sum of them.

Think for a moment about the last time you saw a river—not the cover for Bruce Springsteen's album of the same name, not the image you have of boys throwing rocks by the bank, or the "tugboat, down by the river" (don't you know) in "Mack the Knife," but an actual churning, glistening, splashing river. There is something intrinsically magnificent and captivating about rivers. A psychologist or neuroscientist could set out to study what it is about that particular river that makes it so appealing. Maybe they'll take a bucket of river water and bring it back to the laboratory. But at that very moment, it ceases being a river—it is just a bucket of river water. In the same way, when we take individual rhythms and pitches, maybe a trumpet blast here or a bass line there, out of the musical performance, all we are left with is a bucketful of notes. No movement. No flowing. The magnificent thing we had is no longer alive; scarlet billows, oozing musical life, spill over the ground.

Whatever it is in music that is engaging, emotional, or social, for instance, is not a building block. It is the end product of a number of different elements coming together to make it so. In other words, the elements that coalesce are not in themselves engaging, but they create something engaging through their unique combination, through an epiphenemenon, one that involves, indeed, *requires* you as part of the experience. As David Pettus said, the act of creation is not complete unless and until a person engages with it. Music begins with three atomic elements—frequency, duration, and amplitude—that then become interpreted and remapped by the brain onto attributes such as pitch, melody, harmony, timbre, rhythm, and meter, and those attributes in turn are combined and interpreted as qualities that move us. That latter step of combining and interpreting is dependent on a number of factors, such as our mood state, culture, experience or previous exposure, immediate surroundings (environment), short- and long-term memories, plus some random factors. The stimulus of music then, its gestalt, is engaging, emotional, social, physical, personal, coordinating, and persuasive.

One unique aspect of our brain's processing of pitch is that it produces a multidimensional cognitive structure out of the single dimension of frequency, creating complex tonal relations such as the circle of fifths, octave equivalence, and intervallic consonance (see endnotes). I've come to believe that we find music so mentally, physically, and spiritually engaging because it stimulates such higher dimensional thinking, modeled by six orthogonal dimensions. If the physicists are right, string theory says that we live in a 10-dimensional universe and that all we see of it is its projection, of flattening, onto the four dimensions of length, width, height, and time. This suggests that music may be the one thing in our world that allows us to visit these higher dimensional spaces, and, moreover, have them make intuitive and visceral sense to us.

The ability music has to engage us wholly and completely is one of the reasons it can be such a powerful therapeutic agent. For it is out of a *loss* of the sense of self that we can achieve true healing and reintegration. I began this book with the story of being so engaged with music at the Keystone Corner in San Francisco (listening to Art Blakey and the Jazz Messengers) that I was thrown into a state of experiential fusion. This miraculous transformation and transportation is available to all of us who are open to it, to all who will let it in. It can occur even when we hear a piece of music for the first time, as Richard Powers describes in his novel *Orfeo*:

Young Peter props up on his elbows, ambushed by a memory from the future. The shuffled half scale gathers mass; it sucks up other melodies into gravity. Tunes and countertunes split off and replicate, chasing each other in a cosmic game of tag. At two minutes, a trapdoor opens underneath the boy. The first floor of the house dissolves above a gaping hole. Boy, stereo, speaker boxes, the love seat he sits on: all hang in place, floating on the gusher of sonority pouring into the room.

A memory from the future! Art has great power to change the way we see the world, our friends, ourselves. The power of art is that it can connect us to one another, and to larger truths about what it means to be alive and what it means to be human. While music can increase empathy overall, and on the average, that doesn't mean that it always will in every case. Penicillin doesn't cure every infection, because some bacteria become antibiotic resistant. Sadly, then, there are exceptions. Hitler loved Wagner and neither could be said to have been empathetic individuals. Yet music holds a special place among the arts because, unlike painting or sculpture, it is manifest across time. Literature is manifest across time, as well, but if we are the ones doing the reading, the timing is up to us. If our minds wander while reading, we lose our place and have to go back. Music carries with it its own internal capacity for moving us forward in time without any additional effort. If our minds wander while listening to music, it keeps going, taking us with it. And, unlike these other art forms, our brains can perceive music even when our minds are wandering—one might say that the highest form of immersion in music *is* when we concede control of our thoughts.

The Canadian literary theorist Northrop Frye wrote, "The fundamental purpose of the imagination in ordinary life is to produce, out of the society we live in, a vision of the society we want to live in." Conductor Kent Nagano elaborates: "This is why the arts are and always have been so essential to humanity and why people gather to celebrate festivals and make music, why they surrender to the poetic and get lost in the infinite beauties and the so magically whimsical excitements of nature and life."

Questions about the importance of music are particularly pertinent now, in this time of political, economic, and social stresses. The fine arts are under threat. Their importance is questioned, their contribution to social development is generally undervalued. Nagano continues, "The arts are, by definition, exercises in and manifestations of creative imagination. Healthy societies grow in response to changing needs of their peoples, and changing circumstances of the world. In order for any society to build a world having the values and qualities in which it would like to live, you must first imagine it. Only then can you build it."

Frye goes further and postulates that unless we engage with the arts, we will raise an entire generation of citizens who have not had the opportunity to train their imagination. This will result, he says, "in your inability to live in the kind of world you would wish for. In order to have this world you must effectively imagine what kind of world could possibly exist and then construct the blueprints in your imagination."

Using our imaginations, as we do when we engage with the arts, encourages tolerance. "In the imagination, our own beliefs are also only possibilities, but we can also see the possibilities in the beliefs of others," Nagano says. He continues:

> Bigots and fanatics seldom have any use for the arts, because they are so preoccupied with their beliefs and actions that they can't imagine anything else as possibilities.
>
> What exactly is the kind of person that the business world of today wants? They should be social, communicative, approachable, thoughtful, self-reflexive, value-oriented, disciplined, empathetic, attentive—and capable of making professional, personal and ethical judgments. Naturally, no one is required to play the piano, nor do you have to be a member of an orchestra, or a painter, or dancer. However in order to be valued by our economy today, at some point, a person must have dealt

with existential questions, and thought about themselves and their environment—that which humanist disciplines have at their core. And where does such thinking take place? Almost exclusively in the permanent examination of the arts, of music, literature, philosophy, and painting.

Without imaginative thinking, our social structure would collapse upon itself. Leadership would become increasingly dictatorial. Empathy, like imagination, needs to be developed through experience, through practice. Our children are instinctively hungry for it. Music stimulates, develops, and feeds the imagination.

In this we can all learn from Beethoven. "The basis of his creative mastery," Nagano continues, "lies in his ability to take two or more elements of sharply contrasting character and place them directly against each other. These elements may include an energetic forceful one against a melodic one, a rhythmically active driving one against a singing, lyrical one, the rapid agitated one against a calmer one, or a brightly colored tonality against a darker, more brooding one." Beethoven creates worlds of opposites, of opposing forces that somehow harmoniously work together. It is within his works that questions about societal harmony are negotiated through the metaphor of acoustical harmony.

The German philosopher Arthur Schopenhauer viewed music as not merely a representation of the world, as other arts might be. It does not simply replicate or echo the patterns we observe around us, but is an expression of the very "Will" that drives everything in our corporeal and metaphysical worlds. This distinction elevates music to a plane where it becomes, not a mirror, but a window—a conduit—into the deepest recesses of existence. It's akin to suggesting that music has its own "intentional stance," evoking responses from us because it aligns with fundamental laws of nature. As philosopher (and polymath) Gottfried Leibniz said, *Musica est exercitium arithmeticae occultum nescientis se numerare animi*: "Music is a hidden arithmetic exercise of the soul, which doesn't know that it is counting."

It's a mesmerizing hypothesis, and while one might be tempted to dissect it with the sharp tools of neuroscience and evolutionary psychology, there remains an ineffable allure to the idea that music might indeed tap into something primal, universal, and profoundly mysterious. And that mystery is delivered by music's inherent ambiguity. Ambiguity in music isn't an error in our cognitive apparatus, but a feature that pulls us into a deeper state of introspection. It toys with our evolutionarily ancient pattern-seeking instincts, challenging us to discern meaning in a medium that is entirely a subjective mental construct, such that no single, fixed, and objective meaning exists. According to Schopenhauer, this taps into the fluidity of the "Will," resisting fixedness, and thus engaging our minds and feelings in the deepest and most profound ways.

I think of music as a stream that endlessly branches with each new listening into new tributaries or distributaries. Indeed, listening to an old, well-known performance can yield an even richer set of new branchings, taking the mind to new vistas and cognitive-emotional landscapes, all from the same, original source. Instead of presenting a straightforward narrative, music offers a dynamic interplay of sound, structure, and meaning, continually prompting our brains—at the millisecond level—to adjust and reinterpret. Instead of invoking a single, linear pathway of comprehension, the ambiguities present multiple neural routes of interpretation. This stimulates neuroplasticity, growth of whole new brain pathways, and healing or rerouting of damaged ones.

Ultimately, music can do so many things because it can *mean* so many things. It is intentionally ambiguous. It does not explicitly refer to any concrete object-in-the-world, it refers to metaphorical objects and emotional ones. It is rich with possibilities for personal interpretation. We can choose to like or reject a piece of music; to fall in love with it; to tire of it. It can take on different meanings as our brains change, as we react to the intricacies of life, love, memories, and relationships. And it is among the most intimate of art forms. Visual art, like anything we see, and due to the way our retina evolved, always seems *out there* in the world. Sound, like taste, feels like it is *inside our heads*, inside of *us*. Shut your eyes after viewing a painting and you may get an afterimage for 30 seconds. Stop listening to a piece of music and it can replay in your head for as long as you like. I have never heard of someone who gets paintings stuck in their heads (although I imagine some do), but earworms, those snippets of music that rattle around in your brain, are reported by 90% of the population. Once music gets in there, it stays in there.

Music is, at its heart, both communal and intensely private at the same time. What other human activity can be experienced with tens of thousands of others, and still feel so personal? Earlier I said that in the search for the substantive universals that underlie the power of music we have the harmonic series (related to physical acoustics), and the low-integer ratios of rhythms (related to the physiology of heartbeats and bipedal locomotion). Those are the tangible universals. The more important universal, however, is the intangible: music speaks to us in a language that is only partly understood by others. Your musical semantics and pragmatics are different than mine. Your relationship with music takes place in a secret language, known only to you and her. The secret chord is you, the product of all your life's encounters; every word or note heard, every thought shared or kept to yourself, every dream achieved or struck down. And that chord is on the move, constantly changing, communicated in a language and a dialect you can understand intuitively without ever having to learn it or take lessons in it, a conversation that is continually evolving.

Acknowledgments

THE IMPETUS FOR THIS BOOK WAS BORN OUT OF A SENSE of unease. I had spent 20 years of my professional life performing brain scans (fMRI) of individuals in order to better understand the neuroanatomy of music. We were mapping the musical brain, trying to find which regions are involved in specific aspects of music listening, performing, and composing. I could tell you that pitch is processed *here*, that rhythm is processed *there*; that it all comes together later in the brain in a process called "featural binding." But knowing where something happened was beginning to feel hollow. I recalled the story Dan Dennett tells of the neurosurgeon who had seen thousands of brains but had never seen a *thought*. Had I, or the people in my field, actually learned anything useful, had we seen music in the brain or just identified a bunch of circuits that added up to music somehow (wave hands here)?

So much of life comes down to serendipity. After I'd been in my job as an assistant professor at McGill for about a year, our department hired a new faculty member, Jeffrey S. Mogil, who studied pain. Jeff's work could not have been more different from mine. He studied mice, I studied humans. He studied something most people try to avoid, I studied something that people are drawn toward. But we enjoyed one another's company immensely, played in a band together (Hebb Zeppelin, after our department's founder, Donald Hebb), and we found we had the same sense of humor—when I think of Jeff, I think of all the laughing we do together. Because Jeff is a musician, he was interested in my work. Because I sometimes get aches and pains, like anyone, I was interested in his. And both of us love brainstorming about designing new experiments.

One day, we were talking about studies that showed music's ability to reduce pain (in humans, not in mice). We wondered: what causes the effect? Is it music's ability to distract us, to alter our mood, or something else? In 2009, Laura Mitchell of Glasgow New Caledonia University came and spent her sabbatical year in my lab and we performed studies to investigate this (it turns out it was both). In 2013, the journal *Trends in Cognitive Sciences* asked me to write a review article on the neurochemistry of music. I had a new post-doctoral fellow working for me in my lab, Dr. Mona Lisa Chanda, who was trained as a neurochemist—as it happens—by Jeff Mogil. I was skeptical at first. After all, the great majority of articles I had read were rife with flaws and methodological errors. But Mona and I dug in with an open mind. We were surprised to see that there among all the piles of neuroscientific chaff were some clever experiments with robust findings, the result of careful and rigorous research.

In 2017, Francis Collins, the Director of the U.S. National Institutes of Health (NIH), and Renée Fleming invited me to present those findings at a two-day conference on Music and Brain Health at the Kennedy Center for the Performing Arts, with Charles Limb, Nina Kraus, Sheri Robb, Ani Patel, Ben Folds, and Edwin Outwater. Over the next few years, I was appointed to several expert panels and committees by the NIH and the Global Council on Brain Health to help explore the evidence-based use of music for health and wellness outcomes. It became clear to me then that what had been a field of inquiry mostly marked by pseudoscience and faulty claims has developed into a substantial body of knowledge based on high-quality science. Most of this had not trickled down to the average reader. The time seemed right, then, to write a book that would bring together all we knew, and I started working on it in earnest in January 2019.

During the COVID-19 lockdown, the Kennedy Center asked

Renée Fleming to curate a series of video conferences and performances to help keep the public engaged with arts and science during a time when so many of us were feeling cut off from one another. Renée invited me to one of these virtual events, where we traded ideas about music as medicine, and sang one of my songs together, "Just a Memory."

In March of 2022, Francis Collins—just retired from the NIH—was named Science Advisor to President Biden, and asked me to join a working group to organize an ambitious two-day conference on the science and applications of music for health and wellness. "Music As Medicine: The Science and Clinical Practice" was held December 14–15 in Washington, D.C., sponsored by the National Institutes of Health and the National Endowment for the Arts, and jointly organized by the NIH, NEA, the Renée Fleming Foundation, and the John F. Kennedy Center for the Performing Arts. Francis and Renée co-chaired the meeting. It was a thrilling and intense two days that brought together, in one room, leading scientists, clinicians, funding agency officers, and stellar musicians: Shelly Berg, Jeralyn Glass, Lisa Wong, Grace Leslie, Fred Johnson, and Raul Midón.* By then, I had already handed in the manuscript for this book, and I was grateful when John Glusman, my editor at Norton, allowed me to add some of the newfound wisdom from this important meeting.

One night in 2019, after having started on this book, I was lying in bed listening intently to the newly released *Beethoven: The Complete Piano Sonatas* by Mari Kodama. Mari's interpretations of these pieces utterly transformed the way I thought of them. I find I have very narrow tastes in classical music insofar as interpretations are concerned. I've listened to about 40 versions of the *Pathétique* and most of them leave me flat. One of my favorites has been Barenboim's 2005 live Berlin recording. There is something sublime and

* The event was taped and can be viewed here: https://www.nccih.nih.gov/news/events/music-as-medicine-the-science-and-clinical-practice.

delicate about it that speaks to me. When Mari's 2019 recording came out, my understanding of the piece changed entirely. I have listened to it over and over again, often weeping—sometimes quietly, sometimes profoundly. She brings a majestic quality and an emotional authority. Where Barenboim sometimes plays into the uncertainty of human experience, Mari plays into the passion and constancy of it. I hear a more profound inner strength in her rendering, and a greater power, and—perhaps paradoxically—a commanding sense of unwavering inner calm (a trait she shares with her daughter Karin). I sat there unmoving, unblinking, filled with awe. The genesis of this book is the very genesis of my coming into new awarenesses of music every day, and of its consistent if sometimes ineffable ability to comfort, inspire, and intrigue me.

And so, I am grateful to Mari for her performances, and her friendship, and her guiding hand in helping me to become a better musician. I've learned more about music from Mari, and from playing with Victor Wooten, Rodney Crowell, Rosanne Cash, Bobby McFerrin, Renée Fleming, David Jackson, David Byrne, and Joni Mitchell than from any of the brain scans and scientific conferences. I'm indebted to them, and in great admiration. The late Mike Lankford, one of my favorite authors, gave me valuable advice during the writing of this book. Mostly, I owe an enormous debt of gratitude to the many undergraduate students who have taken my classes over the years and asked the kinds of questions that helped me learn how to explain things better. As J. M. Coetzee wryly noted, those who come to teach often learn more than the students they are charged to instruct; this has been gloriously and richly true for my life in the classroom.

Indispensable to this book at every step, small and large, has been my so-much-more than a research assistant, Lindsay Anne Fleming. Lindsay began as my lab manager in 2001 and has grown into a trusted friend, a confidant, a meticulous researcher, and a fine editor. Lindsay has a knack for helping me to say what I'm thinking when

I can't get the words on the page; for finding the most illustrative musical examples and cultural touchstones; she uncovered wonderful examples of research findings at every turn. She often surprises me with new ways of expressing things that beautifully complement my own deficiencies. She is a treasure, my right hand and my left brain, my literary and scientific soulmate, bringing a steady calm and graceful composure to every interaction. I am immensely grateful also to her family, who lived this book with her for the last five years, often coming up with great ideas that found their way here—David, Kennis, Lua, Grace, and Eliot.

My agents Sarah Chalfant and Rebecca Nagel began working on the book with me when it was still just an idea and have contributed in numerous ways from conception to this very moment. Along the way, their insightful comments and guidance helped me discover what the core of this book would be. I thoroughly love having John Glusman as my editor, and our conversations not only made for a better book, but made me a better writer and a better thinker. Working with the team at Norton, at Allen Lane, and at Cornerstone Books has been a great privilege, particularly indexer Do Mi Stauber, who literally wrote the book on indexing.

I'm also grateful to William Forde Thompson, Jeff Mogil, Nick Garrison, Venetia Butterfield, Anna Argenio, Lew Goldberg, Brian Nova, Vinod Menon, Scott Grafton, Dillon O'Brian, Shelly Berg, Nolan Gasser, Michael Posner, and Jocelyn D'Arcy for reading (and improving) previous drafts of the manuscript, and to Victor Wooten, Rodney Crowell, Michael Thaut, Jasper Rine, Renée Fleming, Stewart Copeland, Rosanne Cash, Kent Nagano, Graham Nash, Kristen Stills, Joni Mitchell, Jeralyn Glass, Tom Brosseau, Paul Felton, David Pettus, and Frank Russo for conversations that guided both the direction and the content of this book; to the team at MIIR Audio Technologies—Paul Moe, Roger Dumas, Aaron Prust, Jon Beck, Gary Katz, Kelly McCollins, and Courtney Jensen—for many helpful discussions; to Heather Lewis and Ronny Goldschmitz who were exemplary personal assistants, and along with the team at the Wylie Agency, for keeping the trains running on time (and knowing what to do when they didn't). *Chapeau* to my student research assistant Bukle Unaldi Kamel and the McGill librarians Emily Kingsland, Katie Lai, Cathy Martin, Nikki Tummon, and Lonnie Weatherby.

I save my greatest appreciation and gratitude to my wife Heather Bortfeld, a brilliant neuroscientist with her own thriving and demanding scientific career. For five years, she's been living with this book as much as I have. Whether she was in the middle of reading the newspaper at breakfast, or while we'd be out walking the dog, she remained an inexhaustible sounding board for my ideas, my reading passages to her, and asking the kinds of questions that allowed me to clarify my own thinking. Every interruption was met with a smile and an insight.

Appendix:
Types of Music Therapy

MUSIC THERAPY ENGAGES A SUITE OF PRACTICES. JUST AS A psychotherapist might be trained in several methods, music therapists have standard techniques from which to draw, and many combine these to cultivate their own personal style. Music therapy techniques fall into two broad categories: passive or active. In passive (or receptive) music therapy, the therapist plays live or recorded music for the client to respond to in words or other artistic responses such as drawing or dance. Active (also called creative or expressive) music therapy involves composition, improvisation, or re-creation (imitation of music created by the therapist). These methods differ from musical interventions, in which a doctor or caregiver may play music in an operating or recovery room or dentist's chair; they differ from most self-directed uses of music, such as for exercise workouts and relaxation. Music therapy is a licensed practice, governed by the American Music Therapy Association (AMTA) and the Certification Board for Music Therapists (CBMT) in the United States, Health & Care Professions Council in the UK, the Canadian Association for Music Therapy (CAMT) in Canada, and others around the world. This summary offers a brief on specific formal practices recognized by these governing bodies.

Neurologic Music Therapy (NMT) stands at the forefront of neurorehabilitation, targeting the improvement of motor, speech, and cognitive functions. Developed by Michael Thaut at Colorado State University, NMT is built on the principle that music engages the entire brain, acting as a catalyst for neuroplastic reorganization of essential pathways. NMT-trained therapists employ rhythm, melody, and harmony to stimulate functional improvements in patients with neurological impairments. Techniques such as rhythmic auditory stimulation, patterned sensory enhancement, and therapeutic singing are integral to NMT. Clinical applications encompass a wide range of disorders, including motor rehabilitation in Parkinson's disease, stroke recovery, and cognitive rehabilitation in traumatic brain injury patients. Thaut and his colleagues provide a robust foundation for understanding rhythmic entrainment and its impact on the motor system; Teppo Särkämö and his team at the University of Helsinki demonstrate the positive effects of music listening on cognitive recovery and mood after a stroke.

Creative music therapy embraces a broad spectrum of approaches that use music as a medium for channeling self-expression, emotional processing, and personal growth. It encompasses diverse techniques such as individual or collaborative songwriting and exploration of different musical instruments. The therapeutic foundation is that self-expression and personal growth can be achieved through emotional processing that occurs during creative acts. The role of the therapist is to guide and facilitate the client's achievement of stated goals. Creative music therapy is very similar to art therapy, with both encouraging individuals to engage in nonverbal expression to tap into inner experiences that may be difficult to put into words. In particular, the arts hold the power to reveal emotional insights that may elude us when we have to describe them using the constraints of language and literal meaning, stimulating us to access metaphorical meanings that ultimately can uncover our true thoughts and feelings. Creative music therapy has been used to treat individuals with a variety of mental health disorders, including depression, anxiety, and schizophrenia. It has also been used with children with autism spectrum disorder and developmental disabilities.

Nordoff-Robbins Music Therapy, named after its founders, Paul Nordoff and Clive Robbins, is a form of creative music therapy. It adopts an improvisational approach that taps into the innate musicality within each of us. Therapist and patient engage in a collaborative process of musical creation, improvisation, and exploration, aiming to stimulate personal growth, emotional expression, and social interaction. This method has showcased successful applications across diverse populations, including children with profound intellectual disabilities and psychiatric patients. Moreover, it has shown promise in working with older adults diagnosed with dementia or Alzheimer's disease.

Orff-Schulwerk is a music education approach named after its co-founder Carl Orff; "Schulwerk" is the German word for "schoolwork." This approach emphasizes the potentially transformative power of rhythm, movement, and play, and aims to promote musical and social development while enhancing communication and self-expression. Orff Music Therapy was developed by Carl's wife, Gertrud, for children with developmental disorders and delays using Orff-Schulwerk as the musical basis. It has shown promise in working with children and adults with developmental disabilities, as well as individuals with mental health disorders. Implementing rhythmic activities and movement, music therapists foster social interaction and communication skills, especially beneficial for children with autism.

Vocal Psychotherapy, pioneered by Diane Austin, is built upon the idea that the voice is intimately connected to the emotions, and that musical vocal expression can be a powerful tool for accessing and processing unconscious feelings. Austin's technique incorporates vocal improvisation, singing, and verbal processing to delve into the patient's emotional world and facilitate healing (similar to Arthur Janov's Primal Scream Therapy, made famous on John Lennon's first solo album, *Plastic Ono Band.*) Vocal Psychotherapy has been employed in treating various populations, including adults with psychiatric disorders and adolescents in foster care. Austin's book, *The Theory and Practice of Vocal Psychotherapy: Songs of the Self*, details the essence of this approach. Vocal Psychotherapy has been used to treat individuals with a variety of mental health disorders, including anxiety, depression, and PTSD. It has also been used to treat individuals with speech and language disorders.

Guided imagery and music entails listening to carefully selected music to induce relaxation, reduce anxiety, and promote introspection through daydreaming-facilitated insights. Typically, the therapist guides the individual through visualization and imagery exercises while integrating the music as a transformative element. It is similar in concept to nonmusical guided imagery conducted with verbal instructions from a therapist. Guided musical imagery has been used to treat anxiety, depression, chronic pain, and post-traumatic stress disorder (PTSD). For instance, a patient with chronic pain can be guided to relax and visualize a peaceful, pain-free state while immersed in carefully curated music.

The Bonny Method of Guided Imagery and Music was developed by Dr. Helen L. Bonny. It is a type of guided imagery practice driven by the belief that Western classical music is uniquely able to evoke profound emotional responses, and that these responses promote personal growth and healing. A specially trained therapist selects music based on the client's mood and energy, then offers suggestions for relaxing the body and focusing the mind. Once the music begins, rather than the therapist making suggestions, the client describes their experience as it unfolds. The therapist facilitates, witnesses, and supports the client's experience. In this respect, it mirrors the experience of many hallucinogenic therapies. After 35–45 minutes, the therapist initiates a return to the present moment, and the client and therapist review the session together. The Bonny Method has been used to treat a variety of conditions, including anxiety, depression, and PTSD. It has also been used with individuals undergoing cancer treatment.

Chinese Five-Element Music Therapy is a therapeutic intervention that merges music with the principles of traditional Chinese medicine. It is based on the belief that the five elements, Wood, Fire, Earth, Metal, and Water, are fundamental to the balance and harmony of the human body and mind. Each element is associated with specific musical tones. By using music that corresponds to the elements, this therapy aims to restore and promote overall well-being by addressing imbalances and disharmony in the individual's energy system. (This is not a scientifically recognized concept.)

Glossary

agraphia. A neurological condition characterized by the impairment of a person's ability to write. It can result from damage to specific brain regions involved in language and motor functions.

alexia. A cognitive disorder that manifests as the inability to read, despite preserved or previously acquired reading abilities. It is often associated with brain injury or neurological conditions affecting language processing.

amusia. A generalized deficit in music perception and recognition. Individuals with amusia may have difficulty discerning pitch, rhythm, or melody and may lack an appreciation for music.

amygdala. A subcortical region that is part of the limbic system, and an important component of emotion processing, including fear, aggression, social interaction, and memory.

anhedonia. The inability to experience pleasure or a diminished interest in activities that were once enjoyable. It is a common symptom in mood disorders such as depression.

anterior cingulate. A region in the frontal lobe that serves as a computational hub for maintaining attention. It is also implicated in the Default Mode Network (DMN), which is active during introspection and mind-wandering.

arcuate fasciculus. A fiber tract connecting speech production and perception areas in the frontal, temporal, and parietal lobes.

auditory cortex. A region of the brain responsible for processing auditory information, including sound perception and interpretation. It plays a crucial role in the perception of pitch, rhythm, and other auditory stimuli.

basal ganglia. A cluster of neurons involved in movement, motor control, decision-making and reward, and timing-related information, located deep beneath the cerebral cortex. Damage to these nuclei can lead to tremors and Parkinson's disease. The basal ganglia comprise the striatum (which includes the putamen and caudate nucleus), globus pallidus, subthalamic nucleus, and substantia nigra.

Broca's area. A region in the frontal lobe of the brain, in the dominant hemisphere (typically the left), that plays a critical role in language production and speech. Damage to this area can result in expressive language deficits, a condition known as Broca's aphasia.

Brodmann Area 47. A small sliver of tissue in the prefrontal cortex that serves as a computational hub for temporal expectations and sequence prediction, playing a role in language, music processing, and working memory.

caudate nucleus (sometimes abbreviated simply as the caudate). One of the major components of the basal ganglia located near the center of the brain, with a role in motor planning, learning, and the formation of procedural (motor-based) memory.

cerebellum. A subcortical part of the brain located at the base of the skull. It helps to maintain a steady gait and to execute voluntary motor actions, and is more generally required for motor control, balance and posture, and, through connections to the frontal lobes, for processing emotion.

contour. An attribute of melodies that describes only the pattern of "ups and downs" (changes in pitch direction) without respect to the size of the interval.

cortex. The outer layer of the brain, responsible for higher-order cognitive functions like thinking, planning, perception, and sensation. It consists of four main lobes: frontal, parietal, occipital, and temporal.

cortisol. Often referred to as the "stress hormone," it plays a crucial role in the body's response to various stressors. It helps regulate metabolism, blood sugar, immune function, libido, and the sleep-wake cycle. Chronic elevation of cortisol levels is associated with long-term stress and may contribute to various health issues.

Default Mode Network (DMN). A network of brain regions that is active when an individual is not focused on the outside world or engaged in specific cognitive tasks. It is associated with introspection, daydreaming, and self-referential thoughts.

diffusion tensor imaging (DTI). A specialized magnetic resonance imaging (MRI) technique used to visualize and analyze the pathways of white matter tracts in the brain.

dopamine. A neurotransmitter, or chemical messenger, that transmits signals in the brain and other areas of the nervous system and plays a crucial role in various physiological functions, including mood regulation, motivation, reward processing, and motor control.

executive network. Also known as the executive control network or the central executive network. A functional brain network involved in higher-order cognitive functions such as decision making, attentional control, working memory, and goal-directed behavior. It plays a key role in coordinating and regulating various cognitive processes to achieve complex tasks and goal-oriented activities. The executive network is associated with regions in the frontal and parietal lobes, and its proper functioning is crucial for adaptive and flexible cognitive control.

experiential fusion. A term coined by psychologist Richard Davidson designating the state of being so fully absorbed in an experience that you are not aware of where you are or what time it is; a conscious state that lacks meta-awareness. It is related to the flow state, but does not necessarily involve the performance of an action.

flow state. A term coined by psychologist Mihaly Csikszentmihalyi for a conscious state of complete immersion and engagement in the performance of an activity, where one loses the sense of time, and one's skills are functioning at peak levels; often referred to as being "in the zone."

frisson. Also known as "chills." A sudden and intense shiver, goosebumps, or tingling sensations often experienced during listening to music, or in response to particularly moving or awe-inspiring stimuli. It is associated with heightened emotional states and is thought to be linked to the release of dopamine in the brain.

frontal lobe. The largest lobe in the human brain, involved in various cognitive functions, including decision-making, problem-solving, motor function, and personality.

functional magnetic resonance imaging (fMRI). A non-invasive neuroimaging technique that measures and maps brain activity by detecting changes in blood flow and oxygenation. It relies on the blood oxygenation level-dependent (BOLD) signal, which indicates increased or decreased neural activity in specific brain regions.

GABA (gamma-aminobutyric acid). A neurotransmitter that serves to inhibit or dampen neuronal activity, playing a crucial role in mood regulation, impulse control, anxiety reduction, and promoting relaxation. GABA-selective neurons produce and release GABA and are concentrated in several key brain regions necessary for inhibitory control, including the prefrontal cortex and basal ganglia. A GABA receptor is a type of protein located on the surface of neurons, the site where the GABA molecule binds.

glutamate. An amino acid and the most abundant excitatory neurotransmitter in the central nervous system. It plays a fundamental role in synaptic transmission, neural plasticity, and learning. While it is essential for normal brain function, excessive glutamate release can be neurotoxic and is implicated in certain neurological disorders and neurodegenerative diseases.

hippocampus. Part of the limbic system, a seahorse-shaped structure located deep within the temporal lobe. It evolved chiefly for spatial geonavigation, and is crucial for the formation of new memories. It is unlikely that old memories are stored there, but it may serve as an index as to where they are stored.

hypothalamus. A small and crucial region in the brain that regulates various physiological processes, including temperature, hunger, thirst, and the sleep-wake cycle. It also controls the release of hormones from the pituitary gland.

inferior colliculus. A structure in the midbrain that plays a central role in the processing of auditory information. It is a key component of the auditory pathway.

insula. A region of the cerebral cortex located deep within the lateral sulcus. It is involved in various functions, including emotional regulation, self-awareness, and processing of visceral sensations.

islands of Calleja. Small structures found in the brain, particularly in the ventral striatum, involved in the modulation of dopamine release and associated with reward-related processes.

key. A system for organizing a piece of music around a central note (the tonic), the key defines which subset of notes will be considered primary to the piece. Most musical pieces (with the exception of 12-tone or atonal music) begin and/or end with a primary (tonic) chord, built from the first note of the scale corresponding to the key. For example, a piece in the key of C major would typically start and/or end with a

C major chord. C major and other chords within the key (D minor, E minor, F major, G major, A minor, and B diminished) would be expected to occur most prominently in the piece, with only occasional, temporary departures using other chords. In the absence of chords, the melody itself implies chords and thus the piece would typically begin and end with notes that are contained in the tonic chord. The key creates harmonic structure, expectations, and allows for violations of those expectations.

limbic system. A complex set of brain structures that includes the amygdala, hippocampus, and hypothalamus, among others. It is responsible for emotions, motivation, memory, and various autonomic functions. Often referred to as the "emotional brain."

loudness. A purely psychological (as opposed to acoustic) construct loosely and nonlinearly related to the amplitude of a signal.

mesolimbic system. A specific pathway within the brain's reward system that is involved in the experience of pleasure and reinforcement. It includes the ventral tegmental area and its projections to the nucleus accumbens, a key structure associated with reward and motivation. Dysregulation of the mesolimbic system is associated with conditions such as addiction and schizophrenia.

meter. The rhythmic groupings of a piece, based on the accent pattern of loud versus soft notes. Common meters are felt in 2, 3, or 4. When meter is notated in the time signature of a piece of music, it defines the rhythmic groupings by which we count a piece. For example, regular march time is 4/4, meaning that a quarter note gets one beat and there are 4 beats to a "measure" of music, counted 1-2-3-4. Waltz time is in 3/4 meter 1-2-3. See also **tactus.**

midbrain. A portion of the brain stem that serves as a relay center for sensory and motor information. It plays a crucial role in functions such as visual and auditory processing.

mode. The word "mode" has two meanings in music. (1) The modality (major or minor mode) of a composition in contemporary music. This is the sense I use in this book. (2) One of the seven Greek modes or scales: Ionian, Dorian, Phrygian, Lydian, Myxolidan, Aeolian, Locrian.

motor cortex. A region near the back of the frontal lobe involved in the planning, control, and execution of voluntary movements.

myelin. A fatty substance that forms a protective sheath around nerve fibers, allowing for faster and more efficient transmission of nerve impulses. It is essential for the proper functioning of the nervous system.

neurochemicals. Chemical substances that play a role in the transmission of signals within the nervous system. Examples include neurotransmitters, hormones, and other molecules that influence neural activity.

neurogenesis. The process by which new neurons, or nerve cells, are generated from neural stem cells in the brain. This phenomenon occurs throughout life, although the extent and significance may vary across different stages of development. Neurogenesis is implicated in learning, memory, and emotional regulation, and it contributes to the overall plasticity of the nervous system.

neuron. The basic structural and functional units of the nervous system. Neurons transmit information through electrical and chemical signals and form the basis of neural networks or circuits.

neuroplasticity. The brain's ability to reorganize itself by forming new neural connections throughout life. This adaptive capacity allows the brain to modify its structure and function in response to learning, experience, and environmental changes. Neuroplasticity plays a crucial role in recovery from injury, skill acquisition, and cognitive development.

nucleus accumbens. Part of the basal ganglia, a major component of the brain's reward system, releasing and modulating levels of dopamine in the brain, working closely with the ventral tegmental area.

occipital lobe. Located at the back of the brain, this lobe is the primary area for processing visual information, including color, shape, and motion. Among blind people, it is neuroplastically repurposed for speech.

olfactory tubercle. A part of the brain that is involved in the processing of olfactory (smell) information. It is located in the ventral striatum.

parietal lobe. Positioned at the top and back of the brain, this lobe is involved in processing spatial orientation, numerical understanding, and certain types of object recognition.

pitch. A purely psychological (as opposed to acoustical) construct, based loosely on the frequency of vibration of a vibrating object.

pituitary gland. A pea-sized gland located at the base of the brain. Often referred to as the "master gland," it secretes hormones that regulate various physiological processes and control the functions of other endocrine glands.

planum temporale. A brain region located in the temporal lobe, just posterior to the auditory cortex. It plays a critical role in language processing, particularly in speech comprehension and production.

pons. A region located in the brain stem that connects different parts of the brain, including the cerebrum and the cerebellum. The pons plays a crucial role in various functions, including relaying signals between different brain regions, regulating breathing, and controlling facial movements.

posterior cingulate cortex. A region of the parietal lobe involved in various cognitive processes, including memory retrieval, attention, and self-reflection, and part of the Default Mode Network.

precuneus. A region of the parietal lobe involved in a range of complex functions, such as visuospatial imagery, episodic memory retrieval, and aspects of consciousness and self-awareness. The precuneus is also a part of the Default Mode Network.

prefrontal cortex. The part of the brain that is the most highly developed in humans, responsible for executive functions such as planning, decision-making, impulse control, working memory, attentional control, and moderating social behaviors.

prolactin. A hormone produced by the pituitary gland that plays a central role in the regulation of various reproductive and metabolic functions. Prolactin is best known for its role in promoting lactation in mammals; it also influences immune system modulation and has effects on metabolism and behavior.

putamen. Part of the striatum of the basal ganglia, a computational hub for controlling motor movements and motor learning.

rhythm. A purely psychological (as opposed to acoustical) construct based on the perception of grouping patterns in successive musical tones. Rhythm is built up of duration, loudness, accent structure, and expectations.

scale. A selection of notes that form the primary alphabet or vocabulary from which a musical piece is written. Different types of scales, such as major, minor, chromatic, or modal, provide different moods and colors to the music, much like different palettes in visual arts.

sensory cortex (somato-sensory cortex). An area positioned in the forward part of the parietal lobe, and responsible for receiving and integrating information from the external senses, vision, taste, olfaction,

touch, and hearing. There is mixed information on whether it also processes vestibular input.

serotonin. A neurotransmitter synthesized in the brain and the gastrointestinal tract that contributes to the regulation of mood, appetite,
sleep, and various cognitive functions. Serotonin is often referred to as
the "feel-good" neurotransmitter, and imbalances in serotonin levels
are associated with conditions such as depression, anxiety, and certain
neurological disorders.

striatum. A subcortical structure in the basal ganglia (in the forebrain)
that includes the caudate nucleus and putamen. It is involved in motor
control, reward processing, and learning.

substantia nigra. A structure located in the midbrain that plays a key role
in motor control. It is particularly important in the production of dopamine, a neurotransmitter that influences movement, reward, and mood.
Dysfunction of the substantia nigra is associated with conditions such as
Parkinson's disease, characterized by motor impairments and tremors.

synapse. A specialized junction between two nerve cells, or between
a nerve cell and a target cell (such as a muscle cell). It is the site where
signals are transmitted from one cell to another. Neurotransmitters are
released at the synapse, allowing communication between neurons. Synapses play a fundamental role in the functioning of the nervous system.

tactus. The basic pulse or beat in a piece of music; the point at which
one would naturally clap hands or tap feet.

temporal lobe. Positioned roughly above the ears, the temporal lobe
houses the auditory cortex and a number of special-purpose circuits that
process auditory information. It is also responsible for the proper formation and storage of memories of all kinds, not just auditory.

thalamus. A central relay station in the brain that processes and relays
sensory information to the cerebral cortex. It is involved in regulating
consciousness, sleep, and alertness.

timbre. The attribute of sound that distinguishes two musical instruments playing the same note, or two voices pronouncing the same word.
It is responsible for "tonal color" and is an amalgam of information
contained in the attack, steady state, and release of a note, comprising
pitch, intensity, duration, and spectro-temporal flux (the ways in which
the sound changes over time).

tympanic membrane. Also known as the eardrum. A thin membrane that separates the outer ear from the middle ear. It vibrates in response to sound waves, transmitting them to the middle ear.

ventral tegmental area. A part of the limbic system, positioned close to the midline on the floor of the midbrain, this region is involved in the release of dopamine, a neurotransmitter associated with feelings of pleasure, reward, and motivation.

visual cortex. A region of the occipital lobe dedicated to processing visual information. It plays a crucial role in visual perception, including the recognition of shapes, colors, and motion.

Wernicke's area. A region located in the left hemisphere of the brain associated with language processing, particularly the comprehension of speech and language interpretation. Damage to Wernicke's area can result in receptive aphasia, where individuals may have difficulty understanding language but may still produce fluent speech that lacks meaningful content.

white matter tracts. Bundles of myelinated nerve fibers in the central nervous system that facilitate communication between different brain regions. They form the structural connections that enable information transfer.

Notes

Chapter 1:
A Musical Species

2   *experiential fusion*: Dahl, C. J., Lutz, A., & Davidson, R. J. (2015). Reconstructing and deconstructing the self: Cognitive mechanisms in meditation practice. *Trends in Cognitive Sciences, 19*(9), 515–523.

3   **33 1/3 rpm sing-alongs:** In the late '50s and early '60s, sing-along records by The Four Roses Society, Mitch Miller ("Sing Along with Mitch"), and others were popular.

3   **"motion" is part of the word "emotion":** "motion, n." OED Online. March 2023. Oxford University Press.

4   **medicine is both a science and an art:** Charon, R. (2021). Knowing, seeing, and telling in medicine. *The Lancet, 398*(10316), 2068–2070.

      Schwartz, T. H. (2024). *Gray Matters: A Biography of Brain Surgery.* Dutton.

      Schwartz goes on to add: "Ideally, we execute every move as pre-visualized. But what about when you're working down a deep narrow hole, and you can only partially see that last tiny bit of tumor because it's obscured by a small piece of bone, or by an unexpected artery that was too small to have been seen on the MRI? Is it worth the risk to pull that last bit out blindly, or do you leave it behind, knowing that it will only grow back and require yet another risky surgery in the future? . . . Is it worth damaging those structures to get the last remaining piece? . . . Decision after decision after decision with so much on the line."

4   **the therapist, the scientist, and the physician must establish a rapport:** Kneebone, R., Houstoun, W., & Houghton, N. (2021). Medicine, magic, and online performance. *The Lancet, 398*(10314), 1868–1869.

5   **Music promotes relaxation:** Chanda, M. L., & Levitin, D. J. (2013). The neurochemistry of music. *Trends in Cognitive Sciences, 17*(4), 179–193.

5   **reduce blood pressure:** Mir, I. A., Chowdhury, M., Islam, R. M., Ling, G. Y., Chowdhury, A. A., Hasan, Z. M., & Higashi, Y. (2021). Relaxing music reduces blood pressure and heart rate among pre-hypertensive young adults: A randomized control trial. *Journal of Clinical Hypertension, 23*(2), 317–322.

5   **make diabetes management easier:** Bacus, I. P., Mahomed, H., Murphy, A. M., Connolly, M., Neylon, O., & O'Gorman, C. (2022). Play, art, music and exercise therapy impact on children with diabetes. *Irish Journal of Medical Science (1971–)*, 1–6.

      Mandel, S. E., Davis, B. A., & Secic, M. (2013). Effects of music therapy and music-assisted relaxation and imagery on health-related outcomes in diabetes education: A feasibility study. *The Diabetes Educator, 39*(4), 568–581.

5   **soothes us when we're depressed:** Aalbers, S., Fusar-Poli, L., Freeman, R. E., Spreen, M., Ket, J. C., Vink, A. C., . . . & Gold, C. (2017). Music therapy for depression. *Cochrane Database of Systematic Reviews, 11.*

5   **energizes us for exercise:** Terry, P. C., Karageorghis, C. I., Curran, M. L., Martin, O. V., & Parsons-Smith, R. L. (2020). Effects of music in exercise and sport: A meta-analytic review. *Psychological Bulletin, 146*(2), 91.

5   **Patients with Parkinson's:** Pereira, A. P. S., Marinho, V., Gupta, D., Magalhães, F., Ayres, C., & Teixeira, S. (2019). Music therapy and dance as gait rehabilitation in patients with Parkinson disease: A review of evidence. *Journal of Geriatric Psychiatry and Neurology, 32*(1), 49–56.

5   **patients with Alzheimer's:** Gallego, M. G., & García, J. G. (2017). Music therapy and Alzheimer's disease: Cognitive, psychological, and behavioural effects. *Neurología (English Edition), 32*(5), 300–308.

Tomaino, C. M. (2014). Music therapy and the brain. In B. Wheeler (ed.), *Music Therapy Handbook* (pp. 40–50). Guilford Press.

5   **change our perception of time:** Droit-Volet, S., Bigand, E., Ramos, D., & Bueno, J. L. O. (2010). Time flies with music whatever its emotional valence. *Acta Psychologica, 135*(2), 226–232.

5   **immersed in a VR game:** Rogers, K., Milo, M., Weber, M., & Nacke, L. E. (2020, November). The potential disconnect between time perception and immersion: Effects of music on VR player experience. In *Proceedings of the Annual Symposium on Computer-Human Interaction in Play* (pp. 414–426).

5   **Default Mode Network:** Sridharan, D., Levitin, D. J., & Menon, V. (2008). A critical role for the right fronto-insular cortex in switching between central-executive and default-mode networks. *Proceedings of the National Academy of Sciences, 105*(34), 12569–12574.

Taruffi, L., Pehrs, C., Skouras, S., & Koelsch, S. (2017). Effects of sad and happy music on mind-wandering and the default mode network. *Scientific Reports, 7*(1), 1–10.

6   **Beliefs about music's power to heal:** Thaut, M. H. (2015). Music as therapy in early history. *Progress in Brain Research, 217*, 143–158.

6   **Our word *shaman*:** Random House Dictionary of the English Language, 2nd edition, unabridged.

7   **to heal spiritual, mental, or physical ills:** Winn, T., Crowe, B. J., & Moreno, J. J. (1989). Shamanism and music therapy: Ancient healing techniques in modern practice. *Music Therapy Perspectives, 7*(1), 67–71.

7   **such as the Inuit *angakok*:** Singh, M. (2018). The cultural evolution of shamanism. *Behavioral and Brain Sciences, 41*.

7   **The shaman, medicine man or medicine woman:** Singh, M. (2018). The cultural evolution of shamanism. Ibid.

7   **The shamanistic tradition:** Tedlock, B. (2005). *The Woman in the Shaman's Body: Reclaiming the Feminine in Religion and Medicine.* Bantam.

7   ***Music is the universal language:*** Longfellow, H. W. (1863). *Outre-mer, a pilgrimage beyond the sea, France, Spain, Italy, Note-book. Drift wood, a collection of essays* (vol. 1, p. 174). Ticknor and Fields.

7   ***Music produces a kind of pleasure:*** Confucius (1967). *Li Chi: Book of Rites. An Encyclopedia of Ancient Ceremonial Usages, Religious Creeds, and Social Institutions* (J. Legge, trans.). University Books. (Original work published c. 551–479 BCE.)

8   **"Music gives soul to the universe":** The earliest known use of this quote comes from Ritter, F. L. (1891). *Music in Its Relation to Intellectual Life* (p. 48). Edward Schuberth & Co. Ritter attributes these words to Plato.

8   **Plato believed that music:** Hamilton, E. & Cairns, H. (1962). *The Collected Dialogues of Plato.* Princeton University Press. Plato, *Republic,* Book 3, Section 401, Paragraph E.

8   **For example, Dorian mode:** Ramis de Pareja, 1482. *Musica Practica.* Bologna.

10 **"Every illness is a musical problem":** Novalis. *Schriften,* vol. iii, ed. R. Samuel et al. (Stuttgart, 1960), p. 310, no. 386. Cited in P. Horden. (2019). *Cultures of Healing: Medieval and After* (p. 295). Routledge.

10   **The neurologist Oliver Sacks:** Greene, S. (2015). The medical humanity of Oliver Sacks: In his own words. SPM Blog. https://participatorymedicine.org/epatients/2015/09/the-medical-humanity-of-oliver-sacks-in-his-own-words.html

10   **One patient, Tony Cicoria:** Sacks, O. (2007, July 16). Bolt from the Blue: Where do sudden intense passions come from? *The New Yorker.*

13   **Mozart makes you smarter:** Rauscher, F. H., Shaw, G. L., & Ky, C. N. (1993). Music and spatial task performance. *Nature, 365*(6447), 611.

13   **studies failed to replicate:** Mozart's music does not make you smarter, study finds. (2010, May 10). *ScienceDaily.*

      Thompson, W. F., Schellenberg, E. G., & Husain, G. (2001). Arousal, mood, and the Mozart effect. *Psychological Science, 12*(3), 248–251.

      Steele, K. M., Bass, K. E., & Crook, M. D. (1999). The mystery of the Mozart effect: Failure to replicate. *Psychological Science, 10*(4), 366–369.

Chapter 2:
If I Only Had a Brain

17   **40 milliseconds:** Isnard, V., Chastres, V., Viaud-Delmon, I., & Suied, C. (2019). The time course of auditory recognition measured with rapid sequences of short natural sounds. *Scientific Reports, 9*(1), 8005.

17   **people with brain lesions:** Stewart, L., von Kriegstein, K., Warren, J. D., & Griffiths, T. D. (2006). Music and the brain: Disorders of musical listening. *Brain, 129*(10), 2533–2553.

18   **This partly involves the thalamus:** Kimura, A. (2020). Cross-modal modulation of cell activity by sound in first-order visual thalamic nucleus. *Journal of Comparative Neurology, 528*(11), 1917–1941.

18   **"I see music as structure":** Pochmursky, C. (writer/director) (2009, January 31). *The Musical Brain* [television broadcast]. In V. Dylyn (producer), Mississauga, ON: Matter of Fact Media Inc. Production in association with CTV Television Inc. and National Geographic Channel.

19   **the *flow state*:** Oppland, M. (2023). 8 Traits of Flow According to Mihaly Csikszentmihalyi. PositivePsychology.com. https://positivepsychology.com/mihaly-csikszentmihalyi-father-of-flow/

22   **no sound in the vacuum:** That doesn't stop filmmakers from adding the sound of a roaring engine or fired photon torpedoes and other crazy noises to make a space scene more exciting. They aren't rendering physical realism, but emotional realism, which is something musicians and composers do as well.

23   **pitch range of spiders' hearing:** Stafstrom, J. A., Menda, G., Nitzany, E. I., Hebets, E. A., & Hoy, R. R. (2020). Ogre-faced, net-casting spiders use auditory cues to detect airborne prey. *Current Biology, 30*(24), 5033–5039.

23   **Can spiders hear:** Image created by Bing AI with instructions from the author.

24   **lateral line of a fish:** Image retrieved from https://www.supercoloring.com/coloring-pages/indo-pacific-king-mackerel-scomberomorus-guttatus. This work is licensed under a Creative Commons Attribution-Share Alike 4.0 License.

24   **we have no direct access:** This formulation was inspired by and is based on Neisser, U. (1967). *Cognitive Psychology.* Appleton-Century-Crofts.

26   **English has 16 tenses:** There are 12 basic tenses and 4 future-in-the-past constructions.

29   **noise pollution:** US EPA (2022, August 11). *Clean Air Act Title IV—Noise Pollution.* https://www.epa.gov/clean-air-act-overview/clean-air-act-title-iv-noise-pollution

30   **thickness of your fingernail:** The average fingernail is 0.4 mm thick.

30   **study of white matter tracts:** Rajan, A., Valla, J. M., Alappatt, J. A., Sharda, M., Shah, A., Ingalhalikar, M., & Singh, N. C. (2019). Wired for musical rhythm?

A diffusion MRI-based study of individual differences in music perception. *Brain Structure and Function, 224*(5), 1711–1722.

31 **100 different neurochemicals:** Nuclear medicine radio tracers track kinetics to estimate regional binding capacity, but they are not able to assay the release of chemicals. See Wikipedia. (n.d.). List of PET radiotracers. See also: Berger, A. (2003). How does it work? Positron emission tomography. *British Medical Journal, 326*(7404), 1449.

Occasionally, researchers obtain access to patients who have had electrodes placed intracranically, for example, to control epileptic seizures. In these cases, voltammetry can be used to track dopamine, norepinephrine, and serotonin. John, C. E., & Jones, S. R. (2006). Fast scan cyclic voltammetry of dopamine and serotonin in mouse brain slices. In A. Michael & L. Borland (eds.), *Electrochemical Methods for Neuroscience* (pp. 49–62). CRC Press.

32 **long-short pattern of notes:** See also: Frank Sinatra Live at the Sands Hotel and Casino, 1966 [video]. YouTube. https://youtu.be/2c7GRYc_rgY

32 **measured jazz beats:** Friberg, A., & Sundström, A. (2002). Swing ratios and ensemble timing in jazz performance: Evidence for a common rhythmic pattern. *Music Perception, 19*(3), 333–349.

32 **subsequent study by Henkjan Honing:** Honing, H., & De Haas, W. B. (2008). Swing once more: Relating timing and tempo in expert jazz drumming. *Music Perception, 25*(5), 471–476.

33 **My student Yuvika Dandiwal:** Dandiwal, Y. (2021). From present to the future: Examining the disproportionate impact of COVID-19 on minorities' mental health and predicting the behaviour of attending in-person events and activities in the future. Undergraduate honors thesis. McGill University.

34 **Contextual variables refer to:** Hargreaves, D. J., Hargreaves, J. J., & North, A. C. (2011). Imagination and creativity in music listening. In Hargreaves, D., Miell, D., & MacDonald, R. (eds.), *Musical Imaginations: Multidisciplinary Perspectives on Creativity, Performance and Perception* (pp. 156–172). Oxford University Press.

34 **In one paper, Yuvika studied:** Dandiwal, Y., Fleming, L., & Levitin, D. J. (2023). Personal and contextual variables predict music consumption during the first COVID-19 lockdown in Canada. *Frontiers in Psychology, 14*, 1116857. doi: 10.3389/fpsyg.2023.1116857

Chapter 3:
Oh, the Shark Bites

36 **Saturday, February 13, 1960:** Not relevant to the story, but astonishing nonetheless, is that as Ella, a Black woman, was singing to a sold-out crowd in Germany, back home in the United States that same day the *New York Times* reported that Black students were protesting segregated lunch counters in South Carolina; one was knocked off a stool by a white student, while 41 Black youth were arrested for "trespassing." Ella, like most Black performers, especially at that time, experienced extreme racism and yet never lost sight of her art.

36 **Ella forgets the words:** Brecht, B., Weill, K., and Blitzstein, M. (1960). Mack the Knife [recorded by E. Fitzgerald feat. the Paul Smith Quartet]. On *Ella in Berlin* [live album]. Verve.

37 **deal with the unexpected:** I'm reminded of the fictional character Stevens, the protagonist of Kazuo Ishiguro's *Remains of the Day*, who performs elegantly under pressure when a potentially devastating event intrudes on his best-laid plans.

37 **Sullenberger describes:** Inc. (2019, March 6). Captain Sully's minute-by-minute description of the Miracle on the Hudson [video]. YouTube. https://www.youtube.com/watch?v=w6EblErBJqw

38   **when Frank Sinatra sang:** oliounidizlove. (2015, December 28). Frank Sinatra
      & Quincy Jones—Mack the Knife (1984) [video]. YouTube. https://www.youtube
      .com/watch?v=eIazK40JKbM

38   **Ella herself reprised pieces:** Smith, S. (2021). Hear Ella Fitzgerald sing
      "Mack the Knife" in an unearthed 1962 recording. *uDiscover Music.* https://www
      .udiscovermusic.com/news/ella-fitzgerald-unearthed-1962-recording/

38   **seldom-sung final stanza:** The final stanza of "Mack the Knife" comes from the
      1931 film adaptation (*Die 3 Groschen-Oper*) directed by G. W. Pabst. It is translated
      here by the author.

39   **we have false memories:** Loftus, E. F. (2022). Tricked by memory. In J. Book-
      wala & N. J. Newton (eds.), *Reflections from Pioneering Women in Psychology* (pp.
      190–201). Cambridge University Press.

40   **Beatles producer George Martin:** G. Martin, personal communication, Sep-
      tember 17, 1993.

43   **These different memory systems:** E.g.,
          Butler, C. R., & Zeman, A. Z. (2008). Recent insights into the impairment of
      memory in epilepsy: Transient epileptic amnesia, accelerated long-term forgetting
      and remote memory impairment. *Brain, 131*(9), 2243–2263.
          Temple, C. M., & Richardson, P. (2004). Developmental amnesia: A new pattern
      of dissociation with intact episodic memory. *Neuropsychologia, 42*(6), 764–781.
          Whiteley, A. M., & Warrington, E. K. (1978). Selective impairment of topo-
      graphical memory: a single case study. *Journal of Neurology, Neurosurgery and Psychi-
      atry, 41*(6), 575–578.
          Vakil, E. (2005). The effect of moderate to severe traumatic brain injury (TBI)
      on different aspects of memory: A selective review. *Journal of Clinical and Experimen-
      tal Neuropsychology, 27*(8), 977–1021.

43   **Even autobiographical memory:** Rosenbaum, R. S., Köhler, S., Schacter, D. L.,
      Moscovitch, M., Westmacott, R., Black, S. E., & Gao, F. (2005). The case of K.C.:
      Contributions of a memory-impaired person to memory theory. *Neuropsychologia,
      43*(7), 989–1021.
          Irish, M., & Piguet, O. (2013). The pivotal role of semantic memory in remem-
      bering the past and imagining the future. *Frontiers in Behavioral Neuroscience, 7,* 27.
          Renoult, L., Davidson, P. S., Palombo, D. J., Moscovitch, M., & Levine, B.
      (2012). Personal semantics: Is it distinct from episodic and semantic memory? An
      event-related potential study of memory for autobiographical facts and repeated
      events. *Neuropsychologia, 50*(5), 1109–1123.

44   **Elizabeth Loftus has shown:** Loftus, E. F. (2022). Tricked by memory.

45   **"Somewhere over the rainbow":** Harburg, H. A. (1939). Over the Rain-
      bow [recorded by J. Garland]. On *The Wizard of Oz* (Original motion picture
      soundtrack) [album]. Decca Records.

47   **Rosanne had been experiencing headaches:** Farley, T. (2012, July 1). Rosanne
      Cash's 10-year ache from Chiari Type I. *Brain and Life.*

47   **she had a "Chiari malformation":** Cash, R. (2010). *Composed: A Memoir.* Viking.

47   **When Rosanne arrived in pre-op:** Cash, R. (2010). *Composed: A Memoir.*

49   **Song recognition remains accurate:** White, B. W. (1960). Recognition of dis-
      torted melodies. *American Journal of Psychology, 73*(1), 100–107.

51   **On-the-fly adaptations:** But here the analogy with recordings or a filing cabi-
      net, although intuitively satisfying, falls short. Say you're working in a large office
      that still puts papers in file drawers, and you need to locate the Penske file. First
      you need to know which cabinet drawer to open and which folder it's in. Many
      organizations file everything in triplicate: one copy in the alphabetical file (under
      "P" for Penske), one in a chronological file (for "Things I was working on in May,
      1994"), and one in a topical file ("R&D Projects: Framistans") or even in a fourth

abeyance or personnel file ("Projects that former employee George Costanza was working on").

None of this helps you if you have only a vague memory of something about a project close to 30 years ago that had something to do with a research project on some new widget that you can't remember the name of. But human memory? That's a horse of a different color.

51   **the name "Jane":** E.g., "Lady Jane" by The Rolling Stones, "Sweet Jane" by The Velvet Underground, "Jane Says" by Jane's Addiction. Or you might retrieve "Lonesome, On'ry and Mean" by Waylon Jennings or "Famous Blue Raincoat" by Leonard Cohen, which have the name Jane in the lyrics—that's how memory works.

51   **Memories are not static:** When we hear the graduation march *Pomp and Circumstance* or "Here Comes the Bride" (pieces by Edward Elgar and Richard Wagner, respectively), we imbue them with meaning based on our memory of their context, completely apart from the choice of notes the composers made. The meaning-conferred-by-context is so powerful that if it were used in the wrong context, it could move you to tears. Imagine being told by the principal that you've just failed eighth grade and need to repeat it. As you walk out, the school band is waiting for you and starts to play the graduation march, an ironic twist of the knife to stir up feelings of regret, self-doubt, and failure.

52   **Even music associated with a distressing:** Sakka, L. S., & Saarikallio, S. (2020). Spontaneous music-evoked autobiographical memories in individuals experiencing depression. *Music and Science*, *3*, 1–15.

The opposite side of remembering things you don't want to is not remembering things you do. We may hear a voice or a song and *not* recognize it, which could either indicate novelty or a failure of memory. The distinction here is often made for us thanks to a graded memory system in which we can detect when something is familiar, even if we can't recall exactly *what* it is (what Doug Hintzman calls *the feeling of knowing*). This *feeling of knowing* can lead to great frustration in people with memory loss. Far more than the occasional tip-of-the-tongue phenomenon we all experience (what was her name again?), a brain ravaged by Alzheimer's or other pathologies might *only* have the feeling of knowing, and never be able to retrieve the actual memory. It's like an itch that you can't scratch. And when the feeling of knowing system breaks down, we can become easily confused into thinking that very familiar things are novel to us. Case studies are filled with accounts of older people with dementia who turn up at a house they haven't lived in for 20 years. Last week, my 86-year-old uncle lost his wife of 50 years. He asked me to call his mother, my grandmother, to tell her about the funeral. I didn't have the heart to tell him that grandma died in 1984.

52   **music-evoked autobiographical memories:** Belfi, A. M., Bai, E., & Stroud, A. (2020). Comparing methods for analyzing music-evoked autobiographical memories. *Perception*, *37*(5), 392–402.

Belfi, A. M., Karlan, B., & Tranel, D. (2016). Music evokes vivid autobiographical memories. *Memory*, *24*(7), 979–989.

52   **dancer Marta González:** Música para Despertar. (2020, October 30). Primera Bailarina—Ballet en Nueva York—Años 60—Música para *Despertar* [video]. YouTube. https://www.youtube.com/watch?v=owb1uWDg3QM

53   **frontal-temporal dementia:** Baird, A., Gelding, R., Brancatisano, O., & Thompson, W. F. (2020). A preliminary exploration of the stability of music- and photo-evoked autobiographical memories in people with Alzheimer's and behavioral variant frontotemporal dementia. *Music and Science*, *3*, 1–15.

55   **The accepted theory:** Cheung, V. K., Harrison, P. M., Meyer, L., Pearce, M. T., Haynes, J. D., & Koelsch, S. (2019). Uncertainty and surprise jointly predict musical pleasure and amygdala, hippocampus, and auditory cortex activity. *Current Biology*, *29*(23), 4084–4092.

Huron, D. (2019). Musical aesthetics: Uncertainty and surprise enhance our enjoyment of music. *Current Biology, 29*(23), R1238–R1240.

Chapter 4:
Look at Me Now

62   **most diverse universities:** McGill University International Student Services. (n.d.). *International Student Body.* https://www.mcgill.ca/internationalstudents/issoffice/international-student-body

64   **They heard the patterns:** There are some complicated statistical methods at play. We didn't give them a list of song titles to choose from, which would be a *recognition* test, rather, we instructed them to name the songs off-the-top-of-their heads, with no other prompt, making it a *free recall* test. What percentage of participants would have to correctly identify a song for us to say that their performance was statistically different from chance? We solved the problem by asking yet a separate group of people to name the songs when played with both rhythm and pitch intact, forming a baseline for a Bayesian *a priori*.

64   **Imagine you're playing *Where's Waldo*:** This is a really silly illustration of cue validity: for people who love "Where's Waldo" but don't have the time, see: https://www.instagram.com/whereswaldothereswaldo

65   **Unless you're Sheldon Cooper:** Sheldon Cooper, a character from the television show *The Big Bang Theory*, did have a breakfast schedule that he followed rigorously. In one episode, he revealed his schedule included oatmeal on Mondays. Years later, Penny made French toast for Sheldon on a Monday, and Leonard expressed concern that Sheldon wouldn't eat it on "Oatmeal day."

68   **Matteo was a 20-year-old musician:** I'm using a pseudonym to protect his identity.
      Piccirilli, M., Sciarma, T., & Luzzi, S. (2000). Modularity of music: Evidence from a case of pure amusia. *Journal of Neurology, Neurosurgery and Psychiatry, 69*(4), 541–545.

69   **stroke in Wernicke's area:** Midorikawa, A., Kawamura, M., & Kezuka, M. (2003). Musical alexia for rhythm notation: A discrepancy between pitch and rhythm. *Neurocase, 9*(3), 232–238.

70   **speakers of American Sign Language:** Newman, A. J., Supalla, T., Hauser, P., Newport, E. L., & Bavelier, D. (2010). Dissociating neural subsystems for grammar by contrasting word order and inflection. *Proceedings of the National Academy of Sciences, 107*(16), 7539–7544.

70   **ability to identify timbre:** Mazzucchi, A., Marchini, C., Budai, R., & Parma, M. (1982). A case of receptive amusia with prominent timbre perception defect. *Journal of Neurology, Neurosurgery and Psychiatry, 45*(7), 644–647.

70   **Oliver Sacks describes a time:** Sacks, O. (2007) *Musicophilia: Tales of Music and the Brain* (p. 105). Knopf.

70   **amateur musician, K.B.:** Steinke, W. R., Cuddy, L. L., & Jakobson, L. S. (2001). Dissociations among functional subsystems governing melody recognition after right-hemisphere damage. *Cognitive Neuropsychology, 18*(5), 411–437.

71   **patient H.M. in textbooks:** A nice overview is here: PBS NewsHour (2016, August 9). Bringing new life to "Patient H.M.," the man who couldn't make memories [video]. YouTube. https://www.youtube.com/watch?v=_7akPs8ptg4

72   **Immanuel Kant:** Kant, I. (1790/2000). *Critique of the Power of Judgment* (P. Guyer & E. Matthews, trans.). Cambridge University Press.

72   **Edmund Husserl:** Husserl, E. (1913/2012). *Ideas: General Introduction to Pure Phenomenology* (W. R. Boyce Gibson, trans.). Routledge.

72   **William James did better:** James, W. (1890/1950). *The Principles of Psychology.* Dover Publications. Page 403.

72 **Martin Heidegger captured the phenomenology:** Heidegger, M. (1927/1962). *Being and Time* (J. Macquarrie & E. Robinson, trans.). Harper & Row.

72 **scientific definition by Mike Posner:** Posner, M. I. (2022). Personal communication, October 15, 2022.

73 **Earl Miller, Ed Awh, and others:** Miller, E. K., & Buschman, T. J. (2015). Working memory capacity: Limits on the bandwidth of cognition. *Daedalus, 144*(1), 112–122.

Fukuda, K., Awh, E., & Vogel, E. K. (2010). Discrete capacity limits in visual working memory. *Current Opinion in Neurobiology, 20*(2), 177–182.

74 **stimulated by memories and ideas:** At times, even as our attention spotlight zooms in and out, we miss things that are right in front of us—our brain fails to perceive information that is present in the environment. *Highlights* is a popular children's magazine often found in doctors' waiting rooms. In the "What's Different?" game, a regular feature since 1946, readers have to spot the differences between two almost identical drawings. This can be difficult because our brains don't know ahead of time on which particular detail to shine the attentional spotlight. Often, we effectively will ourselves to focus on one thing to the exclusion of others, such as when we're trying to work at a loud café, tuning out all the voices. Some pathological conditions, including stroke, traumatic brain injury, and dementia, may cause attentional deficits such as the inability to maintain focus on one thing, or an inability to notice things on one side of our body—hemispatial neglect.

The classic example of selective attention and the attentional filter is the cocktail party. Different sounds bombard your eardrums all at once, but you direct your attention to Marty, the person you're talking to. He may not be the closest talker to you—he could be four seats away and the couple sitting next to you might be louder, but you can filter them out. If that couple you are actively ignoring says your name or some other salient word, your attention will automatically shift to them—this is the evidence your filter is taking everything in. Attention is also multimodal. In a loud room the signal that impinges on your eardrums is likely distorted or deficient, with syllables or words being masked by surrounding conversations or plates crashing in the clearing tray. If you can *see* Marty's lips and paralinguistic movements such as his hand and head gestures, you stand a better chance of understanding him. During the COVID-19 pandemic so many people had difficulty understanding one another during in-person encounters—masks obscured the lips—reinforcing how much we rely on this additional, cross-modal cue.

Chapter 5:
Daydream Believer

77 **dense paperback by Michael Posner:** Ulric Neisser's *Cognitive Psychology*, a dense book written for PhDs, had come out nine years earlier and helped to define the field but was hardly suitable as an undergraduate text. Nevertheless, I read it now, every few years, for inspiration.

78 **radiotracers to track blood flow:** A small amount of radioactive glucose is injected into the veins, and the PET scanner tracks the movement of that radioactive glucose. The amount of radioactivity is considered safe. Anecdotally, one thing scientists have speculated about is that it is more difficult to find people with an "average" level of education to agree to participate in such experiments. Many people with an average education are indiscriminately afraid of *all* radioactivity and they might be more likely to refuse the scans. Well-educated people understand how low the risk is, and so they might be more inclined to say "yes." Very poorly educated people don't even know to be afraid of radioactivity and they also tend to

agree to the experiments. The latter case does raise issues about informed consent and the ethics of using poorly educated people in experiments.

78   **landmark paper published in *Nature*:** Petersen, S. E., Fox, P. T., Posner, M. I., Mintun, M., & Raichle, M. E. (1988). Positron emission tomographic studies of the cortical anatomy of single-word processing. *Nature, 331*(6157), 585–589.

80   **Raichle and colleagues:** Raichle, M. E., MacLeod, A. M., Snyder, A. Z., Powers, W. J., Gusnard, D. A., & Shulman, G. L. (2001). A default mode of brain function. *Proceedings of the National Academy of Sciences, 98*(2), 676–682.

80   **"community structure":** Blake, R. N., Samuel, N., Robert, G., & Laura, D. (2001). Physical resilience in the brain: The effect of white matter disease on brain networks in cognitively normal older adults. *Morbity and Mortality Weekly Report, 50*(7), 120–125.

81   **"functional neighborhood":** Derado, G., Bowman, F. D., Ely, T. D., & Kilts, C. D. (2010). Evaluating functional autocorrelation within spatially distributed neural processing networks. *Statistics and Its Interface, 3*(1), 45.

81   **internally guided thoughts:** Menon, V. (2023). 20 years of the default mode network: A review and synthesis. *Neuron, 111,* 2469–2487.

82   **a region of the brain called the insula:** Sridharan, D., Levitin, D. J., & Menon, V. (2008). A critical role for the right fronto-insular cortex in switching between central-executive and default-mode networks. *Proceedings of the National Academy of Sciences, 105*(34), 12569–12574.

83   **Clive lost this temporal connectivity:** Wilson, B. A., & Wearing, D. (1995). Prisoner of consciousness: A state of just awakening following herpes simplex encephalitis. In R. Campbell & M. A. Conway (eds.), *Broken Memories: Case Studies in Memory Impairment* (pp. 14–30). Blackwell.

83   **the precuneus:** Wilkins, R. W., Hodges, D. A., Laurienti, P. J., Steen, M., & Burdette, J. H. (2014). Network science and the effects of music preference on functional brain connectivity: From Beethoven to Eminem. *Scientific Reports, 4*(1), 1–8.

85   **accompanied by increased connectivity:** Brewer, J. A., Worhunsky, P. D., Gray, J. R., Tang, Y. Y., Weber, J., & Kober, H. (2011). Meditation experience is associated with differences in default mode network activity and connectivity. *Proceedings of the National Academy of Sciences, 108*(50), 20254–20259.

85   **people with schizophrenia:** Wang, D., Zhou, Y., Zhuo, C., Qin, W., Zhu, J., Liu, H., . . . & Yu, C. (2015). Altered functional connectivity of the cingulate subregions in schizophrenia. *Translational Psychiatry, 5*(6), e575–e575.

85   **when they process shared narratives:** Yeshurun, Y., Nguyen, M., & Hasson, U. (2021). The default mode network: Where the idiosyncratic self meets the shared social world. *Nature Reviews Neuroscience, 22*(3), 181–192.

85   **just as consciousness does:** Zadra, A., & Levitin, D. J. (2022). The disintegrated theory of consciousness: Sleep, waking, and meta-awareness. *Behavioral and Brain Sciences, 45* (and work cited therein).

86   **lucid dreamers:** Stephan, A. M., Lecci, S., Cataldi, J., & Siclari, F. (2021). Conscious experiences and high-density EEG patterns predicting subjective sleep depth. *Current Biology, 31*(24), 5487–5500.

87   **jazz musicians in a brain scanner:** Limb, C. J., & Braun, A. R. (2008). Neural substrates of spontaneous musical performance: An fMRI study of jazz improvisation. *PLoS One, 3*(2), e1679.

88   **improvisation in Wernicke's area:** Donnay, G. F., Rankin, S. K., Lopez-Gonzalez, M., Jiradejvong, P., & Limb, C. J. (2014). Neural substrates of interactive musical improvisation: An FMRI study of 'trading fours' in jazz. *PLoS One, 9*(2), e88665.

88   **Different flavors of mindfulness meditation:** Yordanova, J., Kolev, V., Mauro, F., et al. (2020). Common and distinct lateralised patterns of neural coupling

during focused attention, open monitoring and loving kindness meditation. *Scientific Reports*, *10*(1), 7430.

89 **Richard Davidson frames it:** Dahl, C. J., Lutz, A., & Davidson, R. J. (2015). Reconstructing and deconstructing the self: Cognitive mechanisms in meditation practice. *Trends in Cognitive Sciences*, *19*(9), 515–523.

89 **The Dalai Lama:** His Holiness the Dalai Lama (August 31, 2018). Personal communication.

90 **typically called *flow*:** Oppland, M. (2023). 8 traits of flow according to Mihaly Csikszentmihalyi. PositivePsychology.com. https://positivepsychology.com/mihaly -csikszentmihalyi-father-of-flow/

90 **total absorption in the task:** Gillian Sandstrom from the University of Sussex, and Frank Russo from Toronto Metropolitan University, have developed an "Absorption in music" scale, to identify individuals with strong emotional reactions to music. Sandstrom, G. M., & Russo, F. A. (2013). Absorption in music: Development of a scale to identify individuals with strong emotional responses to music. *Psychology of Music*, *41*(2), 216–228.

90 **A peak musical state:** All of these attentional phenomena illustrate that our attention can be outwardly directed or inwardly (self) directed, and that our attention can be allocated voluntarily or involuntarily. Voluntary visual attention involves the most highly evolved dorsal stream in our visual system, whereas involuntary attention involves the evolutionarily older ventral stream.

In our 2007 paper, in which we applied network analysis to music, Vinod and I found that this distinction also applies to auditory perception. Our participants lay in an fMRI scanner listening to short symphonies by the baroque composer William Boyce. We identified two distinct, functional networks during the silent transitions between movements, a time when there was no sound playing. What does the brain *do* during this silence? We hypothesized that some parts were engaged in maintaining attention while others were waiting for the next salient event. That's exactly what we found: a dorsal network connecting the frontal lobes to the parietal was associated with maintaining attention, and updating working memory; a ventral network connecting the frontal and temporal lobes was associated with detecting salient events—that is, waiting to see if our attention needed to be "grabbed" away from the silence. The scans themselves, which we animated in a YouTube video, show the astonishing pattern of peaks of brain activity in these networks occurring in the right hemisphere precisely when the music stops, and very little activity when the music is playing. These are specialized networks attuned to the ebb and flow of music, and our corresponding ebb and flow of attention. Our study showed that movement transitions, rather than being merely perceived as pauses, are an important component of the listening experience.

90 **I asked Donald Fagen:** D. Fagen, personal communication via email, May 22, 2023.

92 **The subjective feeling that consciousness is a unitary thing:** Dennett, D. C., & Kinsbourne, M. (1992). Time and the observer: The where and when of consciousness in the brain. *Behavioral and Brain Sciences*, *15*(2), 183–201.

Dennett, D. C., & Kinsbourne, M. (1992). Escape from the Cartesian theater. *Behavioral and Brain Sciences*, *15*(2), 234–247.

92 **Vinod Menon believes the DMN:** Menon, V. (2023). 20 years of the default mode network: A review and synthesis. *Neuron*, *111*, 2469–2487.

Interlude

95 **Ulric Neisser wrote:** Neisser, U. (1967). *Cognitive Psychology*. Appleton-Century Crofts.

96 **they often interact dynamically:** Gronchi, G., & Giovannelli, F. (2018). Dual process theory of thought and default mode network: A possible neural foundation of fast thinking. *Frontiers in Psychology, 9*, 1237.

van den Berg, B., de Bruin, A. B., Marsman, J. B. C., Lorist, M. M., Schmidt, H. G., Aleman, A., & Snoek, J. W. (2020). Thinking fast or slow? Functional magnetic resonance imaging reveals stronger connectivity when experienced neurologists diagnose ambiguous cases. *Brain Communications, 2*(1), fcaa023.

96　**How do you know "Summertime":** This discussion owes a debt to Neisser's description of the visual system in Neisser, U. (1967). *Cognitive Psychology* (p. 46).

97　**Ella Fitzgerald and Louis Armstrong's slow:** Ella Fitzgerald—Topic. (2018, September 4). Summertime [video]. YouTube. https://www.youtube.com/watch?v=VZRgiuAXRAs

97　**Janis Joplin's earthy and raw:** JanisJoplinVEVO. (2013, January 9). Big Brother & The Holding Company, Janis Joplin—Summertime (official audio) [video]. YouTube. https://www.youtube.com/watch?v=A24JZkgvNv4

97　**Willie Nelson's take on it:** WillieNelsonVEVO. (2016, January 13). Willie Nelson—Summertime (official video) [video]. YouTube. https://www.youtube.com/watch?v=L5xafQXg1yI

97　**Jacob Koller's piano hip-hop arrangement:** Jacob Koller / The Mad Arranger. (2020, July 29). "Summertime" by George Gershwin—Hip-hop jazz piano arrangement by Jacob Koller [video]. YouTube. https://www.youtube.com/watch?v=Gxg0fAr8OY4

98　**Sun Ra and his Arkestra:** Intergalaxtic Music: Rare Sun Ra. (2022, December 25). Sun Ra and his Arkestra—Summertime / Shadow World 6/22/1980 Elaine's, Philadelphia [video]. YouTube. https://www.youtube.com/watch?v=vGAT6sISgl4

99　**Austin Lounge Lizards' bluegrass performance:** Austin Lounge Lizards—Topic. (2018, July 28). Brain Damage [video]. YouTube. https://www.youtube.com/watch?v=KUkhOqM-ffM

Chapter 6:
Music, Movement, and Movement Disorders

102　**Movement disorders and neurological diseases:** Reynolds, E. H., & Kinnier Wilson, J. V. (2014). Neurology and psychiatry in Babylon. *Brain, 137*(9), 2611–2619.

102　**we learn that Goliath:** Berginer, V. M. (2000). Neurological aspects of the David-Goliath battle: Restriction in the giant's visual field. *Israel Medical Association Journal, 2*(9), 725–727.

Samuel, I. (1985). *The Jewish Bible, Tanakh, a new translation of the Holy Scriptures into English.* Jerusalem: Jewish Publication Society (pp. 443–447). Prophets Nevi'im.

102　**his giantism, acromegaly:** Mathew, S. K., & Pandian, J. D. (2010). Newer insights to the neurological diseases among biblical characters of old testament. *Annals of Indian Academy of Neurology, 13*(3), 164.

103　**Scott explains that physical intelligence:** Grafton, S. (2020). *Physical Intelligence: The Science of How the Body and the Mind Guide Each Other through Life.* Pantheon.

103　**learn from that movement:** Grafton, S. *Physical Intelligence.* Ibid.

104　**have yet to replace real-world embodied experience:** Spano, G., Theodorou, A., Reese, G., Carrus, G., Sanesi, G., & Panno, A. (2023). Virtual nature and psychological outcomes: A systematic review. *Journal of Environmental Psychology*, 102044.

White, M. P., Yeo, N. L., Vassiljev, P., Lundstedt, R., Wallergård, M., Albin, M., & Lõhmus, M. (2018). A prescription for "nature"–the potential of using virtual nature in therapeutics. *Neuropsychiatric Disease and Treatment*, 3001–3013.

104 **patients who move the least:** Falck, R. S., Davis, J. C., & Liu-Ambrose, T. (2017). What is the association between sedentary behaviour and cognitive function? A systematic review. *British Journal of Sports Medicine, 51*(10), 800–811.

104 **Older adults who find themselves:** Strahl, A., Kazim, M. A., Kattwinkel, N., Hauskeller, W., Moritz, S., Arlt, S., & Niemeier, A. (2022). Mid-term improvement of cognitive performance after total hip arthroplasty in patients with osteoarthritis of the hip: A prospective cohort study. *The Bone & Joint Journal, 104*(3), 331–340.

104 **three other major movement disorders:** Raglio, A., Giovanazzi, E., Pain, D., Baiardi, P., Imbriani, C., Imbriani, M., & Mora, G. (2016). Active music therapy approach in amyotrophic lateral sclerosis: A randomized-controlled trial. *International Journal of Rehabilitation Research, 39*(4), 365–367.

Ostermann, T., & Schmid, W. (2006). Music therapy in the treatment of multiple sclerosis: A comprehensive literature review. *Expert Review of Neurotherapeutics, 6*(4), 469–477.

105 **"Stay a Little Longer":** Written by Bob Wills and Tommy Duncan (1945).

105 **country singer Mel Tillis:** McArdle, T. (2023, April 9). Mel Tillis, stuttering country star whose music spoke pristinely, dies at 85. *Washington Post.* (You may find Tillis's '60s and '70s TV appearances on *Hee Haw* or *The Glen Campbell Goodtime Hour* painful to watch—as I did—because of the audience reactions. Audiences laughed whenever he tripped on his words, or temporarily went silent while trying to get the words out, perhaps thinking that he was just trying to be funny. This was at a time when performers mimicked physical disabilities, milking them for laughs. Even more unsettling is that his TV hosts also laughed at him, and Tillis . . . well, he gamely laughed along, tolerating being the object of jokes just in order to get along and be part of the show.)

105 **Stuttering (also called stammering):** NIDCD. (2017, March 6). *Stuttering.* https://www.nidcd.nih.gov/health/stuttering

150 **Most childhood stutterers:** NIDCD. (2017, March 6). *Stuttering.* https://www.nidcd.nih.gov/health/stuttering

105 **sign language users also stutter:** Whitebread, G. (2014). A review of stuttering in signed languages. *Multilingual aspects of signed language communication and disorder,* 143–161.

106 **nine syllables per second:** Peelle, J. E., & Davis, M. H. (2012). Neural oscillations carry speech rhythm through to comprehension. *Frontiers in Psychology, 3,* 320.

107 **average speed of four-seam fastball:** Blum, R. (2022, October 6). MLB velocity, shifts set records; average lowest since 1968. *AP News.*

107 **pre-stored motor action plan:** Lashley, K. S. (1951). The problem of serial order in behavior. In L. A. Jeffress (ed.), *Cerebral Mechanisms in Behavior* (pp. 112–131). Wiley.

Rosenbaum, D. A., Cohen, R. G., Jax, S. A., Weiss, D. J., & Van Der Wel, R. (2007). The problem of serial order in behavior: Lashley's legacy. *Human Movement Science, 26*(4), 525–554.

109 **predictive timing mechanisms are impaired:** Falk, S., Müller, T., & Dalla Bella, S. (2015). Non-verbal sensorimotor timing deficits in children and adolescents who stutter. *Frontiers in Psychology, 6,* 847.

110 **arcuate fasciculus:** Cieslak, M., Ingham, R. J., Ingham, J. C., & Grafton, S. T. (2015). Anomalous white matter morphology in adults who stutter. *Journal of Speech, Language, and Hearing Research, 58*(2), 268–277.

110 **initiation and termination of movements:** Max Planck Institute (2017, December 12). Stuttering: Stop signals in the brain prevent fluent speech. https://www.cbs.mpg.de/stuttering-in-the-brain

110 **Stutterers tend to show impairment:** Falk, S., Müller, T., & Dalla Bella, S. (2015). Non-verbal sensorimotor timing deficits in children and adolescents who stutter. *Frontiers in Psychology, 6,* 847.

111 its role in music cognition: Levitin, D. J., & Menon, V. (2005). The neural locus of temporal structure and expectancies in music: Evidence from functional neuroimaging at 3 Tesla. *Music Perception, 22*(3), 563–575.

112 Wayne Shorter was slightly disfluent: Layman, W. (2013, February 20). *Wayne Shorter Quartet: Without a Net, PopMatters*. PopMatters.

Ratliff, B. (2001, June 30). JVC Jazz Festival Review; At 67, Inspiring a Quest for Perfection. *New York Times*.

113 fluency while singing: Falk, S., Schreier, R., & Russo, F. A. (2020). Singing and stuttering. In *The Routledge Companion to Interdisciplinary Studies in Singing, Volume III: Wellbeing* (pp. 50–60). Routledge.

113 reinforce the normal activation patterns: Fiveash, A., Bedoin, N., Gordon, R. L., & Tillmann, B. (2021). Processing rhythm in speech and music: Shared mechanisms and implications for developmental speech and language disorders. *Neuropsychology, 35*(8), 771.

113 temporal planning and execution: Falk, S., Maslow, E., Thum, G., & Hoole, P. (2016). Temporal variability in sung productions of adolescents who stutter. *Journal of Communication Disorders, 62*, 101–114. doi:10.1016/j.jcomdis.2016.05.012

114 no instances of stuttering: Russo, F. A., & Fallah, S. (2018, October 20). Workshop: Let's sing: The potential benefits of choral singing for people who stutter. Toronto: Annual Meeting of the Canadian Stuttering Association.

114 Tourette syndrome: Centers for Disease Control and Prevention. (2022, May 9). *Five Things You May Not Know about Tourette Syndrome*. https://www.cdc.gov/ncbddd/tourette/features/tourette-five-things.html

114 The tics themselves: Leckman, J. F. (2022). The neurobiology of Gilles de la Tourette syndrome and chronic tics. In M. E. Lavoie & A. E. Cavanna (eds.), *International Review of Movement Disorders*, vol. 3, *The Neurobiology of the Gilles De La Tourette Syndrome and Chronic Tics: Part A* (pp. 69–101). Elsevier.

115 Singer Esha Alwani: TED. (2019, April 23). What it's like to have Tourette's—and how music gives me back control | Esha Alwani [video]. YouTube. https://www.youtube.com/watch?v=QtnBMSSk9Ok

115 Billie Eilish describes her Tourette's: Berman, M. R. (2022, June 1). Billie Eilish's Tourette Syndrome. *MedPage Today*.

115 Sound designer Jamie Grace: Jamie Grace. (2020, August 8). Turning my tics into music (Tourette Syndrome) [video]. YouTube. https://www.youtube.com/watch?v=b0bcAz3o6S8

Hornik, S. (2020, August 8). Jamie Grace on Her Uplifting Song "Marching On," life with Tourette's Syndrome and the transformative power of Gospel and Contemporary Christian Music. *GRAMMY*. https://www.grammy.com/news/jamie-grace-her-uplifting-song-marching-life-tourettes-syndrome-and-transformative

115 tics often disappear: Leckman, J. F. (2022). The neurobiology of Gilles de la Tourette syndrome and chronic tics. In M. E. Lavoie & Andrea E. Cavanna (eds.), *International Review of Movement Disorders*, vol. 3, *The Neurobiology of the Gilles De La Tourette Syndrome and Chronic Tics: Part A* (pp. 69–101). Elsevier.

115 Music may be especially effective: Bodeck, S., Lappe, C., & Evers, S. (2015). Tic-reducing effects of music in patients with Tourette's syndrome: Self-reported and objective analysis. *Journal of the Neurological Sciences, 352*(1–2), 41–47.

115 simply *listening* to music: Levitin, D. J. (2006). *This Is Your Brain on Music*. New York: Dutton/Penguin.

115 Aberrant neural oscillations: Bodeck, S., Lappe, C., & Evers, S. (2015). Tic-reducing effects of music in patients with Tourette's syndrome.

116 excessive cortical input: Cothros, N., & Martino, D. (2022). Recent advances in neuroimaging of Tourette syndrome. Chapter 5 in *The Neurobiology of the Gilles De La Tourette Syndrome and Chronic Tics: Part A* (pp. 161–207). Elsevier.

Imaging studies have revealed greater than normal activity in posterior middle cingulate cortex and in bilateral superior parietal cortex and cerebellum, weaker activity in putamen, caudate, and anterior cingulate. I wish I could tell you what the activity in posterior middle cingulate cortex means, but we don't really know what it does, other than that it is part of a control circuit in the DMN. And although the superior parietal cortex is adjacent to areas we know are involved in planning motor activity, we don't have any evidence that this is its function.

117 **Epstein–Barr:** Sollid, L. M. (2022). Epstein-Barr virus as a driver of multiple sclerosis. *Science Immunology, 7*(70), eabo7799.

117 **MS can affect any area:** Khan, F., McPhail, T., Brand, C., Turner-Stokes, L., & Kilpatrick, T. (2006). Multiple sclerosis: Disability profile and quality of life in an Australian community cohort. *International Journal of Rehabilitation Research, 29*(2), 87–96.

117 **pharmacological treatments for RRMS:** Comi, G., Radaelli, M., & Sørensen, P. S. (2017). Evolving concepts in the treatment of relapsing multiple sclerosis. *The Lancet, 389*(10076), 1347–1356.

118 **Her experience was typical:** Farley, T. (2010, September/October). Exene Cervenka's rebellion against the effects of MS. *Brain & Life.* https://www.brainandlife.org/articles/a-punk-rock-icon-takes-on-ms

120 **a recording of footsteps:** Murgia, M., Pili, R., Corona, F., Sors, F., Agostini, T. A., Bernardis, P., . . . & Pau, M. (2018). The use of footstep sounds as rhythmic auditory stimulation for gait rehabilitation in Parkinson's disease: A randomized controlled trial. *Frontiers in Neurology, 9,* 348.

120 **when the music is customized:** Barta, K., Da Silva, C. P., Tseng, S.-C., & Roddey, T. (2019). Does a customized musical song promote a more positive experience vs. rhythmic auditory stimulation when used to enhance walking for people with Parkinson's Disease?. *Music Therapy Today, 15*(1), 119–133.

120 **RAS is successful:** Thaut, M. H., Rice, R. R., Braun Janzen, T., Hurt-Thaut, C. P., & McIntosh, G. C. (2019). Rhythmic auditory stimulation for reduction of falls in Parkinson's disease: A randomized controlled study. *Clinical Rehabilitation, 33*(1), 34–43.

Capato, T. T., de Vries, N. M., IntHout, J., Barbosa, E. R., Nonnekes, J., & Bloem, B. R. (2020). Multimodal balance training supported by rhythmical auditory stimuli in Parkinson's disease: A randomized clinical trial. *Journal of Parkinson's Disease, 10*(1), 333–346.

120 **rehabilitation in stroke patients:** Suh, J. H., Han, S. J., Jeon, S. Y., Kim, H. J., Lee, J. E., Yoon, T. S., & Chong, H. J. (2014). Effect of rhythmic auditory stimulation on gait and balance in hemiplegic stroke patients. *NeuroRehabilitation, 34*(1), 193–199.

Lee, S., Lee, K., & Song, C. (2018). Gait training with bilateral rhythmic auditory stimulation in stroke patients: A randomized controlled trial. *Brain Sciences, 8*(9), 164.

120 **children with cerebral palsy:** Kwak, E. E. (2007). Effect of rhythmic auditory stimulation on gait performance in children with spastic cerebral palsy. *Journal of Music Therapy, 44*(3), 198–216.

120 **traumatic brain injury:** Hurt, C. P., Rice, R. R., McIntosh, G. C., & Thaut, M. H. (1998). Rhythmic auditory stimulation in gait training for patients with traumatic brain injury. *Journal of Music Therapy, 35*(4), 228–241.

120 **people with multiple sclerosis:** Shahraki, M., Sohrabi, M., Torbati, H. T., Nikkhah, K., & NaeimiKia, M. (2017). Effect of rhythmic auditory stimulation on gait kinematic parameters of patients with multiple sclerosis. *Journal of Medicine and Life, 10*(1), 33.

120 **timed, cognitive control:** Devlin, K., Alshaikh, J. T., & Pantelyat, A. (2019). Music therapy and music-based interventions for movement disorders. *Current Neurology and Neuroscience Reports, 19*(11), 1–13.

120 **rhythmic synchronization and entrainment:** Devlin, K., Alshaikh, J. T., & Pantelyat, A. (2019). Music therapy and music-based interventions for movement disorders. Ibid.

121 **predictors of RAS effectiveness:** Bella, S. D., Dotov, D., Bardy, B., & de Cock, V. C. (2018). Individualization of music-based rhythmic auditory cueing in Parkinson's disease. *Annals of the New York Academy of Sciences, 1423*(1), 308–317.

   Cochen De Cock, V., Dotov, D. G., Ihalainen, P., Bégel, V., Galtier, F., Lebrun, C., . . . & Dalla Bella, S. (2018). Rhythmic abilities and musical training in Parkinson's disease: Do they help?. *NPJ Parkinson's Disease, 4*(1), 8.

   Ready, E. A., McGarry, L. M., Rinchon, C., Holmes, J. D., & Grahn, J. A. (2019). Beat perception ability and instructions to synchronize influence gait when walking to music-based auditory cues. *Gait and Posture, 68,* 555–561.

121 **Rhythmic auditory stimulation for MS:** Conklyn, D., Stough, D., Novak, E., Paczak, S., Chemali, K., & Bethoux, F. (2010). A home-based walking program using rhythmic auditory stimulation improves gait performance in patients with multiple sclerosis: A pilot study. *Neurorehabilitation and Neural Repair, 24*(9), 835–842.

121 **Even just imagining movements:** Vinciguerra, C., De Stefano, N., & Federico, A. (2019). Exploring the role of music therapy in multiple sclerosis: Brief updates from research to clinical practice. *Neurological Sciences, 40,* 2277–2285.

121 **musical keyboard training:** Gatti, R., Tettamanti, A., Lambiase, S., Rossi, P., & Comola, M. (2015). Improving hand functional use in subjects with multiple sclerosis using a musical keyboard: A randomized controlled trial. *Physiotherapy Research International, 20*(2), 100–107.

121 **auditory feedback from their movements:** Gatti, R., Tettamanti, A., Lambiase, S., Rossi, P., & Comola, M. (2015). Improving hand functional use in subjects with multiple sclerosis using a musical keyboard. *Physiotherapy Research International, 20*(2), 100–107.

121 **effectiveness of dance therapy:** Van Geel, F., Van Asch, P., Veldkamp, R., & Feys, P. (2020). Effects of a 10-week multimodal dance and art intervention program leading to a public performance in persons with multiple sclerosis: A controlled pilot-trial. *Multiple Sclerosis and Related Disorders, 44,* 102256.

   Young, H. J., Mehta, T. S., Herman, C., Wang, F., & Rimmer, J. H. (2019). The effects of M2M and adapted yoga on physical and psychosocial outcomes in people with multiple sclerosis. *Archives of Physical Medicine and Rehabilitation, 100*(3), 391–400.

121 **dancer Courtney Platt:** Platt, C. (2021, May 1). I was diagnosed with multiple sclerosis after my body went numb while touring for "So You Think You Can Dance." *Women's Health.*

122 **a mood management tool:** Platt, C. (n.d.). MS in Harmony. https://www.msinharmony.com/

122 **she explored music therapy:** Leiber, S. J. (2021, May 27). Interview: Ben Platt & Courtney Platt Talk Music Therapy, MS in Harmony & Finding Joy. *BroadwayWorld.com.*

123 **improve gait in Huntington's:** Thaut, M. H., Miltner, R., Lange, H. W., Hurt, C. P., & Hoemberg, V. (1999). Velocity modulation and rhythmic synchronization of gait in Huntington's disease. *Movement Disorders, 14*(5), 808–819.

123 **higher-order processing deficits:** Beste, C., Schüttke, A., Pfleiderer, B., & Saft, C. (2011). Music perception and movement deterioration in Huntington's disease. *PLoS Currents, 3,* RRN1252.

124 **Huntington speech music therapy:** Brandt, M., Nieuwkamp, M., Kerkdijk, E., & Verschuur, E. (2014). L28 Huntington speech music therapy: A therapy based on the principles of the speech music therapy for aphasia, adjusted for patients with Huntington. *Journal of Neurology, Neurosurgery and Psychiatry, 85,* A91–A92.

124 **Drummer Trey Gray:** Allen, J. (2022). Trey Gray. *Big Bang Distribution.* https://aheaddrumsticks.com/trey-gray.html

Miller, S. (2020, September 8). Trey Gray Beating the Odds. *Granger Magazine.*

124 **Trey believes that his drumming:** Grotto. (2021). Beating the odds of Huntington's Disease. *Grotto Network.* https://grottonetwork.com/navigate-life/health-and-wellness/huntington-disease-story/

124 **measurable improvements in working memory:** Metzler-Baddeley, C., Baddeley, R. J., Canteras, J., Rosser, A., Coulthard, E., & Jones, D. K. (2013). Improved white matter microstructure after a novel drumming training in Huntingtons disease. *Proceedings of the International Society for Magnetic Resonance in Medicine, 21.*

Metzler-Baddeley, C., Cantera, J., Coulthard, E., Rosser, A., Jones, D. K., & Baddeley, R. J. (2014). Improved executive function and callosal white matter microstructure after rhythm exercise in Huntington's disease. *Journal of Huntington's Disease, 3*(3), 273–283.

125 **increased myelination:** Casella, C., Bourbon-Teles, J., Bells, S., Coulthard, E., Parker, G. D., Rosser, A., . . . & Metzler-Baddeley, C. (2020). Drumming motor sequence training induces apparent myelin remodelling in Huntington's disease: A longitudinal diffusion MRI and quantitative magnetization transfer study. *Journal of Huntington's Disease, 9*(3), 303–320.

125 **Woody told his pal Pete Seeger:** Markel, H. (2019, July 14). This genetic brain disorder turned Woody Guthrie's life from songs to suffering. PBS NewsHour.

125 **Americans with Parkinson's doubled:** Dorsey, E., Sherer, T., Okun, M. S., & Bloem, B. R. (2018). The emerging evidence of the Parkinson pandemic. *Journal of Parkinson's Disease, 8*(s1), S3–S8.

125 **Environmental toxins:** Beitz, J. M. (2014). Parkinson's disease: A review. *Frontiers in Bioscience (Scholars Ed.), 6*(1), 65–74.

Chapter 7:
Parkinson's Disease

128 **Parkinson's affects:** National Institute of Neurological Disorders and Stroke. (n.d.). Parkinson's disease: Challenges, progress, and promise. https://www.ninds.nih.gov/current-research/focus-disorders/parkinsons-disease-research/parkinsons-disease-challenges-progress-and-promise. Retrieved March 31, 2024.

Dommershuijsen, L. J., Heshmatollah, A., Darweesh, S. K., Koudstaal, P. J., Ikram, M. A., & Ikram, M. K. (2020). Life expectancy of Parkinsonism patients in the general population. *Parkinsonism & Related Disorders, 77,* 94–99.

Parkinson's Foundation: Statistics. (n.d.). https://www.parkinson.org/understanding-parkinsons/statistics (accessed January 16, 2024)

132 **"my Miles Davis story":** Here Bobby refers to Miles Davis's performance on February 15, 1969, at Club Baron in Nashville, TN.

133 **singing is a repetitive motion:** Schulman, M. (2019, September 1). Linda Ronstadt has found another voice. *The New Yorker.*

133 **Police drummer Stewart Copeland:** Instagram post. https://www.instagram.com/reel/CkhF5p_pOMA/?igshid=MDJmNzVkMjY%3D

134 **music enhances movement synchronization:** Altenmüller, E., & Stewart, L. (2018). Music supported therapy in neurorehabilitation. In V. Dietz and N. S. Ward (eds.), *Oxford Textbook of Neurorehabilitation,* 2nd ed. (pp. 421–432). Oxford University Press.

134 **reported a number of positive outcomes:** de Bruin, N., Doan, J. B., Turnbull, G., Suchowersky, O., Bonfield, S., Hu, B., & Brown, L. A. (2010). Walking with music is a safe and viable tool for gait training in Parkinson's: The effect of a

13-week feasibility study on single and dual task walking. *Parkinson's Disease, 2010,* 1–9. doi:10.4061/2010/483530

134 randomly assigned to music therapy: Pacchetti, C., Mancini, F., Aglieri, R., Fundarò, C., Martignoni, E., & Nappi, G. (2000). Active music therapy in Parkinson's: An integrative method for motor and emotional rehabilitation. *Psychosomatic Medicine, 62*(3), 386–393.

135 improvements in gait and balance: Nuic, D., Vinti, M., Karachi, C., Foulon, P., Van Hamme, A., & Welter, M. L. (2018). The feasibility and positive effects of a customised videogame rehabilitation programme for freezing of gait and falls in Parkinson's patients: A pilot study. *Journal of Neuroengineering and Rehabilitation, 15*(1), 1–11.

135 transcranial alternating currents: Brittain, J. S., Probert-Smith, P., Aziz, T. Z., & Brown, P. (2013). Tremor suppression by rhythmic transcranial current stimulation. *Current Biology, 23*(5), 436–440.

135 music serves as a reward: Chomiak, T., Watts, A., Meyer, N., Pereira, F. V., & Hu, B. (2017). A training approach to improve stepping automaticity while dual-tasking in Parkinson's disease: A prospective pilot study. *Medicine, 96*(5), e5934.

135 stronger effects are from dancing: Hackney, M. E., Kantorovich, S., Levin, R., & Earhart, G. M. (2007). Effects of tango on functional mobility in Parkinson's: A preliminary study. *Journal of Neurologic Physical Therapy, 31*(4), 173–179.

135 wearable sensors, treatment software, and delivery interfaces: Collimore, A. N., Roto Cataldo, A. V., Aiello, A. J., Sloutsky, R., Hutchinson, K. J., Harris, B., . . . & Awad, L. N. (2023). Autonomous control of music to retrain walking after stroke. *Neurorehabilitation and Neural Repair, 37*(5), 255–265.

135 Ambulosono: Hu, B., & Chomiak, T. (2019). Wearable technological platform for multidomain diagnostic and exercise interventions in Parkinson's disease. *International Review of Neurobiology, 147*, 75–93.

135 unpleasant side effects: Sporer, K. A. (1991). Carbidopa-levodopa overdose. *American Journal of Emergency Medicine, 9*(1), 47–48.

136 Music therapist and neuroscientist Elizabeth Stegemöller: Stegemöller, E. L., Radig, H., Hibbing, P., Wingate, J., & Sapienza, C. (2017). Effects of singing on voice, respiratory control and quality of life in persons with Parkinson's disease. *Disability and Rehabilitation, 39*(6), 594–600.

136 effects of an individual singing program: Han, E. Y., Yun, J. Y., Chong, H. J., & Choi, K. G. (2018). Individual therapeutic singing program for vocal quality and depression in Parkinson's disease. *Journal of Movement Disorders, 11*(3), 121.

136 singing interventions had carryover effects: Butala, A., Li, K., Swaminathan, A., Dunlop, S., Salnikova, Y., Ficek, B., . . . & Pantelyat, A. (2022). Parkinsonics: A randomized, blinded, cross-over trial of group singing for motor and nonmotor symptoms in idiopathic parkinson disease. *Parkinson's Disease, 2022.*

136 preliminary evidence for carryover effects: Spina, E., Barone, P., Mosca, L. L., Lombardi, A., Longo, K., Iavarone, A., & Amboni, M. (2016). Music therapy for motor and nonmotor symptoms of Parkinson's disease: A prospective, randomized, controlled, single-blinded study. *Journal of the American Geriatrics Society, 64*(9), e39.

137 a grandfather with Parkinson's: Hamilton, J. (2022, February 7). A brain circuit tied to emotion may lead to better treatments for Parkinson's disease. NPR: *All Things Considered.*

137 Paradoxical kinesia: Ballanger, B., Thobois, S., Baraduc, P., Turner, R. S., Broussolle, E., & Desmurget, M. (2006). "Paradoxical kinesis" is not a hallmark of Parkinson's disease but a general property of the motor system. *Movement Disorders, 21*(9), 1490–1495.

137 playing tennis: Duysens, J., & Nonnekes, J. (2021). Parkinson's kinesia paradoxa is not a paradox. *Movement Disorders, 36*(5), 1115–1118.

137 **soccer:** Asmus, F., Huber, H., Gasser, T., & Schöls, L. (2008). Kick and rush: paradoxical kinesia in Parkinson disease. *Neurology, 71*(9), 695.

137 **boxing, ice skating:** Mercier, B. P. T. (2017). *Ice Skating Is Safe and Skillfully Preserved amongst Some People Living with Parkinson's Disease: Possibility of Neurotherapeutic Intervention.* University of Lethbridge (Canada).

137 **bicycling:** Hamilton, J. (2022, February 7). A brain circuit tied to emotion may lead to better treatments for Parkinson's disease. NPR: *All Things Considered.*

138 **Rapper and filmmaker Walter J. Archey III:** Dogan, S. E. (2019, July 25). I'm trying to be the Stephen Hawking of Parkinson's. *Parkinson's Life.* https://parkinsonslife.eu/walter-archey-tahiti-parkinsons-rapper/

138 **Parkinson's as a younger person:** Dogan, S. E. (2019, July 25). I'm trying to be the Stephen Hawking of Parkinson's. Ibid.

Chapter 8:
Trauma

139 **McGill neuroscientist Michael Meaney:** Price, M. (2009, October). DNA isn't the whole story. *Monitor on Psychology, 40*(9).

140 **contribute to PTSD symptoms:** Pant, U., Frishkopf, M., Park, T., Norris, C. M., & Papathanassoglou, E. (2022). A neurobiological framework for the therapeutic potential of music and sound interventions for post-traumatic stress symptoms in critical illness survivors. *International Journal of Environmental Research and Public Health, 19*(5), 3113.

141 **the ventral tegmental area:** Ning, M., Wen, S., Zhou, P., & Zhang, C. (2022). Ventral tegmental area dopaminergic action in music therapy for post-traumatic stress disorder: A literature review. *Frontiers in Psychology, 13.*

141 **upregulation of genes:** Nair, P. S., Raijas, P., Ahvenainen, M., Philips, A. K., Ukkola-Vuoti, L., & Järvelä, I. (2021). Music-listening regulates human microRNA expression. *Epigenetics, 16*(5), 554–566.

141 **The incidence of PTSD:** U.S. Department of Veteran Affairs (2023, February 3). How common is PTSD in adults? https://www.ptsd.va.gov/understand/common/common_adults.asp

142 **personal theme song:** Bernstein, E. (2023, April 25). You need a personal theme song. *Wall Street Journal.*

143 **Soldiers participating in group drumming:** Bensimon, M., Amir, D., & Wolf, Y. (2008). Drumming through trauma: Music therapy with post-traumatic soldiers. *The Arts in Psychotherapy, 35*(1), 34–48.

144 **United States veterans experiencing PTSD:** Bronson, H., Vaudreuil, R., & Bradt, J. (2018). Music therapy treatment of active duty military: An overview of intensive outpatient and longitudinal care programs. *Music Therapy Perspectives, 36*(2).

144 **participation in community music engagement:** Vetro-Kalseth, D., Vaudreuil, R., & Segall, L. E. (2021). Treatment description and case series report of a phased music therapy group to support veteran reintegration. *Military Psychology, 33*(6), 446–452.

144 **Sam was 52 years old:** Vetro-Kalseth, D., Vaudreuil, R., & Segall, L. E. (2021). Treatment description and case series report of a phased music therapy group to support veteran reintegration. Ibid.

145 **forcibly displaced worldwide:** UNHCR (2022, May 23). Ukraine, other conflicts push forcibly displaced total over 100 million for first time. https://www.unrefugees.org/news/unhcr-ukraine-other-conflicts-push-forcibly-displaced-total-over-100-million-for-first-time/

145 **Refugees and asylum seekers:** Kien, C., Sommer, I., Faustmann, A., Gibson, L., Schneider, M., Krczal, E., . . . & Gartlehner, G. (2019). Prevalence of mental

disorders in young refugees and asylum seekers in European countries: A systematic review. *European Child and Adolescent Psychiatry, 28*, 1295–1310.

    Nesterko, Y., Jäckle, D., Friedrich, M., Holzapfel, L., & Glaesmer, H. (2020). Prevalence of post-traumatic stress disorder, depression and somatisation in recently arrived refugees in Germany: An epidemiological study. *Epidemiology and Psychiatric Sciences, 29*, e40.

145 **STARTTS:** STARTTS. (n.d.). Who we are. https://www.startts.org.au/about-us/

145 **Singing also has physical:** Gick, M. L. (2011). Singing, health and well-being: A health psychologist's review. *Psychomusicology: Music, Mind and Brain, 21*(1–2), 176–207.

    Heydon, R., Fancourt, D., & Cohen, A. J. (eds.). (2020). *The Routledge Companion to Interdisciplinary Studies in Singing, Volume III: Wellbeing.* Routledge.

    Kang, J., Scholp, A., & Jiang, J. J. (2018). A review of the physiological effects and mechanisms of singing. *Journal of Voice, 32*(4), 390–395.

146 **Singing in a group:** Pearce, E., Launay, J., & Dunbar, R. I. (2015). The icebreaker effect: Singing mediates fast social bonding. *Royal Society Open Science, 2*(10), 150221.

146 **a boy from West Africa:** Schmartz, C., Majerus, A., & World Health Organization. (2020). Mateneen (Together)—a music therapy project for and with young refugees and asylum seekers in Luxembourg. *Public Health Panorama, 6*(1), 117–121.

147 **the bridge gives you a release from tension:** I'm indebted to Dillon O'Brian for assistance with this section and the analysis that follows.

151 **artists who hold our attention:** Brooks, D. (2020, October 23). Bruce Springsteen and the Art of Aging Well. *The Atlantic.*

154 **"Dying Speech of an Old Philosopher":** Landor, W. S. (n.d.). Dying speech of an old philosopher. *Poetry Foundation.* https://www.poetryfoundation.org/poems/44562

154 **As Nick Cave observed:** Petrusich, A. (2023, March 23). Nick Cave on the fragility of life. *The New Yorker.*

Chapter 9:
Mental Health

156 **"And as I played":** TED. (2010, March 26). Robert Gupta: Music is medicine, music is sanity [video]. YouTube. https://www.youtube.com/watch?v=C_SBGTJgBGo

157 **Gupta describes the transformation:** TED. (2010, March 26). Robert Gupta: Music is medicine, music is sanity. https://www.ted.com/talks/robert_gupta_music_is_medicine_music_is_sanity

157 **diagnostic criteria for schizophrenia:** American Psychiatric Association. (2013). *Diagnostic and Statistical Manual of Mental Disorders: DSM-5.* Washington, DC: American Psychiatric Association.

157 **Affected individuals display:** Kay, S. R., Fiszbein, A., & Opler, L. A. (1987). The positive and negative syndrome scale (PANSS) for schizophrenia. *Schizophrenia Bulletin, 13*(2), 261–276.

157 **relatively high prevalence:** Velligan, D. I., & Rao, S. (2023). The epidemiology and global burden of schizophrenia. *Journal of Clinical Psychiatry, 84*(1), 45094.

158 *balancing natural selection:* Hedrick, P. W. (2007). Balancing selection. *Current Biology, 17*(7), R230–R231.

158 **five different dopamine receptors:** Baik, J. H. (2013). Dopamine signaling in reward-related behaviors. *Frontiers in Neural Circuits, 7*, 152.

160 *any form of music therapy:* Pedersen, I. N., Bonde, L. O., Hannibal, N. J., Nielsen, J., Aagaard, J., Gold, C., . . . & Nielsen, R. E. (2021). Music therapy vs. music listening for negative symptoms in schizophrenia: Randomized, controlled, assessor- and patient-blinded trial. *Frontiers in Psychiatry*, 2374.

Tseng, P. T., Chen, Y. W., Lin, P. Y., Tu, K. Y., Wang, H. Y., Cheng, Y. S., . . . & Wu, C. K. (2016). Significant treatment effect of adjunct music therapy to standard treatment on the positive, negative, and mood symptoms of schizophrenic patients: A meta-analysis. *BMC Psychiatry, 16*(1), 1–11.

160 **music can improve mood:** Margo, A., Hemsley, D. R., & Slade, P. D. (1981). The effects of varying auditory input on schizophrenic hallucinations. *The British Journal of Psychiatry, 139*(2), 122–127.

Silverman, M. J. (2003). The influence of music on the symptoms of psychosis: A meta-analysis. *Journal of Music Therapy, 40*(1), 27–40.

Thaut, M. H. (1989). The influence of music therapy interventions on self-rated changes in relaxation, affect, and thought in psychiatric prisoner-patients. *Journal of Music Therapy, 26*, 155–166.

160 **A recent Cochrane Review:** Geretsegger, M., Mössler, K. A., Bieleninik, Ł., Chen, X. J., Heldal, T. O., & Gold, C. (2017). Music therapy for people with schizophrenia and schizophrenia-like disorders. *Cochrane Database of Systematic Reviews, 5*(5).

160 **retired security guard:** McInnis, M., & Marks, I. (1990). Audiotape therapy for persistent auditory hallucinations. *The British Journal of Psychiatry, 157*(6), 913–914.

161 **Quincy Jones expressed:** Jones, Q. (2002). *Q: The Autobiography of Quincy Jones.* Three Rivers Press.

161 **Bruce Springsteen has spoken openly:** PBS NewsHour. (2016, December 20). The music is medicine for Bruce Springsteen [video]. https://www.pbs.org/newshour/show/music-medicine-bruce-springsteen#transcript

161 **Guns N' Roses bassist:** McKagan, D. (2023). "This Is the Song" [video]. https://duffonline.com/videos/this-is-the-song/

161 *Tried Lexapro:* McKagan, D. (2023). "This Is the Song" (lyrics).

161 **an open letter to fans:** McKagan, D. (2023, June 12). "Mental Health Awareness— This Is the Song." https://duffonline.com/writing/mental-health-awareness-this -is-the-song/

162 **Tchaikovsky's diaries:** Tchaikovsky, P. I., & Lakond, V. (1945). *The Diaries of Tchaikovsky* (W. Lakond, trans.). W. W. Norton.

162 **Aristotle said:** *"Nullum magnum ingenium sine mixtura dementiae fuit."* Attributed to Aristotle in Seneca the Younger, "On Tranquility of Mind [*De Tranquillitate Animi*]" (17.10). Seneca (2009), *Dialogues and Essays.* (J. Davie, trans). Oxford University Press. (Original work published c. 60 CE.)

162 **Neuroscientist Laura Wesseldijk:** Power, R. A., Steinberg, S., Bjornsdottir, G., Rietveld, C. A., Abdellaoui, A., Nivard, M. M., . . . & Stefansson, K. (2015). Polygenic risk scores for schizophrenia and bipolar disorder predict creativity. *Nature Neuroscience, 18*(7), 953–955.

Wesseldijk, L. W., Lu, Y., Karlsson, R., Ullén, F., & Mosing, M. A. (2023). A comprehensive investigation into the genetic relationship between music engagement and mental health. *Translational Psychiatry, 13*(1), 15.

163 **studies of music and eating disorders:** Coutinho, E., Van Criekinge, T., Hanford, G., Nathan, R., Maden, M., & Hill, R. (2022). Music therapy interventions for eating disorders: Lack of robust evidence and recommendations for future research. *British Journal of Music Therapy, 36*(2), 84–93.

163 **music listening for eating disorders:** Wang, C., & Xiao, R. (2021). Music and art therapy combined with cognitive behavioral therapy to treat adolescent anorexia patients. *American Journal of Translational Research, 13*(6), 6534.

163 **Well-being and psychosocial development:** Chang, E. X., Brooker, J., Hiscock, R., & O'Callaghan, C. (2023). Music-based intervention impacts for people with eating disorders: A narrative synthesis systematic review. *Journal of Music Therapy, 60*(2), 202–231.

163 **The iso principle:** Heiderscheit, A., & Madson, A. (2015). Use of the iso principle as a central method in mood management: A music psychotherapy clinical case study. *Music Therapy Perspectives, 33*(1), 45–52.

164 **Most studies thus far:** Carter, T. E., & Panisch, L. S. (2021). A systematic review of music therapy for psychosocial outcomes of substance use clients. *International Journal of Mental Health and Addiction, 19*, 1551–1568.

Hohmann, L., Bradt, J., Stegemann, T., & Koelsch, S. (2017). Effects of music therapy and music-based interventions in the treatment of substance use disorders: A systematic review. *PLoS One, 12*(11), e0187363.

164 **Music therapy seems to reduce craving:** Pasqualitto, F., Panin, F., Maid-hof, C., Thompson, N., & Fachner, J. (2023). Neuroplastic changes in addiction memory—how music therapy and music-based intervention may reduce craving: A narrative review. *Brain Sciences, 13*(2), 259.

164 **musical negative mood induction:** Woolgar, M., & Tranah, T. (2010). Cognitive vulnerability to depression in young people in secure accommodation: The influence of ethnicity and current suicidal ideation. *Journal of Adolescence, 33*(5), 653–661.

164 **five-element music therapy:** Liu, B., Yu, F., & Shi, J. (2009). Discussing on the progress of musical treatment domestic and abroad. *Journal of Jiangxi University of Traditional Chinese Medicine, 121*, 89–91.

165 ***Letter to You:*** Brooks, D. (2020, October 23). Bruce Springsteen and the Art of Aging Well. *The Atlantic.*

166 **Since 1973, 196 people:** Death Penalty Information Center. Innocence Database. https://deathpenaltyinfo.org/policy-issues/innocence

166 **death row is likely innocent:** Gross, S. R., O'Brien, B., Hu, C., & Kennedy, E. H. (2014). Rate of false conviction of criminal defendants who are sentenced to death. *Proceedings of the National Academy of Sciences, 111*(20), 7230–7235.

166 **Bryan's meeting with Henry:** Stevenson, B. (2014). *Just Mercy: A Story of Justice and Redemption.* New York: Spiegel & Grau, pp. 11–12.

Chapter 10:
Memory Loss, Dementia, Alzheimer's Disease, and Stroke

168 **George was lying beneath:** George is a composite patient based on cases from Dr. Steven Sykes, Dr. Carlos Quintana, and my own observations.

169 **11% of adults will get Alzheimer's:** Alzheimer's Association. (n.d.). *Alzheimer's Disease Facts and Figures.* https://www.alz.org/alzheimers-dementia/facts-figures

169 **more people over age 65:** U.S. Census Bureau. (2022, July 9). *Population Projec-tions.* Census.gov. https://www.census.gov/programs-surveys/popproj.html

169 **more diapers are purchased:** Herships, S. (2016, August 29). There are more adult diapers sold in Japan than baby diapers. *JCIE.* https://www.jcie.org/analysis/in-the-media/there-are-more-adult-diapers-sold-in-japan-than-baby-diapers/

Hymowitz, C., & Coleman-Lochner, L. (2016, February 11). The adult diaper market is about to take off. *Bloomberg.com.* https://www.bloomberg.com/news/articles/2016-02-11/the-adult-diaper-market-is-about-to-take-off#xj4y7vzkg

171 **"I don't have an idea of what I'm going to play":** Chinen, N. (2020, October 21). Keith Jarrett confronts a future without the piano. *New York Times.*

171 **Jarrett tried to play:** Chinen, N. (2020, October 21). Keith Jarrett confronts a future without the piano. Ibid.

171 ***Downbeat* visited Jarrett:** Jackson, M. (2023, March 21). At home with Keith Jarrett. *DownBeat.*

172 **I'm using half the piano:** Jackson, M. (2023, March 21). At home with Keith Jarrett. Ibid.

172 **people worldwide experience a stroke:** Feigin, V. L., Stark, B. A., Johnson, C. O., Roth, G. A., Bisignano, C., Abady, G. G., . . . & Hamidi, S. (2021). Global, regional, and national burden of stroke and its risk factors, 1990–2019: A systematic analysis for the Global Burden of Disease Study 2019. *The Lancet Neurology, 20*(10), 795–820.

172 **saxophonist Sonny Rollins:** S. Rollins, personal communication. June 18, 2018.

173 **man with dementia and his daughter:** Her dad was diagnosed with dementia and "wet brain syndrome," also known as Wernicke-Korsakoff syndrome (WKS) or Wernicke's encephalopathy, a brain disorder caused by a severe deficiency in vitamin B1, a common complication of long-term heavy drinking. Dad (59) is Scott and his daughter is Bailey. She has been chronicling her caretaking journey on TikTok and their story is being made into a documentary called *You're at My House.* https://www.instagram.com/p/CnqQkJvB2-0/

175 **haloperidol, increases mortality:** Maust, D. T., Kim, H. M., Seyfried, L. S., Chiang, C., Kavanagh, J., Schneider, L. S., & Kales, H. C. (2015). Antipsychotics, other psychotropics, and the risk of death in patients with dementia: Number needed to harm. *JAMA Psychiatry, 72*(5), 438–445.

175 **agitation in dementia:** F. Russo, personal communication. May 15, 2023.

176 **Russo's and Mallik's model of the brain's "relaxation network":** Russo, F. A., Mallik, A., Thomson, Z., de Raadt St. James, A., Dupuis, K., & Cohen, D. (2023). Developing a music-based digital therapeutic to help manage the neuropsychiatric symptoms of dementia. *Frontiers in Digital Health, 5,* 1064115.

177 **"personalization is so important":** Russo, F. A., Mallik, A., Thomson, Z., de Raadt St. James, A., Dupuis, K., & Cohen, D. (2023). Developing a music-based digital therapeutic to help manage the neuropsychiatric symptoms of dementia. *Frontiers in Digital Health, 5,* 1064115.

179 **"This little light of mine":** mattys7 (2011, November 20). Gabby Giffords finding voice through music therapy. ABC News part 2/3 [video]. YouTube. https://www.youtube.com/watch?v=tiJ9X_wLSWM&t=209s

179 **the new Gabby Giffords:** CBS Sunday Morning (2015, March 15). Gabby Giffords speaks four years into her recovery [video]. YouTube. https://www.youtube.com/watch?v=4_kTv2t2hlY

179 **restoring language after strokes:** Haro-Martínez, A., Pérez-Araujo, C. M., Sanchez-Caro, J. M., Fuentes, B., & Díez-Tejedor, E. (2021). Melodic intonation therapy for post-stroke non-fluent aphasia: Systematic review and meta-analysis. *Frontiers in Neurology, 12,* 700115.

Schlaug, G., Marchina, S., & Norton, A. (2008). From singing to speaking: Why singing may lead to recovery of expressive language function in patients with Broca's aphasia. *Music Perception, 25*(4), 315–323.

Kim, J. S., Sung, J. E., Kim, J. S., & Sung, J. E. (2022). Treatment efficacy of working memory plus melodic intonation therapy for people with dementia of Alzheimer's type. *Communication Sciences & Disorders, 27*(2), 349–370.

179 **patients can remember music:** Cuddy, L. L., & Duffin, J. (2005). Music, memory, and Alzheimer's disease: Is music recognition spared in dementia, and how can it be assessed? *Medical Hypotheses, 64,* 229–235. PMID: 15607545.

Vanstone, A. D., & Cuddy, L. L. (2010). Musical memory in Alzheimer disease. *Aging, Neuropsychology, and Cognition, 17,* 108–128.

179 **sing along to familiar tunes:** Vanstone, A. D., Sikka, R., Tangness, L., Sham, R., Garcia, A., & Cuddy, L. L. (2012). Episodic and semantic memory for melodies in Alzheimer's disease. *Music Perception, 29,* 501–507.

179 **visiting Tony Bennett in 2021:** Colapinto, J. (2021, February 5). Tony Bennett keeps singing with Alzheimer's. *AARP Magazine.*

181 **everyday recall of Beatles tunes:** Hyman, I. E., & Rubin, D. C. (1990).

Memorabeatlia: A naturalistic study of long-term memory. *Memory and Cognition, 18*(2), 205–214.

181   **"And you know you should be glad"**: Lennon, J. & McCartney, P. (1963). She Loves You [recorded by The Beatles; Single]. Parlophone.

181   **"If you say you love me too"**: Lennon, J. & McCartney, P. (1964). Can't Buy Me Love [recorded by The Beatles]. On *A Hard Day's Night* [album]. Parlophone.

181   **"To help with good Rocky's revival"**: Lennon, J. & McCartney, P. (1968). Rocky Raccoon [recorded by The Beatles]. On *The Beatles* ("The White Album") [album]. Apple.

182   **use their knowledge of scales**: Ericsson, K. A., Krampe, R. T., & Tesch-Römer, C. (1993). The role of deliberate practice in the acquisition of expert performance. *Psychological Review, 100*(3), 363.

182   **encode chord sequences as chunks**: Noice, H., Jeffrey, J., Noice, T., & Chaffin, R. (2008). Memorization by a jazz musician: A case study. *Psychology of Music, 36*(1), 63–79.
      Levitin, D. J. (2002). Memory for musical attributes. In *Foundations of Cognitive Psychology: Core Readings* (pp. 295–310). MIT Press.
      Woody, R. H. (2020). Musicians' use of harmonic cognitive strategies when playing by ear. *Psychology of Music, 48*(5), 674–692.

182   **memorize each individual note**: See for example, Chaffin, R. (2007). Learning Clair de Lune: Retrieval practice and expert memorization. *Music Perception, 24*(4), 377–393.

183   **The guitarist Pat Martino**: Schoof, D. (2014, May 15). Pat Martino discusses relearning to play guitar after a near-fatal brain aneurysm left him with amnesia. *Lehigh Valley Live*. https://www.lehighvalleylive.com/music/2014/05/pat_martino_discusses_relearni.html

183   **Martino learned to play**: Solomon, J. (2016, February 24). How jazz guitarist Pat Martino learned to play again. *Westword*. https://www.westword.com/music/how-jazz-guitarist-pat-martino-learned-to-play-again-7633081

183   **he wrote in his autobiography**: Genzlinger, N. (2021, November 2). Pat Martino, jazz guitarist who overcame amnesia, dies at 77. *New York Times*.

Chapter 11:
Pain

187   **severing the tendons and nerves**: Levitin, D. J. (2018, October 10). Severed. *The New Yorker*.

189   **comedian Joan Rivers died**: Santora, M. (2016, May 12). Settlement reached in Joan Rivers malpractice case. *New York Times*.

188   **played it upside-down**: NeboShaMusic. (n.d.). Paul McCartney teaching a fan how to play All My Loving correctly on guitar [video]. YouTube. https://www.youtube.com/shorts/nzZllaSRnoE

188   **B.B. King changed a guitar string**: peterson dias. (2017, August 17). BB King breaks a guitar string mid-song and handles it like a pro [video]. YouTube. https://www.youtube.com/watch?v=lKf-mU6QrJs

188   **Propofol acts**: Alkire, M. T., Haier, R. J., Barker, S. J., Shah, N. K., Wu, J. C., & Kao, J. Y. (1995). Cerebral metabolism during propofol anesthesia in humans studied with positron emission tomography. *Journal of the American Society of Anesthesiologists, 82*(2), 393–403.

188   **activates GABA receptors**: Sahinovic, M. M., Struys, M. M., & Absalom, A. R. (2018). Clinical pharmacokinetics and pharmacodynamics of propofol. *Clinical Pharmacokinetics, 57*(12), 1539–1558.

188   **hypoxia-inducible factor 1-alpha**: Hsiao, H. T., Liu, Y. Y., Wang, J. C. F., Lin, Y. C., & Liu, Y. C. (2019). The analgesic effect of propofol associated with the

inhibition of hypoxia inducible factor and inflammasome in complex regional pain syndrome. *Journal of Biomedical Science, 26*, 1–11.

188 **produced in the hypothalamus:** Rajan, S., & Rao, S. (2017). Fluid and blood transfusion in pediatric neurosurgery. In *Essentials of Neuroanesthesia* (pp. 643–651). Academic Press.

188 **Pain researcher Jeffrey Mogil:** See also:

Schappert, S. M., & Burt, C. W. (2006). Ambulatory care visits to physician offices, hospital outpatient departments, and emergency departments: United States, 2001–02. *Vital and Health Statistics. Series 13, Data from the National Health Survey,* (159), 1–66.

Simon, L. S. (2012). Relieving pain in America: A blueprint for transforming prevention, care, education, and research. *Journal of Pain and Palliative Care Pharmacotherapy, 26*(2), 197–198.

Smith, B. H., Elliott, A. M., Chambers, W. A., Smith, W. C., Hannaford, P. C., & Penny, K. (2001). The impact of chronic pain in the community. *Family Practice, 18*(3), 292–299.

189 **Well, music, like propofol:** Antioch, I., Furuta, T., Uchikawa, R., Okumura, M., Otogoto, J., Kondo, E., . . . & Tomida, M. (2020). Favorite music mediates pain-related responses in the anterior cingulate cortex and skin pain thresholds. *Journal of Pain Research, 13*, 2729.

Bravo, F., Cross, I., Hopkins, C., Gonzalez, N., Docampo, J., Bruno, C., & Stamatakis, E. A. (2020). Anterior cingulate and medial prefrontal cortex response to systematically controlled tonal dissonance during passive music listening. *Human Brain Mapping, 41*(1), 46–66.

Sridharan, D., Levitin, D. J., & Menon, V. (2008). A critical role for the right fronto-insular cortex in switching between central-executive and default-mode networks. *Proceedings of the National Academy of Sciences, 105*(34), 12569–12574.

189 **music inactivates that HIF-1α:** Wang, J. Z., Li, L., Pan, L. L., & Chen, J. H. (2015). Hypnosis and music interventions (HMIs) inactivate HIF-1: A potential curative efficacy for cancers and hypertension. *Medical Hypotheses, 85*(5), 551–557.

189 **music listening and vasopressin:** Ukkola-Vuoti, L., Oikkonen, J., Onkamo, P., Karma, K., Raijas, P., & Järvelä, I. (2011). Association of the arginine vasopressin receptor 1A (AVPR1A) haplotypes with listening to music. *Journal of Human Genetics, 56*(4), 324–329.

189 **Music listening following surgeries:** Cepeda, M. S., Carr, D. B., Lau, J., & Alvarez, H. (2006). Music for pain relief. *Cochrane Database of Systematic Reviews, 2*, CD004843.

Nilsson, U. (2008). The anxiety-and pain-reducing effects of music interventions: A systematic review. *AORN Journal, 87*(4), 780–807.

189 **even for spinal surgery:** Lin, P. C., Lin, M. L., Huang, L. C., Hsu, H. C., & Lin, C. C. (2011). Music therapy for patients receiving spine surgery. *Journal of Clinical Nursing, 20*(7–8), 960–968.

189 **Prasad Shirvalkar at University of California:** Shirvalkar, P., Prosky, J., Chin, G., Ahmadipour, P., Sani, O. G., Desai, M., . . . & Chang, E. F. (2023). First-in-human prediction of chronic pain state using intracranial neural biomarkers. *Nature Neuroscience,* 1–10.

189 **Dentists figured this out:** Gardner, W. J., & Licklider, J. C. (1959). Auditory analgesia in dental operations. *Journal of the American Dental Association, 59*(6), 1144–1149.

193 **analgesic effects of music persist:** Cepeda, M. S., Carr, D. B., Lau, J., & Alvarez, H. (2006). Music for pain relief. *Cochrane Database of Systematic Reviews,* (2).

Nilsson, U. (2008). The anxiety- and pain-reducing effects of music interventions: A systematic review. *AORN Journal, 87*(4), 780–807.

193   **music reduces self-reported pain:** Garza-Villarreal, E. A., Pando, V., Vuust, P., & Parsons, C. (2017). Music-induced analgesia in chronic pain conditions: A systematic review and meta-analysis. *Pain Physician, 20*(7), 597–610.

193   **the descending pain modulation pathway:** Roy, M., Peretz, I., and Rainville, P. (2008). Emotional valence contributes to music-induced analgesia. *Pain, 134,* 140–147.

194   **self-chosen music significantly reduces pain:** Garza-Villarreal, E. A., Wilson, A. D., Vase, L., Brattico, E., Barrios, F. A., Jensen, T. S. T., . . . & Vuust, P. (2014). Music reduces pain and increases functional mobility in fibromyalgia. *Frontiers in Psychology, 5,* 90.

195   **measure things like dopamine:** We can measure these using radioactive tracers in a PET machine but many undergraduates don't like being injected with radioactivity.

195   **Self-selected music compared:** Dobek, C. E., Beynon, M. E., Bosma, R. L., & Stroman, P. W. (2014). Music modulation of pain perception and pain-related activity in the brain, brain stem, and spinal cord: A functional magnetic resonance imaging study. *Journal of Pain, 15*(10), 1057–1068.

196   **effect of music on pain:** Lunde, S. J., Vuust, P., Garza-Villarreal, E. A., Kirsch, I., Møller, A., & Vase, L. (2022). Music-induced analgesia in healthy participants is associated with expected pain levels but not opioid or dopamine-dependent mechanisms. *Frontiers in Pain Research, 3.*

198   **"saliency matrix":** Mouraux, A., Diukova, A., Lee, M. C., Wise, R. G., & Iannetti, G. D. (2011). A multisensory investigation of the functional significance of the "pain matrix." *Neuroimage, 54*(3), 2237–2249.

198   **Pleasant music accompanying pain:** Bhatara, A. K., Quintin, E. M., Heaton, P., Fombonne, E., & Levitin, D. J. (2009). The effect of music on social attribution in adolescents with autism spectrum disorders. *Child Neuropsychology, 15*(4), 375–396.

   Boltz, M. G. (2004). The cognitive processing of film and musical soundtracks. *Memory & Cognition, 32,* 1194–1205.

   Cohen, A. J. (2001). Music as a source of emotion in film. In P. N. Juslin & J. A. Sloboda (eds.), *Music and Emotion: Theory and Research* (pp. 249–272). Oxford University Press.

   Hoeckner, B., Wyatt, E. W., Decety, J., & Nusbaum, H. (2011). Film music influences how viewers relate to movie characters. *Psychology of Aesthetics, Creativity, and the Arts, 5*(2), 146.

199   **increase NK (natural killer) cell count:** Hasegawa, Y., Kubota, N., Inagaki, T., & Shinagawa, N. (2001). Music therapy induced alternations in natural killer cell count and function. *Nihon Ronen Igakkai zasshi. Japanese Journal of Geriatrics, 38*(2), 201–204.

   Wachi, M., Koyama, M., Utsuyama, M., Bittman, B. B., Kitagawa, M., & Hirokawa, K. (2007). Recreational music-making modulates natural killer cell activity, cytokines, and mood states in corporate employees. *Medical Science Monitor, 13*(2), CR57–CR70.

199   **optimal musical characteristics:** Martin-Saavedra, J. S., Vergara-Mendez, L. D., Pradilla, I., Velez-van-Meerbeke, A., & Talero-Gutierrez, C. (2018). Standardizing music characteristics for the management of pain: A systematic review and meta-analysis of clinical trials. *Complementary Therapies in Medicine, 41,* 81–89.

Chapter 12:
Neurodevelopmental Disorders

200   **Neurodevelopmental Disorders:** I have previously published peer-reviewed articles on this topic. For the interested reader:

   Ng, R., Lai, P., Levitin, D. J., & Bellugi, U. (2013). Musicality correlates with sociability and emotionality in Williams Syndrome. *Journal of Mental Health Research in Intellectual Disabilities, (6)*4, 268–279.

Levitin, D. J., & Bellugi, U. (2006). Rhythm, timbre and hyperacusis in Williams-Beuren Syndrome. In C. Morris, H. Lenhoff, and P. Wang (eds.), *Williams-Beuren Syndrome: Research and Clinical Perspectives* (pp. 343–358). Johns Hopkins University Press.

Levitin, D. J. (2005). Musical behavior in a neurogenetic developmental disorder: Evidence from Williams syndrome. *Annals of the New York Academy of Sciences, 1060*(27), 325–334. doi:10.1196/annals.1360.027

Levitin, D. J., Cole, K., Lincoln, A., & Bellugi, U. (2005). Aversion, awareness and attraction: Understanding hyperacusis in Williams Syndrome. *Journal of Child Psychiatry, Psychology and Allied Disciplines, 46*(5), 514–523. doi:10.1111/j.1469-7610.2004.00376.x

Levitin, D. J., Cole, K., Chiles, M., Lai, Z., Lincoln, A., & Bellugi, U. (2004). Characterizing the musical phenotype in individuals with Williams syndrome. *Child Neuropsychology, 10*(4), 223–247.

Levitin, D. J., Menon, V., Schmitt, J. E., Eliez, S., White, C., Glover, G., Kadis, J., Korenberg, J. R., Bellugi, U., & Reiss, A. L. (2003). Neural correlates of auditory perception in Williams Syndrome: An fMRI study. *NeuroImage, 18*, 74–82.

Levitin, D. J., & Bellugi, U. (1998). Musical abilities in individuals with Williams Syndrome. *Music Perception, 15*(4), 357–389.

204   **rules for writing lyrics:** Hammerstein II, O. (1985). *The complete lyrics of Oscar Hammerstein II.* Hal Leonard.

208   **music camp in the Berkshires:** The Williams Syndrome Music and Art Camp at Belvoir Terrace.

209   **hold a regular job:** Howlin, P., & Udwin, O. (2006). Outcome in adult life for people with Williams syndrome—results from a survey of 239 families. *Journal of Intellectual Disability Research, 50*(2), 151–160.

212   **Crystal, Neil, Lisa, and Clark:** Not their real names.

215   **play a musical game:** At the time I was conducting a similar study of time perception, and I had already obtained a blanket IRB (Independent Review Board) ethical approval to conduct music-based studies with a wide range of people, including minors and individuals with disabilities. We obtained informed consent from each parent before administering the test, and verbal consent from each minor, and we stopped every few trials and asked the children if they wanted to continue so that they knew they had a choice. Several of our participations *did* opt to discontinue the study early, and that reinforced for us that in general they understood the instruction.

220   **When we performed the subtraction analysis:** Levitin, D. J., Menon, V., Schmitt, J. E., Eliez, S., White, C., Glover, G., Kadis, J., Korenberg, J. R., Bellugi, U., & Reiss, A. L. (2003). Neural correlates of auditory perception in Williams Syndrome: An fMRI study. *NeuroImage, 18*, 74–82.

220   **findings of Al Galaburda:** Galaburda, A. M., Holinger, D. P., Bellugi, U., & Sherman, G. F. (2002). Williams syndrome: Neuronal size and neuronal-packing density in primary visual cortex. *Archives of Neurology, 59*(9), 1461–1467.

Thompson, P. M., Lee, A. D., Dutton, R. A., Geaga, J. A., Hayashi, K. M., Eckert, M. A., . . . & Reiss, A. L. (2005). Abnormal cortical complexity and thickness profiles mapped in Williams syndrome. *Journal of Neuroscience, 25*(16), 4146–4158.

224   **"could hum and sing many tunes accurately":** Kanner, L. (1943). Autistic disturbances of affective contact. *Nervous Child, 2*(3), 217–250.

224   **emotion processing deficits in ASD:** Quintin, E. M., Bhatara, A., Poissant, H., Fombonne, E., & Levitin, D. J. (2011). Emotion perception in music in

high-functioning adolescents with autism spectrum disorders. *Journal of Autism and Developmental Disorders, 41,* 1240–1255.

224   **ASD showed reduced physiological activity:** Stephenson, K. G., Quintin, E. M., & South, M. (2016). Age-related differences in response to music-evoked emotion among children and adolescents with autism spectrum disorders. *Journal of Autism and Developmental Disorders, 46,* 1142–1151.

225   **from utterly robotic to wholly expressive:** Bhatara, A., Quintin, E. M., Levy, B., Bellugi, U., Fombonne, E., & Levitin, D. J. (2010). Perception of emotion in musical performance in adolescents with autism spectrum disorders. *Autism Research, 3*(5), 214–225.

225   **pitch perception ability:** Bhatara, A. (2009). Commentary on "Why does music therapy help in autism?" by N. Khetrapal. *Empirical Musicology Review,* 4(1).

Chapter 13:
Learning How to Fly

232   **Older adults are better able:** Levitin, D. J. (2020). *Successful Aging: A Neuroscientist Explores the Power and Potential of Our Lives.* Dutton.

232   **making music in his sixties:** John Glusman, personal communication, October 3, 2023.

232   **Concert pianist Albert Frantz:** TEDx Talks (2010, October 5). TEDxPannonia—Albert Frantz—Finding our hidden dreams [video]. YouTube. https://www.youtube.com/watch?v=WwhDJi3bWrw

233   **The critic Joel Selvin:** Cited in Zanes, W. (2023, May 19). What we can learn from Bruce Springsteen's Great Left Turn. *New York Times.*

235   **stronger preliteracy skills:** Bolduc, J. (2009). Effects of a music programme on kindergartners' phonological awareness skills 1. *International Journal of Music Education, 27*(1), 37–47.

Moreno, S., Friesen, D., & Bialystok, E. (2011). Effect of music training on promoting preliteracy skills: Preliminary causal evidence. *Music Perception, 29*(2), 165–172.

Moritz, C., Yampolsky, S., Papadelis, G., Thomson, J., & Wolf, M. (2013). Links between early rhythm skills, musical training, and phonological awareness. *Reading and Writing, 26,* 739–769.

235   **tests of reading comprehension:** Corrigall, K. A., & Trainor, L. J. (2011). Associations between length of music training and reading skills in children. *Music Perception, 29*(2), 147–155.

235   **memory for words:** Franklin, M. S., Sledge Moore, K., Yip, C.-Y., Jonides, J., Rattray, K., & Moher, J. (2008). The effects of musical training on verbal memory. *Psychology of Music, 36*(3), 353–365.

235   **verbal fluency:** Zuk, J., Benjamin, C., Kenyon, A., & Gaab, N. (2014). Behavioral and neural correlates of executive functioning in musicians and non-musicians. *PLoS One, 9*(6), e99868.

235   **maintaining attention:** Strait, D. L., Kraus, N., Parbery-Clark, A., & Ashley, R. (2010). Musical experience shapes topdown auditory mechanisms: Evidence from masking and auditory attention performance. *Hearing Research, 261*(1–2), 22–29.

235   **associated with intelligence:** Bugos, J., & Groner, A. (2008). The effects of instrumental training on non-verbal reasoning in eighth-grade students. *Research Perspectives in Music Education, 12*(1), 14–19.

Swaminathan, S., Schellenberg, E. G., & Khalil, S. (2017). Revisiting the association between music lessons and intelligence: Training effects or music aptitude? *Intelligence, 62,* 119–124.

235   **visuospatial abilities:** Sluming, V., Brooks, J., Howard, M., Downes, J. J., &

Roberts, N. (2007). Broca's area supports enhanced visuospatial cognition in orchestral musicians. *Journal of Neuroscience, 27*(14), 3799–3806.

235 **processing speed:** Bugos, J., & Mostafa, W. (2011). Musical training enhances information processing speed. *Bulletin of the Council for Research in Music Education,* (187), 7–18.

Jentzsch, I., Mkrtchian, A., & Kansal, N. (2014). Improved effectiveness of performance monitoring in amateur instrumental musicians. *Neuropsychologia, 52,* 117–124.

235 **executive control:** Jentzsch, I., Mkrtchian, A., & Kansal, N. (2014). Improved effectiveness of performance monitoring in amateur instrumental musicians. *Neuropsychologia, 52,* 117–124.

Medina, D., & Barraza, P. (2019). Efficiency of attentional networks in musicians and non-musicians. *Heliyon, 5*(3), e01315.

235 **attention and vigilance:** Kaganovich, N., Kim, J., Herring, C., Schumaker, J., MacPherson, M., & Weber-Fox, C. (2013). Musicians show general enhancement of complex sound encoding and better inhibition of irrelevant auditory change in music: An ERP study. *European Journal of Neuroscience, 37*(8), 1295–1307.

Rodrigues, A. C., Loureiro, M. A., & Caramelli, P. (2013). Long-term musical training may improve different forms of visual attention ability. *Brain and Cognition, 82*(3), 229–235.

Román-Caballero, R., Martín-Arévalo, E., & Lupiáñez, J. (2021). Attentional networks functioning and vigilance in expert musicians and non-musicians. *Psychological Research, 85,* 1121–1135.

235 **episodic and working memory:** Talamini, F., Altoè, G., Carretti, B., & Grassi, M. (2017). Musicians have better memory than nonmusicians: A meta-analysis. *PLoS One, 12*(10), e0186773.

235 **better at impulse control:** Brown, E. D., Blumenthal, M. A., & Allen, A. A. (2022). The sound of self-regulation: Music program relates to an advantage for children at risk. *Early Childhood Research Quarterly, 60,* 126–136.

Hennessy, S. L., Sachs, M. E., Ilari, B., & Habibi, A. (2019). Effects of music training on inhibitory control and associated neural networks in school-aged children: A longitudinal study. *Frontiers in Neuroscience,* 1080.

236 **inferred causation from correlational designs:** Schellenberg, E. G. (2020). Correlation = causation? Music training, psychology, and neuroscience. *Psychology of Aesthetics, Creativity, and the Arts, 14*(4), 475.

236 **due to preexisting differences:** Woodworth, R. S., & Thorndike, E. L. (1901). The influence of improvement in one mental function upon the efficiency of other functions (I). *Psychological Review,* 8, 247–261.

Melby-Lervåg, M., Redick, T. S., & Hulme, C. (2016). Working memory training does not improve performance on measures of intelligence or other measures of "far transfer" evidence from a meta-analytic review. *Perspectives on Psychological Science, 11*(4), 512–534.

Sala, G., & Gobet, F. (2019). Cognitive training does not enhance general cognition. *Trends in Cognitive Sciences, 23*(1), 9–20.

Wan, C. Y., & Schlaug, G. (2010). Music making as a tool for promoting brain plasticity across the life span. *The Neuroscientist, 16*(5), 566–577.

236 **experiments with randomized control groups:** For a review, see Román-Caballero, R., Vadillo, M. A., Trainor, L. J., & Lupiáñez, J. (2022). Please don't stop the music: A meta-analysis of the cognitive and academic benefits of instrumental musical training in childhood and adolescence. *Educational Research Review, 35,* 100436.

236 **musical instrument training is associated:** Rogers, F., & Metzler-Baddeley, C. (2024). The effects of musical instrument training on fluid intelligence and executive functions in healthy older adults: A systematic review and meta-analysis. *Brain and Cognition, 175,* 106137.

237 10,000 hours: Ericsson, K. A., Krampe, R. T., & Tesch-Römer, C. (1993). The role of deliberate practice in the acquisition of expert performance. *Psychological Review, 100*(3), 363.

Reader, you may be thinking that becoming a surgeon requires more training: 4 years for a bachelor's work, 4 years of medical school, and typically 5 years as a resident (in Canada, students can enter medical school without having a bachelor's degree). But surgeons aren't beginning their education until age 18, and musicians typically start around age 5 or 6. That means musicians have already been training 12 years by the time surgeons get started.

237 profound brain plasticity: Münte, T. F., Altenmüller, E., & Jäncke, L. (2002). The musician's brain as a model of neuroplasticity. *Nature Reviews Neuroscience, 3*(6), 473–478.

237 musicians are happier: Eveleth, R. (2013, September 5). Artists might not make much, but they're happier with their jobs than you. *Smithsonian Magazine.*

Creech, A., Hallam, S., Varvarigou, M., McQueen, H., & Gaunt, H. (2013). Active music making: A route to enhanced subjective well-being among older people. *Perspectives in Public Health, 133*(1), 36–43.

Wheatley, D., & Bickerton, C. (2017). Subjective well-being and engagement in arts, culture and sport. *Journal of Cultural Economics, 41,* 23–45.

237 much smaller declines in processing speed: Mansens, D., Deeg, D., & Comijs, H. (2018). The association between singing and/or playing a musical instrument and cognitive functions in older adults. *Aging and Mental Health, 22*(8), 964–971.

237 nonverbal auditory memory: Hanna-Pladdy, B., & Gajewski, B. (2012). Recent and past musical activity predicts cognitive aging variability: Direct comparison with general lifestyle activities. *Frontiers in Human Neuroscience, 6,* 198.

Parbery-Clark, A., Strait, D. L., Anderson, S., Hittner, E., & Kraus, N. (2011). Musical experience and the aging auditory system: Implications for cognitive abilities and hearing speech in noise. *PLoS One, 6*(5), e18082.

237 verbal memory: Hanna-Pladdy, B., & MacKay, A. (2011). The relation between instrumental musical activity and cognitive aging. *Neuropsychology, 25*(3), 378.

237 mild cognitive impairment: Geda, Y. E., Topazian, H. M., Lewis, R. A., Roberts, R. O., Knopman, D. S., Pankratz, V. S., . . . & Ivnik, R. J. (2011). Engaging in cognitive activities, aging, and mild cognitive impairment: A population-based study. *Journal of Neuropsychiatry and Clinical Neurosciences, 23*(2), 149–154.

Wilson, R. S., Boyle, P. A., Yang, J., James, B. D., & Bennett, D. A. (2015). Early life instruction in foreign language and music and incidence of mild cognitive impairment. *Neuropsychology, 29*(2), 292.

237 dementia: Balbag, M. A., Pedersen, N. L., & Gatz, M. (2014). Playing a musical instrument as a protective factor against dementia and cognitive impairment: A population-based twin study. *International Journal of Alzheimer's Disease, 2014,* 836748.

238 amateur musicianship may confer: Rogenmoser, L., Kernbach, J., Schlaug, G., & Gaser, C. (2018). Keeping brains young with making music. *Brain Structure and Function, 223*(1), 297–305.

Musgrave, G. (2023). Music and wellbeing vs. musicians' wellbeing: Examining the paradox of music-making positively impacting wellbeing, but musicians suffering from poor mental health. *Cultural Trends, 32*(3), 280–295.

238 One study analyzed death records of over 13,000 professional musicians: Kenny, D. T., & Asher, A. (2016). Life expectancy and cause of death in popular musicians: Is the popular musician lifestyle the road to ruin? *Medical Problems of Performing Artists, 31*(1), 37–44.

238 a question of lumping or splitting: Simpson, G. G. (1945). *The Principles of Classification and a Classification of Mammals* (vol. 85). American Museum of Natural History.

239 **many different kinds of cancer:** Stratton, M. R., Campbell, P. J., & Futreal, P. A. (2009). *Nature, 458*(7239), 719–724.

National Cancer Institute. The Cancer Genome Atlas Program (TCGA), https://www.cancer.gov/ccg/research/genome-sequencing/tcga

239 **George Musgrave of University of London:** Musgrave, G. (2023). Music and wellbeing vs. musicians' wellbeing: examining the paradox of music-making positively impacting wellbeing, but musicians suffering from poor mental health. *Cultural Trends, 32*(3), 280–295.

240 **As Adam Gopnik has said:** Gopnik, A. (2011, June 20). Life studies. *The New Yorker.*

240 **David Remnick interviewed Bruce Springsteen:** Remnick, D. (Host). (2016, November 25). Bruce Springsteen talks with David Remnick (no. 58) [audio podcast episode]. *The New Yorker Radio Hour.* WNYC Studios and *The New Yorker.* https://www.newyorker.com/podcast/the-new-yorker-radio-hour/episode-58-bruce-springsteen-talks-with-david-remnick

244 **apply this technique to musicians:** Rogenmoser, L., Kernbach, J., Schlaug, G., & Gaser, C. (2018). Keeping brains young with making music. *Brain Structure and Function, 223*(1), 297–305.

245 **stave off cognitive and motor decline:** Worschech, F., James, C. E., Jünemann, K., Sinke, C., Krüger, T. H., Scholz, D. S., . . . & Altenmüller, E. (2023). Fine motor control improves in older adults after one year of piano lessons: Analysis of individual development and its coupling with cognition and brain structure. *European Journal of Neuroscience, 57*(12), 2040–2061.

245 **piano lessons group showed improved:** Worschech, F., James, C. E., Jünemann, K., Sinke, C., Krüger, T. H., Scholz, D. S., . . . & Altenmüller, E. (2023). Fine motor control improves in older adults after one year of piano lessons: Analysis of individual development and its coupling with cognition and brain structure. *European Journal of Neuroscience.*

246 **keyboard lessons to children:** Hyde, K. L., Lerch, J., Norton, A., Forgeard, M., Winner, E., Evans, A. C., & Schlaug, G. (2009). Musical training shapes structural brain development. *Journal of Neuroscience, 29*(10), 3019–3025.

246 **cross-domain skill transfer:** See also: Altenmüller, E., & Furuya, S. (2017). Apollos gift and curse: Making music as a model for adaptive and maladaptive plasticity. *e-Neuroforum, 23*(2), 57–75.

248 **the putamen:** See also: Eimontaite, I., Schindler, I., De Marco, M., Duzzi, D., Venneri, A., & Goel, V. (2019). Left amygdala and putamen activation modulate emotion driven decisions in the iterated prisoner's dilemma game. *Frontiers in Neuroscience, 13*, 741.

248 **During the initial two stages:** Worschech, F., James, C. E., Jünemann, K., Sinke, C., Krüger, T. H., Scholz, D. S., . . . & Altenmüller, E. (2023). Fine motor control improves in older adults after one year of piano lessons: Analysis of individual development and its coupling with cognition and brain structure. *European Journal of Neuroscience, 57*(12), 2040–2061.

Penhune, V. B., & Steele, C. J. (2012). Parallel contributions of cerebellar, striatal and M1 mechanisms to motor sequence learning. *Behavioural brain research, 226*(2), 579–591.

Doyon, J., Gabitov, E., Vahdat, S., Lungu, O., & Boutin, A. (2018). Current issues related to motor sequence learning in humans. *Current Opinion in Behavioral Sciences, 20*, 89–97.

248 **classroom instruction *about* music:** Moreno, S., Bialystok, E., Barac, R., Schellenberg, E. G., Cepeda, N. J., & Chau, T. (2011). Short-term music training enhances verbal intelligence and executive function. *Psychological Science, 22*(11), 1425–1433.

Chapter 14:
Music in Everyday Life

250 **They tend to be hummed:** Of the 4500 species of animals on our planet that sing or hum, humans are the only species that live on the ground—all the other singing and humming species live in the water or in trees. Ethnomusicologist Joseph Jordania notes that far more people in the world hum than sing, and cites the overwhelming evidence that humming for us is an expression of well-being, comfort, and enjoyment. Even our primary means of conveying agreement, the nonverbal *mm-hmm*, is a hum.

251 **premature infants in a neonatal unit:** Graff-Radford, M. (2021, January 12). *Humming Your Way to Relaxation*. Mayo Clinic Connect. https://connect .mayoclinic.org/blog/living-with-mild-cognitive-impairment-mci/newsfeed -post/humming-your-way-to-relaxation

  Loewy, J., Stewart, K., Dassler, A. M., Telsey, A., & Homel, P. (2013). The effects of music therapy on vital signs, feeding, and sleep in premature infants. *Pediatrics, 131*(5), 902–918.

251 **Steinbeck writes:** Steinbeck, J. (1952). *East of Eden* (p. 120). Viking.

  Not all songs or humming are comforting of course. Around the world, people have complained of hearing a mysterious "global hum," often attributed to industrial and urban noise pollution. Because low frequencies can travel many hundreds of miles, the exact source is difficult to pinpoint. One theory is that it is the low frequency vibration of ventilation fans on large buildings, diesel-electric generators, or trucks traveling on roads. Such sounds are viscerally unsettling at best, frightening at the worst. I sometimes sit in my front yard early in the morning, around 5:30 am, before anyone else is about, before there is traffic, or trucks or busses in my relatively quiet part of California. But on some days, especially cold ones when the thicker air can more readily transmit low frequencies, I hear it. Evolutionarily, sounds that are so low that they vibrate the very ground on which we walk signaled danger—a mastodon or an avalanche approaching. All my cognition and awareness of the source can't affect the way it disturbs me.

252 **we conducted a controlled experiment:** *SONOS and Apple Music*. (2016). The homes. https://musicmakesithome.com/tagged/sonoshomes. (Full disclosure: I received financial compensation from Sonos and Apple Music via a consulting contract with Weber-Shandwick to help design the study and analyze the data.)

  Stein, J. (2016, February 11). Study shows music brings my family closer together (as long as I'm not the one choosing the music). *Time*.

252 **During the music weeks:** *SONOS and Apple Music*. (2016). The homes.

253 **Heavy metal:** Olsen, K. N., Terry, J., & Thompson, W. F. (2022). Psychosocial risks and benefits of exposure to heavy metal music with aggressive themes: Current theory and evidence. *Current Psychology*, 1–18.

253 **heavy metal fans often feel:** Thompson, W. F., Geeves, A. M., & Olsen, K. N. (2019). Who enjoys listening to violent music and why? *Psychology of Popular Media Culture, 8*(3), 218–232. doi:10.1037/ppm0000184

254 **they'd all want to get with your grandma:** I'm grateful to my friend the comedian Gary Mule Deer for this line.

254 **neurosurgeon Katrina Firlick:** Firlik, K. (2006). *Another Day in the Frontal Lobe: A Brain Surgeon Exposes Life on the Inside.* Random House.

255 **self-chosen meditation music:** Firlik, K. (2006). *Another Day in the Frontal Lobe.*

255 **Pomodoro method:** The Pomodoro Technique is a time management method developed by Francesco Cirillo in the late 1980s. It is a simple yet effective approach to improve productivity and focus by breaking tasks into intervals of focused work and short breaks. Here's how the technique typically works:

1. Choose a task: Select a task or project you want to work on.
2. Set a timer: Set a timer for a specific duration, traditionally 25 minutes, known as one "Pomodoro" interval. You can use a kitchen timer, a smartphone app, or any other timer of your choice.
3. Work on the task: Focus solely on the chosen task during the Pomodoro interval. Avoid any distractions or interruptions and work with deep concentration.
4. Take a short break: When the timer goes off after 25 minutes, take a short break, typically 5 minutes. Use this time to relax, stretch, get a drink, or do anything unrelated to work.
5. Repeat the cycle: Once the break is over, start another Pomodoro interval and continue working on the task. Each interval represents one Pomodoro.
6. Long break: After completing four Pomodoro intervals (four focused work sessions), take a longer break (15–30 minutes). This break allows for more substantial rest and rejuvenation.
7. Track progress: Keep a record of completed Pomodoros and breaks. You can use a simple tally system on paper or use digital tools specifically designed for Pomodoro Technique tracking.
8. The idea behind the Pomodoro Technique is that the frequent breaks help prevent burnout and maintain mental freshness, while the time constraints provide a sense of urgency and enhance focus during the work periods. When tasks are broken into smaller, manageable intervals, it can be easier to stay motivated and maintain productivity throughout the day.
9. It's worth noting that while the traditional Pomodoro Technique uses 25-minute work intervals, you can adjust the duration based on your preferences and work style. The key is to maintain a consistent ratio of work and break times.
10. Many people find the Pomodoro Technique helpful for managing their time, avoiding procrastination, and increasing their overall productivity. However, it may not be suitable for everyone, so feel free to adapt or modify the method to best suit your needs and work habits.

256 **people are people and sometimes:** Caillat, C. & Swift, T. (2021). Breathe (Taylor's Version) [recorded by T. Swift feat. C. Caillat]. On *Fearless (Taylor's Version)* [album]. Republic Records.

257 **music in a large classroom:** Boyle, D. (2011). Exploring a university teacher's approach to incorporating music in a cognition psychology course (Order no. NR78600). Available from Dissertations & Theses @ McGill University. (1034470902).

The Lys Blues Awards are an annual recognition in Quebec, Canada, dedicated to honoring outstanding achievements in the field of blues music. The three awards are bestowed by the Trois-Rivières en Blues organization, also known for organizing the annual Trois-Rivières en Blues festival, a significant event in the Quebec blues scene. Dale Boyle, an award-winning guitarist, won "Folk/Blues Artist of the Year" in 2005 and 2006, and "Songwriter of the Year" in 2008. He has won 10 other music awards as well.

257 **journalist Madeleine Davies observed:** Davies, Madeleine (2013, May 31). The best/most ridiculous college a cappella team names in the country. *Jezebel*.

258 **adults who took music lessons:** Marie, D., Müller, C. A., Altenmüller, E., Van De Ville, D., Jünemann, K., Scholz, D. S., . . . & James, C. E. (2023). Music interventions in 132 healthy older adults enhance cerebellar grey matter and auditory working memory, despite general brain atrophy. *Neuroimage: Reports*, *3*(2), 100166.

258 **lifelong musicianship improves the ability:** Zhang, L., Wang, X., Alain, C., &

Du, Y. (2023). Successful aging of musicians: Preservation of sensorimotor regions aids audiovisual speech-in-noise perception. *Science Advances, 9*(17), eadg7056.

258 **Blondie drummer Clem Burke:** Smith, M., Draper, S., & Potter, C. (2008). Physiological demands of rock drumming: A case study. British Association of Sport and Exercise Sciences Conference (London, UK).

258 **"Influence of Music on Speed":** Ayres, L. P. (1911). The influence of music on speed in the six day bicycle race. *American Physical Education Review, 16*(5), 321–324.

259 **increasing adherence to exercise:** Clark, I. N., Baker, F. A., & Taylor, N. F. (2016). The modulating effects of music listening on health-related exercise and physical activity in adults: A systematic review and narrative synthesis. *Nordic Journal of Music Therapy, 25*(1), 76–104.

259 **push ourselves harder:** Clark, I. N., Baker, F. A., & Taylor, N. F. (2016). The modulating effects of music listening on health-related exercise and physical activity in adults: A systematic review and narrative synthesis. *Nordic Journal of Music Therapy, 25*(1), 76–104.

Hutchinson, J. C., Jones, L., Vitti, S. N., Moore, A., Dalton, P. C., & O'Neil, B. J. (2018). The influence of self-selected music on affect-regulated exercise intensity and remembered pleasure during treadmill running. *Sport, Exercise, and Performance Psychology, 7*(1), 80.

Hutchinson, J. C., Karageorghis, C. I., & Black, J. D. (2017). The diabeates project: Perceptual, affective and psychophysiological effects of music and music-video in a clinical exercise setting. *Canadian Journal of Diabetes, 41*(1), 90–96.

Terry, P. C., Karageorghis, C. I., Curran, M. L., Martin, O. V., & Parsons-Smith, R. L. (2020). Effects of music in exercise and sport: A meta-analytic review. *Psychological Bulletin, 146*(2), 91.

259 **Music can distract occasional exercisers:** Hutchinson, J. C., & Karageorghis, C. I. (2013). Moderating influence of dominant attentional style and exercise intensity on responses to asynchronous music. *Journal of Sport and Exercise Psychology, 35*(6), 625–643, as cited in Terry, P. C., Karageorghis, C. I., Curran, M. L., Martin, O. V., & Parsons-Smith, R. L. (2020). Effects of music in exercise and sport: A meta-analytic review. *Psychological Bulletin, 146*(2), 91.

259 **a lack of enjoyment of exercise:** Terry, P. C., Karageorghis, C. I., Curran, M. L., Martin, O. V., & Parsons-Smith, R. L. (2020). Effects of music in exercise and sport: A meta-analytic review. *Psychological Bulletin, 146*(2), 91.

Burgess, E., Hassmén, P., & Pumpa, K. L. (2017). Determinants of adherence to lifestyle intervention in adults with obesity: A systematic review. *Clinical Obesity, 7*(3), 123–135.

259 **boost performance:** For a review, see Karageorghis, C. I. (2020). Music-related interventions in the exercise domain: A theory-based approach. In G. Tenenbaum & R. C. Eklund (eds.), *Handbook of Sport Psychology*, 4th edition (pp. 929–949). Wiley.

259 **prolong male orgasm:** Chia, M., & Abrams, D. (2009). *The Multi-Orgasmic Man: Sexual Secrets Every Man Should Know.* Harper Collins.

259 **improve desire, arousal, and orgasm in women:** Mohammadi, E., Abdi-Shahshahani, M., Noroozi, M., Mohammadi, A. Z., & Beigi, M. (2023). Improving sexual dysfunction through guided imagery music (GIM): A clinical trial study. *Journal of Education and Health Promotion, 12*(1), 442.

Mohammadi, E., Shahshahani, M. A., Noroozi, M., & Beigi, M. (2024). The effect of guided imagery and music on the level of sexual satisfaction of women of reproductive age: A parallel cluster-randomized trial. *Journal of Midwifery & Reproductive Health, 12*(1).

259 **reduce chemotherapy-induced nausea:** Hurt, A. (2023, June 4). Cancer patients turn to music therapy for nausea relief. *Discover.*

259 **substantial reduction in nausea:** Kiernan, J. M., & Vallerand, A. H. (2023).

Mitigation of chemotherapy-induced nausea using adjunct music listening: A pilot study. *Clinical Nursing Research*, *32*(3), 469–477.

259 **undergraduates and older adults with dementia sang together:** Harris, P. B., & Caporella, C. A. (2019). Making a university community more dementia friendly through participation in an intergenerational choir. *Dementia*, *18*(7–8), 2556–2575.

261 **synchronization of brain activity:** Abrams, D. A., Ryali, S., Chen, T., Chordia, P., Khouzam, A., Levitin, D. J., & Menon, V. (2013). Intersubject synchronization of brain responses during natural music listening. *European Journal of Neuroscience*, *37*(9), 1458–1469.

261 **As Pink Floyd sang:** R. Waters, "Take Up Thy Stethoscope and Walk." Pink Floyd, *The Piper at the Gates of Dawn* [album]. EMI Columbia.

262 **Tony Award–winning *Avenue Q*:** Whitty, J., Marx, J., & Lopez, R. (2003). *Avenue Q: The Musical*. Music Theatre International.

263 **music for self-medication:** Baltazar, M., & Saarikallio, S. (2016). Toward a better understanding and conceptualization of affect self-regulation through music: A critical, integrative literature review. *Psychology of Music*, *44*(6), 1500–1521.

Stewart, J., Garrido, S., Hense, C., & McFerran, K. (2019). Music use for mood regulation: Self-awareness and conscious listening choices in young people with tendencies to depression. *Frontiers in Psychology*, *10*, 1199.

264 **varied exercise is more healthful:** Baz-Valle, E., Schoenfeld, B. J., Torres-Unda, J., Santos-Concejero, J., & Balsalobre-Fernández, C. (2019). The effects of exercise variation in muscle thickness, maximal strength and motivation in resistance trained men. *PLoS One*, *14*(12), e0226989.

Rauch, J. T., Ugrinowitsch, C., Barakat, C. I., Alvarez, M. R., Brummert, D. L., Aube, D. W., . . . & De Souza, E. O. (2020). Auto-regulated exercise selection training regimen produces small increases in lean body mass and maximal strength adaptations in strength-trained individuals. *The Journal of Strength and Conditioning Research*, *34*(4), 1133–1140.

264 **President Calvin Coolidge:** Gregory A. Kimble; Norman Garmezy; Edward Zigler (1974). *Principles of General Psychology* (4th ed.). Ronald Press Company. Page 249.

Wilson, J. R., Kuehn, R. E., & Beach, F. A. (1963). Modification in the sexual behavior of male rats produced by changing the stimulus female. *Journal of Comparative and Physiological Psychology*, *56*(3), 636–644.

Hughes, S. M., Aung, T., Harrison, M. A., LaFayette, J. N., & Gallup, G. G. (2021). Experimental evidence for sex differences in sexual variety preferences: Support for the Coolidge effect in humans. *Archives of Sexual Behavior*, *50*, 495–509.

267 **discovered *more* new music:** Dandiwal, Y., Fleming, L., & Levitin, D. J. (2023). Personal and contextual variables predict music consumption during the first COVID-19 lockdown in Canada. *Frontiers in Psychology*, *14*, 2162.

270 **Experiments in my lab:** Martin, L. J., Hathaway, G., Isbester, K., Mirali, S., Acland, E. L., Niederstrasser, N., Levitin, D. J., . . . & Mogil, J. S. (2015). Reducing social stress elicits emotional contagion of pain in mouse and human strangers. *Current Biology*, *25*(3), 326–332.

274 **The Nazis used music:** Eyre, M. (2023, May 18). The man who saved the music of the Nazi camps. *Wall Street Journal*.

275 **As Rosanne Cash writes:** Cash, R. (2010). *Composed: A Memoir*. New York: Viking.

Chapter 15:
Fate Knocking on Your Door

276 **As Thomas Mann wrote:** Mann, T. E. (1944, May). What Is German? *The Atlantic*.

276 **Chick's use of extended chords:** Drotos, R. (2018, November 3). The evolution

of jazz ballads. *Keyboard Improv.* https://keyboardimprov.com/the-evolution-of
-jazz-ballads/

278 McCoy Tyner accidentally hitting a wrong note: Klemp, N., McDermott, R.,
Raley, J., Thibeault, M., Powell, K., & Levitin, D. J. (2008). Plans, takes, and mistakes. Outlines. *Critical Practice Studies, 10*(1), 4–21.

Klemp, N., McDermott, R., Duque, J., Thibeault, M., Powell, K., & Levitin, D.
J. (2016). Plans, takes, and mis-takes. *Éducation et didactique, 3*, 105–120.

278 Shawn Camp singing: Rune Fløtre (2013, May 9). Shawn Camp—Off To Join The
World [video]. YouTube. https://www.youtube.com/watch?v=ZAX7IBDn-W8

278 Clare and the Reasons singing: Frogstand (2011, September 23). Clare and
the Reasons—Pluto [video]. YouTube. https://www.youtube.com/watch?v=8qdj
FsgW1cc

278 Jeff Silbar singing: Studio City Sound (2011, August 2). Jeff Silbar—Wind
beneath My Wings—Studio City Sound Live [video]. YouTube. https://www
.youtube.com/watch?v=379kwTDdGuA

The best explanation I've seen:

How to Build Chords from Scales

Basic chords are constructed by taking any scale tone as the starting note, skipping
a note and adding the next note of the scale, skipping a note, and adding the one
after that. For example, in the key of C major, the scale tones are C–D–E–F–G–
A–B–C, so a C chord is C–E–G. A three-note chord like this is called a triad.
Similarly, a D chord is D–F–A.

Because of the unequal spacing of notes on the keyboard (white and black keys)
the "distance" between those first two notes of the triad or chord, C and E, is
four piano keys (four half steps in the scale—C, C-sharp, D, D-sharp, E), but the
distance between D and F is only three piano keys; this causes the chords to have
a different tonal quality. Notice that in both chords, the "distance" between the
second two notes also changes, from three half steps in the C chord to four in the
D chord. (You can also count guitar or bass frets and get the same result.) When
the larger interval is in the bottom of the chord, we call it a *major* chord; when the
smaller is in the bottom, we call it a *minor* chord. Thus, the C chord we built is
called *C major* and the D is *D minor.*

We use Roman numerals as a shorthand for chord names, based on their position in the scale. In the key of C, C major is the I chord. D minor is the ii chord
(lowercase Roman numeral for a minor chord). If we change keys, everything
simply shifts up. In the key of D, D major is I, E minor is ii, and so on. This allows
musicians to talk about the chord progression in the abstract—a chord progression
of I–ii–V–I sounds the same regardless of starting note.

So what is G7? We simply continue our pattern and add a fourth note to the chord
we already have. After G–B–D, we skip a note (E) and then add F, creating a four-note
chord, the G7 (or V chord, because we're on the fifth scale degree). The B–F interval
inside G7 is the unstable tritone I mentioned; it is what makes us want to resolve to
something. The B wants to move up to C (to bring us back home to the key we are in),
and the F wants to resolve *down* to the E that is a defining part to the C major chord.

The full list of chords in the key of C major gives us C major, D minor, E minor,
F major, G major, A minor, and B diminished, or I–ii–iii–IV–V–vi–viio. The B
chord, the vii chord, is neither major nor minor: the "distance" between the first
two notes, B and D, is three half steps just as in a minor chord; but the distance
between the D and the F is *also* only three half steps, giving us a kind of double
minor chord (a minor third above the minor third) that we call a *diminished* chord.

290 This openness to interpretation: Not all music is composed to be ambiguous; in
many cases the opposite is true. Earlier, I mentioned Carl Stalling, who composed
music to reinforce animated depictions of characters engaged in specific actions,

such as tiptoeing up a flight of stairs, chasing each other around a tree, or walking along a path while drunk. Such explicitly *depictive* music as *Peter and the Wolf* or Grofé's *Grand Canyon Suite* are an exception to the rule of ambiguity. And yet they operate within a cultural tradition in which certain musical gestures tend to be interpreted in particular ways by experienced audiences. The same musical tropes that work for us may not work at all for someone from a different time and place.

294 **Bernstein's facial expressions:** aguniaaaaak. (2013, December 7). Ludwig van Beethoven Symphony No. 5 in C minor, Op. 67 - Leonard Bernstein [video]. YouTube: https://www.youtube.com/watch?v=1lHOYvIhLxo

296 **Billie Holiday's "My Man":** gemurin (2007, April 10). Billie Holiday, My Man [video]. YouTube. https://www.youtube.com/watch?v=IQlehVpcAes&t=111s

296 **Leonard Bernstein noticed:** Bernstein, L. (1976). *The Unanswered Question: Six Talks at Harvard*. Harvard University Press. Page 7.

297 **Music's universality may derive:** Bernstein, L. (1976). *The Unanswered Question: Six Talks at Harvard*. Page 9.

297 **Henkjan Honing and his research group:** Winkler, I., Háden, G. P., Ladinig, O., Sziller, I., & Honing, H. (2009). Newborn infants detect the beat in music. *Proceedings of the National Academy of Sciences, 106*(7), 2468–2471.

298 **Henkjan employed an alternative paradigm:** Háden, G. P., Bouwer, F. L., Honing, H., & Winkler, I. (2024). Beat processing in newborn infants cannot be explained by statistical learning based on transition probabilities. *Cognition, 243*, 105670.

298 **Subsequent work with macaques:** Honing, H., Bouwer, F. L., Prado, L., & Merchant, H. (2018). Rhesus monkeys (*Macaca mulatta*) sense isochrony in rhythm, but not the beat: Additional support for the gradual audiomotor evolution hypothesis. *Frontiers in Neuroscience, 12*, 475.

298 **offering an evolutionary advantage to humans:** See:
Levitin, D. J. (2008). *The World in Six Songs: How the Musical Brain Created Human Nature*. Dutton/Penguin Random House.
Levitin, D. J. (2021). Knowledge songs as an evolutionary adaptation to facilitate information transmission through music. *Behavioral and Brain Sciences, 44*, e105. doi:10.1017/S0140525X20001090
Savage, P. E., Loui, P., Tarr, B., Schachner, A., Glowacki, L., Mithen, S., & Fitch, W. T. (2021). Music as a coevolved system for social bonding. *Behavioral and Brain Sciences, 44*, e59.

301 **Harrison supports this lyrically:** In rehearsals, Harrison wasn't sure what to write next. He had "attracts me like . . . " But then what? Lennon advises him "Just say whatever comes into your head each time—'*attracts me like a cauliflower*'—until you get the words." "Yeah," Harrison responds, "but I've been through this one for, like, the last six months. I mean, just that line, I couldn't think of anything."

301 **comfortable, partial resolution:** In the mediant of the subdominant chord.

301 **"For No One" by Lennon and McCartney:** The original recording from *Revolver* is in the key of B, but that's a difficult key to play on the piano, and in actual performance, McCartney plays it in C (and slowed it down ~5.9% for the recording, dropping the key one semitone).

304 **irregular pattern of enharmonic waves:** Unlike harmonic waves, which are integer multiples of a fundamental frequency (like those produced by musical instruments), enharmonic waves consist of a complex mix of frequencies that do not have a simple mathematical relationship to each other. When you hit a table, the sound produced is the result of a complex interaction of vibrations in the material of the table. These vibrations are irregular and produce a broad spectrum of frequencies. The sound is more akin to noise than to a musical note because it lacks a clear pitch or harmonic structure. Sounds like banging on a table, clapping

hands, or white noise are often described as "unpitched" or "nonmusical" sounds because they don't have a definable pitch that can be easily notated in music.

304   **overtones or harmonics:** All harmonics are overtones, but not all overtones are harmonics. An overtone is any sound with a frequency higher

306   **going up the overtone series:** You don't get all the notes of our scale perfectly in tune because of some compromises that were made for equal-tempered tuning beginning in the eighteenth century, but you get close enough that the tones are easily recognizable. The overtone series is a natural phenomenon whereby a vibrating body (like a string, drum, or air column) produces a fundamental frequency and a series of higher frequencies (overtones or harmonics). These overtones are whole-number multiples of the fundamental frequency. For example, if the fundamental is 100 Hz, the overtones would be 200 Hz, 300 Hz, 400 Hz, etc.

The Western chromatic scale consists of 12 tones. These tones are derived from the pitches found in the overtone series, but the precise frequencies vary in different systems. In our current tuning system of equal temperament (adopted in the early eighteenth century) the octave is divided into 12 equal parts (semitones). This division does not perfectly match the natural harmonics (as seen in the overtone series), but it allows for consistent interval sizes across keys, making it highly versatile for modulation and transposition.

The idea of equal temperament dates back to ancient Chinese music theory with scholars like Zhu Zaiyu (1536–1611), a Ming dynasty prince, who is often credited with a form of equal temperament. He published his findings in "Fusion of Music and Calendar" (《律曆融通》) in 1584, where he described a 12-tone equal temperament system. In Europe, the concept evolved over time, with various theorists and musicians experimenting with it. One of the earliest mentions in Europe was by Vincenzo Galilei, father of Galileo Galilei, in the late sixteenth century. However, it wasn't widely adopted in Europe until much later.

308   **The 60,000-year-old bone flute:** Wong, K. (1997). Neanderthal notes. *Scientific American, 277*(3), 28–29.

314   **As Leonard Bernstein mused:** Bernstein, L. (1976). *The Unanswered Question: Six Talks at Harvard.* Page 79.

312   **Philippe Lalitte:** Lalitte, P. (2022). Évolution de la durée et du tempo dans les enregistrements du Sacre du printemps d'Igor Stravinski de 1929 à 2019. Une cartographie des styles d'interprétation. *Revue musicale OICRM, 9*(1), 1–26.

313   **a single Chopin etude:** Repp, B. H. (1998). A microcosm of musical expression. I. Quantitative analysis of pianists' timing in the initial measures of Chopin's Etude in E major. *The Journal of the Acoustical Society of America, 104*(2), 1085–1100.

313   **study of dynamic variations:** Repp, B. H. (1999). A microcosm of musical expression: II. Quantitative analysis of pianists' dynamics in the initial measures of Chopin's Etude in E major. *The Journal of the Acoustical Society of America, 105*(3), 1972–1988.

314   **create *super-expressive* versions:** This is the experiment first mentioned in Chapter 12 with respect to individuals with ASD being *unable* to recognize variations in emotional expressivity.

315   **Joseph Polisi . . . tells this story:** Personal communication. Polisi heard the story from William Schuman, Billy Rose's assistant in the production of this Broadway review.

Chapter 16:
Music Medicine, Mystery, and Possibility

317   **having a "transcendent experience":** Levitin, D. J., & Fleming, L. (2023). Transcendent experiences in music listening. Unpublished research report.

317   **The Therapeutic Music Capacities Model:** Brancatisano, O., Baird, A., &

Thompson, W. F. (2020). Why is music therapeutic for neurological disorders? The Therapeutic Music Capacities Model. *Neuroscience & Biobehavioral Reviews, 112,* 600–615.

319   and those attributes in turn are combined: I referred earlier to music's multi-dimensional qualities, comprising pitch, rhythm, timbre, loudness, and so on. Even pitch, all by itself, is multidimensional. This is not simply metaphor. We may be tempted to think of pitch as moving from low to high, across a line, constituting a single dimension based on frequency. But this ignores all that our magnificent brains *do* with that pitch information. Pitches have relationships with one another, and from those relationships come chords, harmony, timbre, and the tonal structures of music.

When we hear two notes an octave apart (one with double the frequency of the other), our brains perceive them as far more similar than two adjacent tones of a scale. Geometrically, then, pitch needs to be represented in two dimensions, with pitches wrapped around a circle so that as we move from C to C# to D, we eventually end up back at C again (the bottom circle in this figure). At the same time, we can detect that the two C's are different in some other quality that Roger Shepard calls pitch height and at the same time, they are perceptually more similar than adjacent notes on the circle. When one takes this into account, we end up with a three-dimensional model of pitch perception: traversing around the helical coil called "chroma" we move upward from C, to C#, to D, etc., and so that when we end up a the next C (C′ in the figure) the perceptual path we have traveled (a) is a greater distance than if we had simply moved straight up (path b).

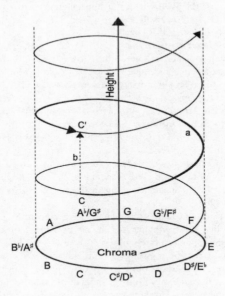

Roger Shepard's three-dimensional model of pitch.

When one takes into account interval relations (e.g., the pleasing consonance of the fifth, a ratio of 3:2) and then key relations that many listeners are sensitive to (the key of C shares all its notes with A minor, and the key of C major sounds harmonically close to G major because they only have one note different), one ends up with a fourth and fifth dimension for pitch. Roger argues that we ultimately need

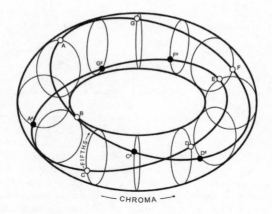

Shepard's five-dimensional model, projected onto the
two dimensions of the page.

six dimensions to account for all of the psychoacoustic data of how people actually hear musical pitch (I've shown only the 5D model he drew, collapsed onto the 2D of this page, because the 6D model is much more difficult to visualize and interpret).

319   in his novel *Orfeo*: Powers, R. (2014). *Orfeo: A Novel.* W. W. Norton. Used with the generous permission of Richard Powers.

320   music can increase empathy: Langford, D. J., Crager, S. E., Shehzad, Z., Smith, S. B., Sotocinal, S. G., Levenstadt, J. S., . . . & Mogil, J. S. (2006). Social modulation of pain as evidence for empathy in mice. *Science, 312*(5782), 1967–1970.

321   opportunity to train their imagination: Frye, N. (2002). *The Educated Imagination.* House of Anansi.

321   engage with the arts: K. Nagano, personal communication, December 22, 2021.

323   According to Schopenhauer: Schopenhauer, A. (2010). *The World as Will and Representation.* Edited and translated by J. Norman, A. Welchman, and C. Janaway. Cambridge University Press.

The German physicist Werner Heisenberg famously noted that the act of observing changes the observation. See: Heisenberg, W. (1927). Über den anschaulichen Inhalt der quantentheoretischen Kinematik und Mechanik. *Zeitschrift für Physik, 43*(3–4), 172–198.

It may be a stretch to extend the Heisenberg Uncertainty Principle. One hundred years earlier, Schopenhauer had already noted that a human observer cannot directly apprehend, or accurately observe, themselves.

Science, and scientific inquiry, can only advance when we discover aspects of reality that do not fall under our existing concepts of reality. Schopenhauer, who was by all accounts not a very funny guy, did have a theory of humor that converges with this. In every case, laughter arises from the sudden recognition of an incongruity between a concept you are holding in mind and some real object (or other concept) that is unexpectedly presented. Laughter itself is simply the expression of this incongruity. Music has much in common with humor, and indeed, avid listeners often burst into laughter while hearing what might be intended as a very serious piece of music! How does this happen?

Consider the structure of a joke. For the joke to be effective, the narrator has to lead you down a path and you have to be willing to follow, and the path has to be sensible and more or less logicially coherent within itself. But at the very last minute, the comic says something that causes you to realize you're not over *here* where you thought you were, but you are over *there*, somewhere you hadn't seen, and in fact, the path had been leading you there all along. This is what it is to truly follow an innovative musical piece. Whether the particular sonic trickery has to do with rhythm, melody, harmony, timbre or some combination doesn't matter. When Beethoven begins his *First Symphony* on what sounds like the dominant (V7) chord in the key of G, but we discover rather quickly that we are in C, he has set up an incongruity and its resolution. Even if you do not know the Key of C from, say, the Key of H, you will know that something's up, that a musical rug has been pulled out from under your feet. Another example is Bob Welch's song *Hypnotized* (from the aptly named album *Mystery to Me* by Fleetwood Mac). Drummer Mick Fleetwood launches the song with a drum pattern, and we can easily, subconsciously, feel the 1-2-3-4 beat as he lets it develop. When the guitar comes in, it enters at an entirely unexpected moment, causing us to completely recontextualize where we are in the time-space of the piece.

Appendix:
Types of Music Therapy

331 **Neurologic Music Therapy (NMT):** Thaut, M. H. (2010). Neurologic music therapy in cognitive rehabilitation. *Music Perception, 27*(4), 281–285.

Thaut, M. H., McIntosh, G. C., & Hoemberg, V. (2015). Neurobiological foundations of neurologic music therapy: Rhythmic entrainment and the motor system. *Frontiers in Psychology, 5*, 1185.

332 **Särkämö and his team:** Särkämö, T., Tervaniemi, M., Laitinen, S., Forsblom, A., Soinila, S., Mikkonen, M., . . . & Hietanen, M. (2008). Music listening enhances cognitive recovery and mood after middle cerebral artery stroke. *Brain, 131*(3), 866–876.

332 **Nordoff-Robbins Music Therapy:** Nordoff, P., Robbins, C., & Marcus, D. (2007). *Creative Music Therapy: A Guide to Fostering Clinical Musicianship.* Barcelona Publishers.

333 **dementia or Alzheimer's disease:** Wang, Y. C. (2021). The potential of Nordoff-Robbins music therapy techniques in music therapy practice for people living with dementia: A literature review. Masters thesis. Lesley University.

333 **Orff Music Therapy:** Voigt, M. (2013). Orff music therapy: History, principles and further development. *Approaches: Music Therapy & Special Music Education, 5*(2), 97–104.

333 **Vocal Psychotherapy:** Austin, D. S. (1996). The role of improvised music in psychodynamic music therapy with adults. *Music Therapy, 14*(1), 29–43.

Austin, D. (2007). Music therapy with adolescents in foster care. In V. A. Camilleri (ed.), *Healing the Inner City Child: Creative Arts Therapies with At-Risk Youth* (pp. 92–103). Jessica Kinglsey Publishers.

Austin, D. (2008). *The Theory and Practice of Vocal Psychotherapy: Songs of the Self.* Jessica Kingsley Publishers.

334 **The Bonny Method of Guided Imagery and Music:** Bonny, H. L. (1976). Music and psychotherapy: A handbook and guide accompanied by eight music tapes to be used by practitioners of Guided Imagery and Music. Doctoral dissertation, Union Institute and University.

Index

Page numbers in *italics* indicate illustrations.

A&R managers, 205–6
Abrams, Dan, 270–71
absolute pitch, 44, 208
a cappella groups, 257–58
accent structure, 39, 45, 55, 63, 341
acromegaly, 102
ACTH (adrenocorticotrophic hormone), *197,* 197n
active (creative, expressive) music therapy, 331
Adagio for Strings (Barber), 44
Adderley, Cannonball, 266
adrenaline, 114, 193, 196
African American Vernacular English (AAVE), 294–95
African music, 32, 33, 307, *308*
"Against All Odds" (Collins), 60n
aging
    cognitive decline and, 169–70, 172–73
    learning an instrument and, 231–33, 237–38, 245–46, 258
    social meaning and, 260
agitation, 174–76
agraphia, 68, 335
AI (artificial intelligence), 177–78, 267, 268–70
alexia, 68, 335
"Alfie" (Bacharach), 58
algorithmic music, 291
Alive Inside, 11
Alphabet song, 55
Altenmüller, Eckart, 244–45
Alwani, Esha, 115
Alzheimer's disease, 52, 168–69, 173, 179–80
    See also dementia

ambiguity, 284–85, 289–90, 293, 323
Ambulosono, 135
amusias, 67–69, 335
amygdala, *28,* 29, 103–4, 137, 140, 174, *176,* 220, 335
ancient music, 6–7, 8–9, 192, 308
anesthesiology, 187
aneurysm, 183–86
anhedonia, 68, 335
anterior cingulate, 88, *176,* 188, 189, 192–93, 335
Anthony, Marc, 339
Anthropologist on Mars, An (Sacks), 212
Antony and the Johnsons, 257
anxiety
    dementia and, 174–76
    eating disorders and, 163, 164
    movement disorders and, 104
    music therapy and, 165, 175, 193, 259
    prisoners and, 166
    professional musicians and, 238
    trauma and, 141, 145
aphasia, 124, 172
Apollo, 8
Arabic music, 31, 310
Arcade Fire, 256
Archey, Walter J., III, 138
arcuate fasciculus, 110, *110,* 335
Aristotle, 98, 162
Arlen, Harold, 15n
Armstrong, Louis, 32, 36, 37, 38, 39, 97, 289
Art Blakey and the Jazz Messengers, 32, 319
arterial dilation, 199
artificial intelligence (AI), 177–78, 267, 268–70
"Artist's Choice" (Mitchell), 183–85, *185*
art therapy, 332

ASD (autism spectrum disorder), 222–25,
    246–48, 292
Asleep at the Wheel, 32
Association, The, 263
Atkins, Chet, 278
attention
    consciousness and, 71, 75–76, 83, 85–86,
        91–92
    conscious sedation and, 188
    cue validity and, 64–67
    dopamine and, 158
    Default Mode Network and, 81–83, 85,
        91–92
    definitions of, 71–73
    emotion and, 43
    filter, 64–65, 86
    functional networks for, 75, 86
    meditation and, 88
    memory and, 43, 63–65, 71
    multitasking and, 73
    music therapy and, 91–93
    musical elements and, 63–64, 65–66,
        349–50, 273
    musicianship and, 235–236
    pain and, 188–89
    pattern matching and, 60–63
    PET studies and, 79
    research methods for, 349–50
    selective, 73–74, 75, 350–51
    sleep and, 85–86
    voluntary vs. involuntary, 74–75, 83–84,
        86, 353
    workplace environment and, 255
auditory cortex, 16, *17, 28,* 120, 248, 297,
    335, 342
auditory feedback, 109
auditory objects, 61
auditory pathway, 16–18, *17,* 338, 342
Austin, Diane, 333
Austin Lounge Lizards, 99
authenticity, 315–16
autism spectrum disorder (ASD), 222–25,
    246–48, 292
autobiographical memory, 18, 42–43, 44,
    52–53, 71, 81, 82–83, 87, 349
Avenue Q, 262
Awh, Ed, 73
Ayers, Nathaniel, 156–57
Ayres, Leonard, 258–59

"Baa, Baa, Black Sheep," 62
"Baby Shark," 55
Bach, Johann Sebastian, 219, 227, 232, 296

Bacharach, Burt, 58
Bailey, Steve, 188
Balanchine, George, 316
balancing natural selection, 158
Balinese music, 307, *307*
"Ballad of Mack the Knife, The" (Weill),
    36–37, 38, 60, 146, 152, 180
Barber, Samuel, 44
Barenboim, Daniel, 90, 204
Barney the Dinosaur, 55
basal ganglia, *28*
    defined, 335–36
    GABA and, 338
    movement disorders and, 110–11
    Parkinson's disease and, 128, 134
    startle reflex and, 29
    Tourette syndrome and, 115, 116
    See also specific components
Bavelier, Daphne, 70
Beatles, The
    chords and, 303, *303,* 306, 308, 309, *309*
    false memories and, 39–40
    keys and, 379
    lyrics and, 181, 301–2
    melody and, 300–302, *301,* 303, *303,
        309, 309*
    memory for Beatles songs, 181
    musical transcendence and, 278
    rhythm and, 298–99, *299,* 301–2
    timbre and, 35
Becker, Walter, 149
Beethoven, Ludwig van
    attention and, 62, 64
    chords and, 383
    Congress of Vienna and, 273, 274
    imagination and, 322
    learning an instrument and, 249
    melody and, 280–81, *281,* 301
    musical meaning and, 280–83, *281, 282,
        283,* 287, 293–94, 296
    performance interpretation and, 32
    Williams syndrome research and, 219
Belfi, Amy, 52
Believers, The, 149
Bellugi, Ursula, 11, 201–3, 208, 211, 212,
    216–17, 218
"Be My Baby" (Ronettes), 262
Bennett, Susan, 179–80
Bennett, Tony, 179–80, 238
Berg, Shelly, 58, 182
Berlin, Irving, 204, 227
Bernstein, Leonard
    emotion and, 294

Bernstein, Leonard (*continued*)
  linguistic analogies and, 291
  on melody, 280–81, 296
  on musical meaning, 293
  on pragmatics, 310–11
  on universality of music, 297
Berry, Chuck, 184
Beyoncé, 263
Bhatara, Anjali, 32, 224, 225, 313–14
Bhramari, 250–51
biblical references, 7, 102
Big Brother and the Holding Company, 97
bipolar disorder, 162
Bird. *See* Parker, Charlie
Bird Gets the Worm (Parker), 108
Blink (Gladwell), 96
Blitzstein, Marc, 36, 38
Blondie, 258
blood oxygenation level-dependent
  (BOLD) signal, 337
blue note, 295–96, 301, 306, 308
Blue Öyster Cult, 253
blues, 31, 182, *230*, 287, 295–96
Bolero (Ravel), 9, 286
Bonham, John, 203
Bonny, Helen L., 334
Bonny Method of Guided Imagery and
  Music, 334
"Boogie On Reggae Woman" (Wonder),
  278
Boomwhackers, 146
Bowie, David, 203
Boyce, William, 353
Boyle, Dale, 257
bradykinesia, 137
Brahms, Johannes, 55, 250
Brahms's "Lullaby," 55, 250
brain. *See* neuroanatomy
"Brain Damage" (Pink Floyd), 99
brain stem, *17, 28, 29*
Brantley, Timothy, 12
Breau, Lenny, 278
Brendel, Alfred, 315
Broca's area, 336
Brodmann Area, 27–29, *28,* 69, 83, 111,
  189, 336
Brosseau, Tom, 35, 267, 277
Buddhism, 89
Buena Vista Social Club, 277
Bugs Bunny, 288
Buhl, J. D., 149–53
Buncke, Greg, 188
Burke, Clem, 258

Burroughs, William, 291
Burton, Gary, 276–77
"Butch and Butch" (Nelson), 171

Calapinto, John, 179–80
Camp, Shawn, 278
Campbell, Glen, 52, 179, 238
Campbell, Joseph, 148
cancer, 259
"Can't Buy Me Love" (Lennon and McCa-
  rtney), 298–99, *299,* 303, *303,* 306,
  308, *309*
Carey, Susan, 77
Carpenter, Karen, 286
Carrell, Steve, 291
Carter, James Lewis, 233
cartoons, 288, 378–79
Casals, Pablo, 244
Cash, Johnny, 34–35, 222, 277
Cash, Rosanne, 46–47, 192, 231, 244, 275
caudate nucleus, 30, 116, 248, 336, 356
Cave, Nick, 154
ceiling effects, 217
Centerfield (Fogerty), 200
central executive (goal-directed attention),
  74, 75, 86, 90, 96
central nervous system (CNS), 116, 117
cerebellar vermis, 111, *111*
cerebellum, *17, 28*
  Chiari malformation and, 47–48
  defined, 336
  functions, 16
  learning an instrument and, 248
  movement disorders and, 120, 123, 128
  rhythm and, 297
  rhythmic auditory stimulation and, 120
  sound perception and, 16
  startle response and, 29
  stuttering and, 108, 111, *111*
  Tourette syndrome and, 356
  Williams syndrome and, 220
cerebral cortex, 16, 342
cerebral palsy (CP), 120
Cervenka, Exene, 116, 117–18
Cetera, Peter, 203
Champs, The, 286
Chanda, Mona Lisa, 190, 326
Chapman, Tracy, 147–48
Chekhov, Anton, 222–23
"Cherish" (The Association), 262
Chiari malformation, 47–48
children's songs, 55
Chinen, Nate, 171

Chinese five-element music therapy, 164, 334

Chinese music, 310, 380

cholecystokinin (CCK), 197n

Chomiak, Taylor, 135

Chomsky, Noam, 26, 200–201

Chopin, Frédéric, 32, 227, 228–29, *229*, 313–14

chords
 chunking and, 49, 182
 construction of, 378
 cultural variations in, 310
 extended, 276
 frequencies and, 303–5
 harmonic progression, 203
 humor and, 383
 keys and, 339
 modulation and, 284
 musical meaning and, 287, 301, 302–3, 306, 309, *309*
 musical memory and, 182
 universality of music and, 305, 310

chunking, 49, 56, 181–82

Cicoria, Tony, 10–11

cingulate cortex, 175

Cirillo, Francesco, 374

Clare and the Reasons, 35, 277

Clark, Guy, 228

classroom environment, 256–57

Clement, Alan, 243

Cobain, Kurt, 308

cochlea, 24

cochlear nucleus, *17*

Cocker, Joe, 148

Cognition: An Introduction (Posner), 77

cognitive behavioral therapy (CBT), 142, 163, 164–65

cognitive impairment, 173, 174
 See also dementia

cognitive neuroscience, 26, 95

cognitive psychology, 77–78

cognitive transfer, 235–37, 246

cold pressor task, 194–95, 270

collaborative filtering, 6

college a cappella, 257–58

Collins, Francis, 326

Collins, Phil, 60n

Coltrane, John, 266

compassion (loving-kindness) meditation, 88

Condon, Sean Michael, 274

conflict resolution, 271–75

Confucius, 7

Congress of Vienna, 273–74

consciousness, 26, 52
 attention and, 75–76, 82, 83, 85, 91–92
 constructive processes and, 101
 Default Mode Network and, 85, 92
 experiential fusion and, 2
 meta-awareness and, 89

conscious sedation, 187–88

context
 attention and, 62
 musical meaning and, 278–79, 289–90, 349
 musical memory and, 51–52
 musical preferences and, 34
 pain and, 198

contour, 16, 48, 306, 336

Coolidge effect, 263–64

Copeland, Stewart, 133

Copland, Aaron, 296

Corea, Chick, 276–77

corpus callosum, *28,* 30, 33

cortex, 16, 336

corticospinal tracts, *28*

cortico-striato-thalamo-cortical circuit (CSTC), 116, 120

cortisol, 113, 165, 166, 176, 193, 196, 197n, 336

Creative Forces, 144

creative music therapy, 332

Creedence Clearwater Revival, 200, 257

Croce, Jim, 262

Crosby, Stills & Nash, 203

Crowell, Rodney, 34, 35, 142, 227n, 244

"Crystal Silence" (Corea and Burton), 276–77

Csikszentmihalyi, Mihaly, 19, 337

cues, 51, 58, 348
 cross-modal cues, 351
 defined, 51
 musical memory cues, 38
 nonmusical cues, 66
 nonverbal cues, 69
 olfactory cues, 57
 pre-conscious cues, 71
 rhythmic cues, 114
 spectral-temporal cues, 18
 visual cues, 52, 128

cue validity, 64–67, 350

Curry, Steph, 90

Curtin, Hoyt, 288

cytokines, 198

Dadaists, 290

Dalai Lama, 88, 89

dancing
  embodied cognition and, 171
  multiple sclerosis and, 121–22
  musicality and, 204, 206
  music therapy for pain and, 190, 195
  music therapy for Parkinson's disease
    and, 135
  neuroprotection and, 18
  playlist design and, 263
  rhythm and, 286–87, 318
  swing and, 32
  Tourette's and, 115
  Williams syndrome and, 11
Dandiwal, Yuvika, 33–34, 57
Darin, Bobby, 36
David, Hal, 204
David (biblical king) 7, 102
Davidson, Richard, 2, 89, 337
Davies, Madeleine, 257
Davis, Miles
  classroom environment and, 257
  empathy and, 171
  improvisation and, 132–33, 221
  musicality and, 90, 192
  novelty-seeking and, 265–66
  performance interpretation and, 97
  timbre and, 203
"Daydream Believer" (Stewart), 77
Days Like This (Morrison), 44
death row inmates, 166–67
Debussy, Claude, 184
"Deck the Halls," 62, 63
Deep Forest, 184
Default Mode Network (DMN), 5, 80–84, 86
  anterior cingulate and, 335
  attention and, 81–82
  central executive and, 90
  consciousness and, 92
  constructive processes and, 96
  defined, 336
  dementia and, 175–76
  discovery of, 79, 80–81
  expectations and, 91
  group drumming and, 143
  meditation and, 95
  memory and, 82–83
  synchronization of, 85
  workplace music and, 255–56
defense mechanisms, 140
déjà vu, 58
dementia
  aging and, 169–70, 172–73
  attention and, 351

  cultural narratives on, 173–74
  experience of, 168–69
  frontal temporal, 53
  feeling of knowing and, 349
  learning an instrument and, 237–38
  movement and, 104
  musical memory and, 52–53, 180
  music therapy for, 169, 174, 175–77, 179,
    260–61, 333
  Parkinson's disease and, 128
  social meaning and, 260–61
  Wernicke-Korsakoff syndrome, 364
Dennett, Daniel, 91, 104, 325
Depeche Mode, 97, 205
depictive music, 378–79
depression, 5, 7, 8, 20, 52, 104, 136, 142,
    143–44, 161–62, 163–66, 174, 193,
    238, 240, 260
Derado, Gordana, 81
Derek and the Dominos, 267
Determined (Sapolsky), 147
Dharma, Buck, 253
dialects, 294–95
differential diagnosis, 116, 117–18
diffusion tensor imaging (DTI), 27, 336–37
Dimmock, Jonathan, 271, 273
disc jockeys, 205
disco, 287
discovery, 56, 263
distraction, 193–94, 195
Divje Babe cave flute, 308
"Don't Worry, Be Happy" (McFerrin), 126
Doobie Brothers, 190–91
dopamine
  act of discovery and, 101
  chorea and, 123
  defined, 337
  frisson and, 337
  function, 158
  islands of Calleja and, 338
  measurement of, 195
  music therapy for mental health condi-
    tions and, 166
  music therapy for trauma and, 140, 141
  pain and, 193, 196
  Parkinson's disease and, 128, 135
  physical health and, 259
  pleasure, reward and, 27–28, 140–41,
    166, 176, 259, 261
  receptors, 158–60
  schizophrenia and, 158–60
  substance misuse and, 164
  substantia nigra and, 342

Tourette's and, 114
ventral tegmental area, 342
dorsolateral prefrontal cortex (DLPFC), 87, 193
dorsomedial prefrontal cortex (DMPFC), *176*
Down syndrome, 209, 216, 217, 292
drug treatments
  for agitation, 175
  for cancer, 259
  conscious sedation, 187–88
  dynamic nature of, 101
  half-life and, 190
  for Huntington's disease, 123
  individual differences and, 19–20
  for mental health conditions, 142, 163, 165, 175
  for multiple sclerosis, 117
  opioids, 177, 190
  for pain, 188, 196
  for Parkinson's disease, 135–36
  placebo effect and, 196
  for schizophrenia, 157, 158
  trauma and, 140, 141, 142
drumming, 6, 112, 115, 124–25, 133, 143, 164, 212–13, 258, 297
  See also playing an instrument
Dryer, Henry, 11, 52
Duan, Marie, 313
duration, 16, 17, 341, 342
"Dying Speech of an Old Philosopher" (Landor), 154
Dylan, Bob, 64–65, 149
dynamics, 232
dysphagia, 178

eardrum (tympanic membrane), 22–24, 342
earworms, 161, 324
East of Eden (Steinbeck), 251
eating disorders, 163–64
1812 Overture (Tchaikovsky), 213, 219
Eilish, Billie, 115
elastin, 201, 210
Elgar, Edward, 349
Ellington, Duke, 32, 184, 185
embodied cognition, 103, 124, 171, 249
emotion
  anhedonia, 68, 335
  attention and, 43
  autism spectrum disorder and, 223–25
  cerebellum and, 16, *28, 336*
  empathy and, 274
  frisson and, *28,* 138, 337

learning an instrument and, 228, 234, 249
limbic system and, *28,* 339
meditation and, 87–88
memory and, 43
memory loss and, 174
movement and, 3–4
musicality and, 192, 204, 206
musical meaning and, 290–91, 293, 296, 302, 311
musical memory and, 44, 51–52
music therapy and, 134, 137–38
music therapy for mental health conditions and, 166
music therapy for trauma and, 143–45
Parkinson's disease and, 136–38
research on, 312
social meaning and, 260
songwriting and, 145, 149, 153
transcendence and, 275
Williams syndrome and, 220, 221
  See also expressivity
empathy, 84, 88, 171, 270, 274–75
"End, The" (Lennon and McCartney), 278
endogenous opioids, 190, 197n
endorphins, 145–46
enharmonic waves, 304, 379
entorhinal cortex, *176*
epigenetics, 141, 225–26
epinephrine, 199
episodic memory, 42, 82–83
Epstein-Barr virus (EBV), 117
Esalen Institute, 142, 272
Escher, M. C., *87*
Evans, Bill, 227, 266
everyday music, 250–70
  home environment, 250–52
  physical health and, 258–59
  self-selected, 262–63
  smart ecosystems for, 258–70
  social meaning and, 260–62
  workplace environment, 253–58
evolution, 179
  brain language/music centers and, 179
  low sounds and, 374
  memory and, 41, 43, 58, 97
  movement and, 103
  neuroanatomy and, 29, 41, 97, 120
  rhythm and, 298
  schizophrenia and, 157–58
  sound perception and, 21–22, 27–28
  stimulus generalization and, 97
  trauma and, 140
  variety and, 264

executive network, 337
exercise, 5, 258–59, 263
expectations
    Brodmann Area 47 and, 336
    Default Mode Network and, 91
    flow state and, 90–91
    individual differences and, 100
    key and, 339
    musical memory and, 56–57
    music therapy for pain and, 196
    placebo effect and, 196
    rhythm and, 298, 341
    selective attention and, 75
experiential fusion, 2–3, 19, 89, 319, 337
    See also flow state
experimental music, 290–91
exposure therapy, 142
expressivity, 10
    autism spectrum disorder and, 225
    improvisation and, 87
    learning an instrument and, 228, 232,
        234–35
    musical memory development and, 56
    music therapy for trauma and, 144, 145
    See also emotion

Fagen, Donald, 90–91, 149
    See also Steely Dan
false memories, 39–40, 44
family, 251–53
Fanconi, Guido, 208
"Fast Car" (Chapman), 147–48
featural binding, 325
feedback loops, 109, 170–71
feeling of knowing, 349
fermata, 281, 281
Fiedel Michel, 119
Fifth Symphony (Beethoven), 280–83, 281,
    282, 283, 293–94, 296
"Fire" (Springsteen), 300, 300
Firlick, Katrina, 254–55
First Piano Concerto (Tchaikovsky), 298,
    299
First Symphony (Beethoven), 285, 287, 383
fish, lateral line of (for hearing), 24
Fitzgerald, Ella
    flow state and, 90
    modulation and, 60
    musical memory and, 36–38, 39, 46, 51,
        52, 180
    performance interpretation and, 97
    racism and, 347
Fleetwood, Mick, 383

Fleetwood Mac, 383
Fleming, Renée, 244, 326–27
Flight of the Bumblebee (Rimsky-Korsakov),
    108
Flintstone, Fred, 288
flow state, 19, 37, 87, 90, 171, 319, 337
    See also experiential fusion
fMRI (functional magnetic resonance
    imaging), 81, 83, 120, 218–20, 337
focused attention meditation, 88
Fogerty, John, 200, 261–62
Foreigner, 263
"For No One" (Lennon and McCartney),
    301–2, 301, 379
Fox, Peter, 78
Frantz, Albert, 232–33
free recall tests, 350
frequencies, 9, 24, 303–8, 310, 379
"Frère Jacques/Are You Sleeping," 55, 62,
    64
Friberg, Anders, 32
frisson, 29, 90, 138, 311, 337
frontal lobe, 28, 337
Frye, Northrop, 320, 321
"Fusion of Music and Calendar" (Zhu), 380

GABA (gamma-aminobutyric acid), 116,
    188, 193, 337–38
gait training, 119–20
Galaburda, Al, 211, 220
Galilei, Galileo, 212
Galilei, Vincenzo, 380
Gardner, Howard, 202
"Gather In Your Promises" (Buhl), 151
Gatti, Roberto, 121
Geertz, Clifford, 297
genetics, 202–7, 208, 221, 225–26
genius, 162
genotypes, 202
Georgia Satellites, 234
Gestalt psychology, 56, 60–61
Getz, Stan, 203
Giffords, Gabby, 178–79
Gigi, 285
Gillespie, Dizzy, 95n
Gladwell, Malcolm, 96
Glee, 257
globus pallidus, 336
glutamate, 193, 338
goal-directed attention (central executive),
    74, 75, 86, 90
Goliath, 102
González, Marta, 52

Goodman, Benny, 32
Gopnik, Adam, 240, 241
Grace, Jamie, 115
Grafton, Scott, 103–4, 111, 137, 231
Grand Canyon Suite (Grofé), 379
Grateful Dead, 90
Gray, Trey, 124
gray matter, 29
Greek modes, 339
Greicius, Mike, 81
Grice, Paul, 292
grief, 53
Grofé, Ferde, 379
group music therapy, 143, 144, 145–46,
    163, 261
guided imagery and music (GIM), 163,
    333–34
guitar diagrams
    blues scales, *230*
    reading, 230
Gupta, Robert Vijay, 156–57
Guthrie, Woody, 125
Guy, Buddy, 233

hair cells, 23, 24, *24*
Halle, Morris, 200
"Halo" (Beyoncé), 263
Hammerstein, Oscar, II, 204–5
Hampton, Lionel, 155
Han, Eun Young, 136
Hancock, Herbie, 185
Handel, George Frederick, 290
"Happy Birthday," 63
Harburg, Yip, 15n, 45
Hardenberg, Philipp Friedrich Freiherr von
    (Novalis), 10
harmonic (overtone) series, 304–8, 310, 380
harmonics, 304
Harmonics (Ptolemy), 8–9
harmony, 17, 203, 305, 322
    See also chords
Harrison, Christopher, 257
Harrison, George, 300–301, *300*, 379
Harvest (Young), 185
Hawaiian Pidgin English, 295
Haydn, Joseph, 100
"Headed for the Fall" (Levitin), 154
heavy metal, 253
Hebrew chants, 308
Heidegger, Martin, 72
Heifetz, Jascha, 25, 233
Heisenberg, Werner, 382
hemophilia, 207

Hendrix, Jimi, 35, 242, 267
Henry, Joe, 142
"Here Comes the Bride" (Wagner), 63,
    349
Hertz, Heinrich, 304
Heschl's gyrus, *28*
Heuser, Frank, 207
HIF-1α (hypoxia-inducible factor 1-alpha),
    188, 189
Hintzman, Doug, 349
hippocampus, *28, 176*
    defined, 338
    dopamine and, 159
    learning an instrument and, 248
    memory and, 40–41, 58, 71, 174, 338
    movement and, 103–4
    movement disorders and, 110
    trauma and, 140
Hippocrates, 8
Hirst, Damien, 290
"Hokey Pokey, The," 55
Holiday, Billie, 296, 308
"Honey Pie" (Lennon and McCartney),
    309, *309*
Honing, Henkjan, 32–33, 297–98
Hooker, John Lee, 105
Houston, Whitney, 242
"How to Grow a Woman from the
    Ground" (Brosseau), 277
HPA (hypothalamic-pituitary-adrenal)
    system, 140
Hu, Bin, 135
humming, 250–51, 373
humor, 382–83
Huntington's disease, 123–25
Huntington speech music therapy (HSMT),
    124
Husserl, Edmund, 72
Huxley, Parthenon, 216, 217, 266
Hyde, Krista, 246–47
hypnosis, 196
Hypnotized (Welch), 383
hypothalamus, 338

"If I Only Had a Brain" (Arlen and Har-
    burg), 15
"I'll Have to Say I Love You in a Song"
    (Croce), 262
imagination, 320–22
"Imagine" (Lennon), 222
immune system, 199
immunoglobulin A (IgA), 198
"Impossible Elephant," *87*

improvisation
  embodied cognition and, 171
  learning an instrument and, 232
  mastery and, 37
  medicine and, 4
  meditative states and, 87–88
  musical memory and, 36–37, 51, 181
  music therapy for mental health conditions and, 164
  Nordoff-Robbins Music Therapy and, 332
  performance and, 129–31, 132–33, 170–72
  repetition and, 289
  song recognition and, 49
  stroke and, 170–72
  Williams syndrome and, 217–18
"In a Sentimental Mood" (Ellington), 3
Indian music, 31, 32, 310
individual differences
  attention and, 62, 64, 66
  curated playlists and, 5–6
  drug treatments and, 19–20
  musical preferences and, 33–34, 62, 98
  music therapy and, 99–101, 120, 165, 176–78
  scalability and, 177–78, 268
  smart ecosystems and, 268–70
  See also self-selected music
inferior colliculus, 16, *17*, 338
inferior frontal gyrus, 88, 110, *111*
inferior parietal lobule, *28,* 88
instruments. *See* learning an instrument; playing an instrument; timbre
insula, 194
insulae, 82, 88, 338
intelligence, 103, 202, 209–10, 213, 235
intensity, 342
"Interlude" ("A Night in Tunisia") (Gillespie), 95
interoception, 88
"In the Mood" (Miller), 32
intonation, 286
involuntary attention, 74, 86
"In Walked Bud" (Monk), 3
"I Remember It Well" (Lerner), 285
Ishiguro, Kazuo, 347
islands of Calleja, 338
iso principle, 163–64
Israeli-Palestinian conflict, 272–73
"It's Alright with Me," 171
"I Walk the Line" (Cash), 34–35, 222, 277

"I Walk the Line (Revisited)" (Crowell), 34
"I Want to Know What Love Is" (Foreigner), 263

Jackendoff, Ray, 291
Jackson, Michael, 241
Jakobson, Lorna, 70–71
jamais vu, 58
James, Etta, 184
James, William, 72–73
Janov, Arthur, 333
Japanese music, 307, *307*
Jarrett, Akiko, 171
Jarrett, Keith, 170–73, 183
Jars, The, 149
jazz, 32–33, 49, 289, 291
  See also specific artists
Jennings, Will, 148
John, Elton, 267
"Johnny B. Goode" (Berry), 184
Jones, James Earl, 105, 112
Jones, Philly Joe, 221
Jones, Quincy, 161
Jones, Spike, 214
Joplin, Janis, 97, 242
Jordan, Louis, 184
"Just a Memory" (Levitin), 327
"Just Like Someone Who Loves You" (Levitin & Buhl), 153
Just Mercy (Stevenson), 166–67

Kadis, Jay, 219
Kahneman, Daniel, 95–96
Kanner, Leo, 223–24
Kant, Immanuel, 72
Kasten, Fritz, 112
Keele, Steve, 128
"Keep Your Hands to Yourself" (Georgia Satellites), 234
keyboard diagrams
  Beethoven's Fifth Symphony, *281, 282, 283*
  "Can't Buy Me Love," *303, 309*
  Chopin's Prelude in E Minor, *229*
  "Honey Pie," *309*
  Japanese and Balinese pentatonic scales, *307*
  musical meaning and, *279*
  reading, *281*
  "The Star Spangled Banner," *309*
keys
  defined, 338–39

modulation, 60, 263, 283–84
  musical meaning and, 291, 293
  playlist design and, 263
Khergiani, Mikheil, 251
"Kind of Blue" (Davis), 266
King, B. B., 188, 227, 229–31, *230*
Kinsbourne, Marcel, 91
Kirwan, Danny, 149
Kissinger, Henry, 246
Klein, Howie, 12
Knox, Daniel, 35
Kodama, Mari, 228, 315, 327
Koller, Jacob, 97
Köln Concerts, The (Jarrett), 170
Korenberg, Julie, 211
Koulis, Theodoro, 194

Lady Gaga, 267
Lalitte, Philippe, 312
Lamar, Kendrick, 113
Landor, Walter Savage, 154
language
  generative nature of, 288–89
  neuroanatomy and, 179
  tonal perception and, 68, 69
  Williams syndrome and, 200–201, 202,
    209, 213
  See also linguistic analogies for music
lateral line, 23, *24*
lateral sulcus, 338
Latin music, 33
L-dopa, 135–36
learning an instrument, 227–38
  aging and, 231–33, 237–38, 245–46, 258
  authenticity and, 315
  cognitive transfer and, 235–37, 244, 246
  embodied cognition and, 249
  emotion and, 228, 234, 249
  expressivity and, 228, 232, 234–35
  goals and, 227–28
  guitar diagrams and, 229–30, *230*
  mistakes and, 233–34
  motor action plans and sequences and,
    246, 248
  musical memory and, 231
  musical notation and, 228–29, *229*
  physical health and, 258
  pragmatics and, 247–48
  stages of, 248
  YouTube and, 231
Led Zeppelin, 253
Leibniz, Gottfried, 322
Leigh, Brennen, 32, 142

Lennon, John, 222, 303, 333, 379
  See also Beatles, The
Lerdahl, Fred, 291
Letter to You (Springsteen), 165
Levitin, Joseph, 153–55
Levy, Bianca, 313
Lewisohn, Mark, 40
Limb, Charles, 87, 326
limbic system, 28, *28,* 335, 339
linguistic analogies for music
  phonemes, 279–80, *279*
  pragmatics, 291–94, 295–96, 299–300,
    310–11, 315
  semantics, 284–86, 288
  syntax, 280–82, *281, 282,* 285–86,
    288–89, 294–95
  See also pragmatics
listening to music
  brain synchronization and, 85, 270–71
  conflict resolution and, 271–72, 273–75
  as constructive process, 99–100
  Default Mode Network and, 82, 83, 84,
    91
  empathy and, 85, 270
  endogenous opioids and, 190
  experiential fusion and, 2–3, 90
  family and, 251–53
  learning, 248–49
  memory and, 18–19, 49, 50, 55–56, 58
  musicality and, 206
  music therapy for mental health condi-
    tions and, 166
  music therapy for multiple sclerosis and,
    119
  music therapy for pain and, 189–90, 193
  music therapy for trauma and, 143
  neuroanatomy and, 18
  social meaning and, 260, 261–62
  song recognition, 49, 50, 96–99
  sound perception development and,
    26–27
  Tourette syndrome and, 115–16
  transitions and, 353
  workplace environment and, 253–58
  See also musical preferences; playlists
Loeb, Lisa, 142
Loftus, Elizabeth, 44
Lone Ranger theme (*William Tell* Overture)
    (Rossini), 62, 65
Longfellow, Henry Wadsworth, 7
Lopez, Steve, 156–57
loudness, 16, 17, 339, 341
"Love Affair" (Spektor), 59

"Love Me Tender" (Presley), 114
"Love the One You're With" (Stills), 149
loving-kindness (compassion) meditation, 88
lucid dreaming, 86, *87*
lullabies, 250, 286
lyrics
    amusias and, 71
    Default Mode Network and, 84
    dementia and, 180
    musicality and, 204–5
    musical meaning and, 300, 301–2, 379
    musical memory and, 180–81
    musical preferences and, 34
    See also songwriting
Lys Blues Awards, 375

"Mack the Knife" (Weill), 36–37, 38, 60, 146, 152, 180
Madonna, 205
"Maggie May" (Stewart), 287
Mahler, Gustav, 74, 276, 277
major triad, 308–9, *309*
Making of "Sgt. Pepper," The, 40
Mallik, Adiel, 175–76, *176*, 177, 189–90
"Man in a Case" (Chekhov), 223
Mann, Thomas, 276
maqams, 310
Marr, Johnny, 146
Martin, George, 40
Martin, Loren, 270
Martin, Steve, 184, 291
Martino, Pat, 183
mastery, 37
McCartney, Paul, 40, 188, 205, 287n, 301–3, 309
    See also Beatles, The
McFerrin, Bobby, 126–27, 129–32, 134, 194, 244
McKagan, Duff, 161
Meaney, Michael, 139
medial prefrontal cortex (MPFC), 87
medicine
    improvisation and, 4, 344
    individual differences and, 19–20
    as science and art, 3, 4
    See also drug treatments
meditative states, 76, 85, 87–89
melakartas, 310
melatonin, 198
Melodic Intonation Therapy (MIT), 178–79
melody
    attention and, 64

auditory pathway and, 16–17
contour and, 336
cultural variations in, 310
key and, 339
lullabies and, 250
modulation and, 60
motifs, 282–83, *282, 283*
musicality and, 203
musical meaning and, 288, 300–302, *300, 301, 303, 303*
musical memory development and, 56
musical preferences and, 34
tone deafness and, 67
memory, 1–2
    amusias and, 71
    autobiographical, 42–43, 52–53, 81, 82–83, 87, 349
    caudate nucleus and, 336
    cues and, 51, 58, 348
    Default Mode Network and, 82–83
    distortion of, 39–40, 44
    distributed nature of, 40
    episodic, 42, 82–83
    hippocampus and, 40–41, 174, 338
    movement and, 104
    multiple sclerosis and, 122
    neuroanatomy and, 18–19
    Parkinson's disease and, 125
    procedural, 42, 46
    rehabilitation of, 48
    sensory, 41–42, 46
    state-dependent retrieval, 52
    temporal lobe and, 342
    types of, 41–43
    unreliability of, 40
    See also musical memory
memory loss, 66, 67, 122, 183, 349
    See also dementia
Menon, Vinod, 81, 82, 85, 92, 111, 189, 270–71, 353
mental chronometry, 78
mental health, 156–67
    depression, 104, 143–44, 161–62, 163–66, 238
    drug treatments for, 142, 163, 165, 175
    earworms and, 161
    eating disorders, 163–64
    music therapy for, 160, 163–66, 175
    panic disorder, 161
    professional musicians and, 162, 238–42
    schizophrenia, 156–61, 162, 163
    substance misuse, 164
    See also anxiety

mesolimbic system, 160, 339
Messiah (Handel), 290
meta-awareness, 88–89, 90
metazoa, 21
meter, 339
Meyer, Leonard, 291
microtones, 31
midbrain, 339
Middle Eastern music, 308
Miller, Bob, 243
Miller, Earl, 73
Miller, Glenn, 32
Miller, Steve, 191
Milošević, Slobodan, 274
mindfulness meditation, 87–88
Mingus, Charles, 185
minor third, 306–7
Mintun, Mark, 78
mistakes, 107–8, 233–34
Mitchell, Joni, 35, 154–55, 183–85, *185*
Mitchell, Laura, 194, 326
mnemonic devices, 45
"Moanin'" (Art Blakey and the Jazz Messengers), 32
mode, 339
modulation, 60, 263, 283–84
Mogil, Jeffrey, 188, 196, 270, 325
Molaison, Henry, 71
Monkees, The, 76n, 267
Montaigne, Michel de, 296, 297
mood, 194, 195
    iso principle and, 163–64, 240
    maqams and, 310
    memory and, 44, 143
    movement disorders and, 124, 129
    music and changes in, 5, 9, 14, 19, 52,
        100, 105, 122, 140, 141, 160, 166, 194,
        195, 249, 259, 261, 263, 276
    music perception and, 319
    neuroanatomy and, 15, 159
    Williams syndrome and, 220, 221
Moon, Keith, 162
Moonlight Sonata (Beethoven), 249
Moreno, Sylvain, 248–49
morphology, 279
Morrison, Van, 44
Morrow, Maegan, 178–79
Mortals, The, 243
motifs, 282–83, *282, 283*
motor action plans and sequences
    learning an instrument and, 246, 248
    movement and, 106–8, 110, *110*,
        128–29

motor cortex, 16, *28, 29*, 30, 110, 135, 245,
    246, 248, 339
Mott the Hoople, 267
movement
    cognition and, 103–4
    motor action plans and sequences,
        106–8, 110, *110*, 128–29
    music therapy for pain and, 195
    speech and, 106–7
    See also movement disorders
movement disorders, 102–3
    ancient evidence of, 102
    auditory feedback and, 109
    cognitive impact of, 104
    Huntington's disease, 123–25
    motor action plans and sequences and,
        108, 110, *110*, 113, 128–29
    multiple sclerosis, 116–22
    muscle memory and, 108
    predictive timing and, 109
    stuttering, 105, 108, 109, 110–12,
        113–14, 355
    Tourette syndrome, 114–16, 356
    See also Parkinson's disease
Mozart, Wolfgang Amadeus, 13, 219, 227
Muldaur, Clare, 277
Mull, Martin, 170
multimodal nature of music
    constructive processes and, 95
    musical memory and, 46, 51, 52, 57–58
    neural connectivity and, 30–31
    neuroanatomy and, 18, 19, 236–37
    playing an instrument and, 19, 236–37
    See also musical elements
multiple intelligences theory, 202
multiple sclerosis (MS), 116–22
multitasking, 73
mu-opioids, 176–77
muscle memory, 108
Musgrave, George, 239
musical ability, 204, 207
musical elements, 9
    AI and, 177–78
    amusias and, 68, 69–70
    attention and, 63–64, 65–66, 349–50
    combined, 300–303, *300, 301*
    constructive processes and, 98
    gestalt of, 317–19, 381
    musicality and, 203, 204
    musical memory and, 55, 97, 180–81
    musical preferences and, 34–35
    musical transcendence and, 317–19
    music as constructive process and, 99

musical elements (*continued*)
  music therapy for pain and, 199
  notes as phonemes and, 279, *279*
  as parts of speech, 286, 295–96
  research and, 199
  song recognition and, 96–98
  songwriting and, 204–5
musical instruments, range of, *279*
musical interventions, 331
musicality, 69, 90, 192, 201–7, 224
musical meaning, 276–316
  ambiguity and, 284–85, 289–90, 293
  authenticity and, 315–16
  chords and, 287, 301, 302–3, 306, 309,
    309
  combined musical elements and,
    300–303, *300, 301*
  context and, 278–79, 289–90, 349
  cultural variations and, 310
  depictive music and, 378–79
  emotion and, 290–91, 293, 296, 302, 311
  harmonic series and, 304–8, 380
  lyrics and, 300, 301–2, 379
  melody and, 288, 300–302, *300, 301,*
    303, *303*
  modulation and, 283–84
  motifs and, 282–83, *282, 283*
  performance interpretation and, 311–15
  preferences and, 276–78
  rhythm and, 282, *282,* 286–87, 297–300,
    299, 300, 302
  surprise and, 276–77
  timbre and, 286, 288
  transcendence and, 277–78
  universality and, 296–97, 303, 305, 307,
    308, 310
  See also linguistic analogies for music
musical memory
  chunking and, 49, 56, 181–82
  context and, 51–52
  dementia and, 52–53, 180
  development of, 53–56
  dynamic nature of, 51–52, 58–59
  emotion and, 51–52
  expectations and, 56–57
  forgetting and, 39
  higher-order representations and, 48–49
  improvisation and, 36–37, 51, 181
  learning an instrument and, 231
  lyrics and, 180–81
  mastery and, 36–38
  multimodal nature of music and, 46, 51,
    52, 57–58

  musical elements and, 55, 97, 180–81
  pattern matching and, 61
  pattern prediction and, 56
  performance and, 36–39, 58, 107–8
  personality and, 56
  repetition and, 49–50, 181
  stimulus generalization and, 96–98
  unique nature of, 44–46
  Williams syndrome and, 201
musical negative mood induction proce-
    dures, 164
musical notation
  Beethoven's Fifth Symphony, *281, 282*
  "Can't Buy Me Love," *299, 303, 309*
  Chopin's Prelude in E Minor, *229*
  "Fire," *300*
  "For No One," *301*
  "Honey Pie," *309*
  reading, 228–29, 281
  "Something," *300*
  "The Star Spangled Banner," *309*
  Tchaikovsky's First Piano Concerto, *299*
  "This Land Is Your Land," *299*
  "When the Saints Go Marching In," *299*
musical preferences
  context and, 34
  cultural variations in, 31
  individual differences and, 33–34, 98
  learning and, 55
  meaning and, 276–78
  musical elements and, 34–35
  naming and, 33
  transcendence and, 317
  workplace environment and, 253–54
musical transcendence, 5, 317–24
  ambiguity and, 323
  frisson and, 90
  imagination and, 320–22
  intimacy and, 323–24
  musical elements and, 317–19
  musical meaning and, 277–78
  music therapy and, 319
  mystery and, 322–23, 382–83
  time-bound nature and, 320
  universality and, 324
music as constructive process, 24–26, 95,
    96–101
music business, 238–43, 266
Music for the Young, Op. 39, No. 3: Mamma
    (Tchaikovsky), 162
music theory. *See* musical elements
music therapy
  AI and, 177–78

ancient origins of, 6–7, 8–9, 192
for aneurysm damage, 183–86, *185*
autism spectrum disorder and, 225
Default Mode Network and, 91
for dementia, 169, 174, 175–77, 179,
    260–61
distraction and, 19, 193–94
dosage and, 190
dynamic nature of, 101
emotion and, 137–38
epigenetics and, 141
group format, 143, 144, 145–46, 163,
    261
for Huntington's disease, 123–24
individual differences and, 99–101
licensing of, 331
Melodic Intonation Therapy, 178–79
for mental health conditions, 160,
    163–66, 175
movement disorders and, 104, 120
for multiple sclerosis, 119–20, 122
musicality and, 192
musical transcendence and, 319
music as constructive process and, 96,
    98, 101
for pain, 189, 190, 192–96, *197*, 198–99
for Parkinson's disease, 120, 129, 133–36,
    137–38
passive vs. active, 331
for relaxation, 176–77
scalability and, 177
songwriting as, 146–55
sound baths, 190
for stroke, 120, 178, 179
for stuttering, 113–14
for trauma, 140–41, 142–46, 149–55
treatment persistence and, 193
types overview, 331–34
Muzak, 254
myelin, 29, 116–17, 339–40
"My Favorite Things" (Tyner), 278
"My Man" (Holiday), 296

Nagano, Karin Kei, 35, 320–22
Nagano, Kent, 320–22
NASA, 184
Nash, Graham, 185, 186
Nazi Germany, 274
Neisser, Ulric, 95
Nelson, Oliver, 171
Nelson, Willie, 97
network analysis, 81, 353
neural connectivity

learning an instrument and, 246, 248
meditation and, 85, 88
multimodal nature of music and, 30–31
music therapy for mental health condi-
    tions and, 166
music therapy for movement disorders
    and, 120
music therapy for stroke and, 178
research methods for, 27
reward center and, 176–77
sound perception development and, 53
trauma and, 140
neural synchrony, 85
neuroanatomy, 15–35
auditory pathway, 16–18, *17*, 338, 342
brain regions, 27–29, *28*
cognitive transfer and, 235–37, 244, 246
constructive processes and, 95
group synchronization and, 270–71
learning an instrument and, 244–46, 248
memory and, 18–19
multimodal nature of music and, 18, 19,
    236–37
musical impact on, 4–5
pain and, 188–89, 192–93, 195–96, 198
placebo effect and, *197*
playing an instrument and, 18, 19,
    236–37
redundancies in, 120
relaxation network, 175–76, *176*
reward center, 176–77
rhythm and, 297–98
sensory processing, 15–16
stuttering and, 110–11, *110, 111*
trauma impact on, 139–40
visual cortex, 18, *28*
Williams syndrome and, 220
See also neural connectivity; neuroplasti-
    city; *specific regions*
neurochemicals, 31, 196, 340
See also neurotransmitters
neurochemical tags, 58
neurogenesis, 140–41, 340
Neurologic Music Therapy (NMT),
    331–32
neurons, 340
neuroplasticity
deafness and, 16
defined, 340
learning an instrument and, 231, 237, 246
musical transcendence and, 323
music therapy for mental health condi-
    tions and, 166

neuroplasticity (*continued*)
  music therapy for stroke and, 178
  music therapy for trauma and, 141
  music therapy for traumatic brain injury
    and, 179
  Neurologic Music Therapy and, 331
  professional musicians and, 30–31
neurotransmitters, 114, 164, 337–38, 341,
    342, 347
  See also dopamine
Newton, Elizabeth, 61–62
Newton, Isaac, 20
Nietzsche, Friedrich, 10
Nitzsche, Jack, 148
NK (natural killer) cells, 199
NMT (Neurologic Music Therapy),
    331–32
noise pollution, 29
Nordoff, Paul, 332
Nordoff-Robbins Music Therapy, 332–33
norepinephrine, 199
notation-performance differences, 32
note order, 280–82, *281, 282*
Novalis, 10
novelty-seeking, 265
nucleus accumbens, 28, *28,* 340
Nutcracker, The (Tchaikovsky), 162

Obama, Barack, 273
occipital lobe, 16, 18, 25, *28,* 128, 340
octave, 31
Octet (Stravinsky), 296
"Ode to Joy" (Beethoven), 62, 64
Officer and a Gentleman, An, 148
"Off to Join the World" (Camp), 278
"Oh, What a Beautiful Morning" (Rodgers
    and Hammerstein), 205
older adults. *See* aging
olfactory tubercle, 340
opening acts, 267
open monitoring meditation, 88
opioid receptors, 193
orbitofrontal cortex, *176,* 189
Orfeo (Powers), 319–20
Orff, Carl, 333
Orff, Gertrud, 333
Orff-Schulwerk, 333
Osbourne, Ozzy, 105
"Our House" (Nash), 185
Outwater, Edwin, 283
"Over the Rainbow" (Harburg), 45, 261
overtone (harmonic) series, 304–8, 310, 380
overtones, 304–5

Owsley, Will, 266
"Owsley" (Owsley), 266
oxytocin, 221, 250

pain, 187–99
  drug treatments for, 188, 196
  music therapy for, 189, 190, 192–96, *197,*
    198–99
  neuroanatomy and, 188–89, 192–93,
    195–96, 198
  placebo effect and, 196, *197*
panic disorder, 161
paradoxical kinesia, 137–38
parasympathetic nervous system, 166
parietal cortex, *176, 356*
parietal lobe (cortex), 16, *28,* 70, 88, 110,
    128, *176,* 271, 340
Parker, Charlie (Bird), 108
Parkinson's disease, 126–38
  emotion and, 136–38
  music therapy for, 120, 129, 133–36,
    137–38
  neuroanatomy and, 110, 111
  symptoms of, 125, 127–28
passive (receptive) music therapy, 331
Patel, Ani, 225
Pathétique Sonata (Beethoven), 249
patterned sensory enhancement, 331
pattern matching, 60–63, 97
pattern prediction, 27–28, 56, 75
Penrose triangle, *87*
pentatonic scales, 31, 307, *307,* 308, *308,* 310
Pentatonix, 203
Peraza, Armando, 54
performance
  anxiety, 240
  autism spectrum disorder and, 222–23
  flow state and, 89–90, 171
  healing intent and, 191
  improvisation and, 129–31, 132–33,
    170–72
  interpretation in, 32, 49, 97, 98–99,
    311–15
  mental health and, 242
  musical memory and, 36–39, 58, 107–8
  music therapy for pain and, 190
  opening acts, 267
  relaxation and, 113
  stuttering and, 105, 112
  See also professional musicians
"Personal Jesus" (Depeche Mode), 97
PET (positron emission tomography) scan-
    ning, 78–79, 351

Peter and the Wolf (Prokofiev), 286, 379
Petersen, Steve, 78
Pettus, David, 275, 318–19
pharmaceutical interventions. *See* drug
    treatments
phenotypes, 202–7
Phish, 90
phonemes, 279
physical health, 258–59
physical intelligence, 103
physical therapy (PT), 134–35, 162, 178
Piaf, Edith, 184
Piano Variations (Copland), 296
Pierce, Billy, 3, 133
Pink Floyd, 99, 261
pitch
    amusia and, 67–68, 69
    attention and, 63, 64
    auditory pathway and, 16–17
    autism spectrum disorder and, 225
    defined, 340
    dimensional models of, 381–82, *381, 382*
    musicality and, 208
    timbre and, 342
    See also melody
Pitch Perfect films, 257–58
pituitary gland, 340
placebo effect, 196, *197*
Plant, Robert, 253
planum temporale, 208
Plastic Ono Band (Lennon), 333
Plato, 8
Platt, Courtney, 121–22
play, 235
playing an instrument
    as constructive process, 25–26
    dementia and, 179
    elision and, 106
    movement and, 107–8
    multimodal nature of music and, 19,
        236–37
    music therapy for Huntington's disease
        and, 124–25
    music therapy for multiple sclerosis and,
        121
    music therapy for Parkinson's disease
        and, 136
    music therapy for trauma and, 143
    neuroanatomy and, 18, 19, 236–37
    relaxation and, 111–13
    social meaning and, 261
    stuttering and, 108, 112
    timbre and, 203

    Tourette syndrome and, 115–16
    Williams syndrome and, 210–11
    See also learning an instrument; singing
playlists
    designing, 263–64
    programming ability and, 205
    progressive, 163–64
    relationship, 262–63
    smart technology for, 5–6, 268–70
    See also self-selected music
"Pluto" (Clare and the Reasons), 277
Pointer Sisters, The, 300
Polisi, Joseph, 316
Pollan, Michael, 263
polyrhythms, 33
Pomodoro Technique, 255, 374–75
Pomp and Circumstance (Elgar), 349
pons, 340–41
popular music, 66, 114, 181–82, 239–40
    See also specific artists
Porter, Cole, 316
Posner, Michael, 27, 72, 77–78
posterior cingulate cortex, 30, 81, 85, *176*
Powers, Richard, 319–20
pragmatics
    learning an instrument and, 247–48
    musical meaning and, 291–94, 295–96,
        299–300, 310–11, 315
precuneus, 81, 83–84, 87, 176
prediction. *See* expectations; pattern pre-
    diction
predictive timing, 109
prefrontal cortex, *28*
    decision fatigue and, 255
    Default Mode Network and, 81, 83, 85,
        88
    defined, 341
    dopamine and, 158, 159
    GABA and, 338
    improvisation and, 87
    learning an instrument and, 248
    meditation and, 85, 88
    movement disorders and, 110, 111, 120,
        123
    pain and, 193
    trauma and, 140
    See also specific components
Prelude in E Minor (Chopin), 227, 228–29,
    229
pre-motor cortex, 110
Presley, Elvis, 105, 114
Pretenders, 205
Primal Scream Therapy, 333

primary auditory cortex, *17, 28*, 248
Prince, 35
procedural memory, 42, 46
professional musicians, 238–43
    finances and, 242–43
    mental health and, 162, 238–42
    musical memory and, 34, 181–83
    neuroplasticity and, 30–31
    training of, 237, 371
    See also performance
programming music, 205
progressive playlists, 163–64
Prokofiev, Sergei, 286, 379
prolactin, 140, 250, 341
propofol, 187–88
proprioception, 109
Ptolemy, 8–9
PTSD (post-traumatic stress disorder), 140,
    141–42, 143, 144, 145, 152
Punch Brothers, 267
Pussycat Dolls, 267
putamen, 30, 116, 175, *176*, 245–46, 248,
    341, 356

Queen, 62, 267
Quincy Jones Big Band, 38
Quintin, Eve-Marie, 224–25

ragas, 310
Raichle, Marcus, 78, 79, 80–81, 85
Ramones, 205
Rapsodie espagnole (Ravel), 296
RAS (rhythmic auditory stimulation),
    119–21, 123, 134, 137, 331
Rascal Flatts, 267
Ravel, Maurice, 9, 286, 296
Reagan, Ronald, 43
receptive aphasia, 343
refugees, 145–46
Reiss, Allan, 211, 218
relationship playlists, 262–63
relaxation
    music therapy for, 176–77
    neuroanatomy and, 175–76, *176*
    playing an instrument and, 111–13
relaxation network, 175–76, *176*
Remains of the Day (Ishiguro), 347
Remnick, David, 240–41
Rentfrow, Jason, 6
repetition
    autism spectrum disorder and, 222
    generative nature of, 289
    movement disorders and, 110, *110*

musical memory and, 49–50, 181
music therapy for dementia and, 261
music therapy for stuttering and, 114
Parkinson's disease and, 133
Repp, Bruno, 32, 312–13
research, 11–14
    bias and, 62–63
    correlation vs. causation and, 235–36
    current advances, 8
    emotion and, 312
    informed consent and, 351, 369
    Bobby McFerrin's contributions, 127
    music therapy for pain and, 190, 194–96,
        198, 199
    placebo effect and, 196
    scientific method and, 12–14, 79–80
    substance misuse and, 164
    See also Williams syndrome
research methods
    attention and, 349–50
    brain lesions and, 17
    brain mapping, 27
    cold pressor task, 194–95, 270
    diffusion tensor imaging (DTI), 27,
        336–37
    early brain scanning, 78–79
    fMRI, 81, 83, 120, 218–20, 337
    network analysis, 81, 353
    neuroimaging, 17
    neurotransmitters and, 347
    pain and, 194–95
    performance interpretation and, 313–14
    PET (positron emission tomography)
        scanning, 78–79, 351
    Williams syndrome and, 215–17
Resonance Project, The, 271, 273
Resurrection (Second) Symphony (Mahler),
    276, 277
reward center and pathways, 141, 166,
    176–77
Reyes, Carlos, 182, 190–92
rhythm
    amusias and, 70
    attention and, 62–63, 64
    cultural variations in, 32, 310
    defined, 341
    expectations and, 298, 341
    humor and, 383
    meter and, 339
    musicality and, 203
    musical meaning and, 282, *282*, 286–87,
        297–300, *299, 300*, 302
    musical memory and, 38, 56

musical preferences and, 34
music therapy for multiple sclerosis and,
    119–21, 331
music therapy for stuttering and, 114
neuroanatomy and, 297–98
Orff-Schulwerk and, 333
predictive timing and, 109
ratios, 32–33
rhythmatism, defined, 133
swing, 32–33, 38
tactus and, 342
trance states and, 9
universality of music and, 297, 310
Williams syndrome and, 209, 215
rhythmic auditory stimulation (RAS),
    119–21, 123, 134, 137, 331
Riddle, Nelson, 204
"Rikki Don't Lose That Number" (Steely
    Dan), 291
Rimsky-Korsakov, Nikolai, 108
Rio, Danny, 286
Rite of Spring, The (Stravinsky), 312
"River: The Joni Letters" (Hancock), 185
Rivers, Joan, 187
Robbins, Clive, 332
Rock, Irv, 67
"Rock-a-bye, Baby," 250
Rock Band, 270
Rollins, Sonny, 172, 173
Romeo Void, 243
Ronettes, The, 262
Ronstadt, Linda, 133, 134
Rose, Billy, 315–16
Ross, Lee, 61–62, 215, 271
Rossini, Gioachino, 62
Rubin, Rick, 206, 287n
Rubinstein, Arthur, 204, 233, 241
Rundman, Dawn, 243
Rundman, Jonathan, 243
Russell Garcia Orchestra, 97
Russo, Frank, 114, 175–76, *176,* 177

Sacks, Oliver, 10–11, 70, 71, 211
Sage, Jimmy, 216, 217
Sainte-Marie, Buffy, 148
Santana, Carlos, 54, 133
Sapolsky, Robert, 147
sarcasm, 292
Särkämö, Teppo, 332
"Saturday Night Fish Fry" (Jordan), 184
Saul (biblical king), 7
"Save Your Tears" (The Weeknd), 97
scalability, 177–78

scales
    chord construction and, 378
    cultural variations in, 31, 310
    defined, 341
    frequencies and, 306
    Greek modes and, 8, 339
    key and, 338
    linguistic analogies and, 279–80, *279*
    musical meaning and, 295–96
    musical memory and, 181–82
    pentatonic, 307, *307,* 308, *308,* 310
    universality of music and, 305
scatting, 37
Scenes de Ballet (Stravinsky), 315–16
schizophrenia, 85, 156–61, 162, 163
Schopenhauer, Arthur, 322, 323, 382
Schwartz, Theodore, 4, 344
secret chord, the, 324
Seinfeld, Jerry, 291
selective attention, 73–74, 75, 350–51
self-reflection, 83–84, 85
self-selected music
    cancer treatment and, 259
    everyday music and, 262–63
    finding new music, 267
    music therapy for aneurysm damage and,
        183–84, *185*
    music therapy for pain and, 195–96
    variety and, 263–66
    workplace environment and, 255, 256
    See also individual differences
self-soothing, 251
Selvin, Joel, 233–34
semantic memory, 42, 46
semantics, 284–86, 288
sensory cortex (somato-sensory cortex), *28,*
    194, 341
sensory memory, 41–42, 46
sensory-motor integration, 109
sensory processing, 15–16
serial music, 290
serotonin, 140, 221, 341
    transporter gene, *SLC6A4,* 341
Service for the Treatment and Rehabilita-
    tion of Torture and Trauma Survivors
    (STARTTS), 145–46
Seven Lively Arts, The, 315–16
Seventh Symphony (Beethoven), 273
sexual abuse, 150–52
sexual activity, 259
Sgt. Pepper (*Sgt. Pepper's Lonely Hearts Club
    Band,* making of), 40
Shakespeare, William, 74

shamanism, 6–7, 9, 192
Shamblin, Allen, 142
Shazam, 98
Sheeran, Ed, 113
Shepard, Roger, 86, *87*, 381–82, *381*, *382*
Shirvalkar, Prasad, 189
Shorter, Wayne, 112, 185
short fibers, *28*
Shulman, Gordon, 79, 80
sign language, 16, 70, 105, 201
Silbar, Jeff, 278
silence, 251
Silver, Horace, 291
Simon, Carly, 105
Simon, Paul, 241
Sinatra, Frank, 38, 308
Singh, Nandini, 30
singing
    harmonic series and, 308
    human potential and, 167
    lyrics and, 204–5
    Melodic Intonation Therapy, 178–79
    musical meaning and, 296
    music therapy for dementia and, 169,
        260–61
    music therapy for Parkinson's disease
        and, 136
    music therapy for trauma and, 145–46
    Parkinson's disease and, 129, 131–32, 136
    social meaning and, 260
    stuttering and, 105, 113
    Vocal Psychotherapy and, 333
    Williams syndrome and and, 201,
        212–13
"Sioux City Sue New" (Jarrett), 171
SLC6A4 (serotonin transporter gene), 221
Sleater-Kinney, 256
sleep, 85–86, 193
sleepwalking, 86
smart technology, 5–6, 268–70
Smiths, The, 146
Soloist, The (Lopez), 156
"Something" (Harrison), 300–301, *300*,
    302, 379
"Somewhere Over the Rainbow" (Har-
    burg), 45, 261
"Song for My Father" (Silver), 291
song recognition, 49, 50, 96–99
songwriting
    aspirational nature of, 222
    conflict resolution and, 272–73
    musical elements and, 204–5

    musical meaning and, 300
    as music therapy, 146–55
    music therapy for mental health condi-
        tions and, 164
    music therapy for trauma and, 142–43,
        144–45
    See also lyrics
Sonos and Arora, 257
Sorrentino, Gilbert, 291
sound baths, 190
sound perception
    auditory pathway, 16–18, *17*, 338, 342
    constructive nature of, 24–25
    early development of, 26–27
    evolution of, 21–22, 27–28
    frequencies and, 24
    individual differences and, 19–20
    learning an instrument and, 248
    movement and, 20–21, 22–24, *24*,
        303–4
So You Think You Can Dance, 121–22
spectro-temporal flux, 342
speech, 105–7, 108, 136, 306–7
speech music therapy, 124
Spektor, Regina, 58–59
spiders, range of hearing, 23, *23*
Springsteen, Bruce
    on mental health, 161, 165, 240–41
    mistakes and, 233–34
    musical meaning and, 300, *300*
    on trauma, 151
Stalling, Carl, 18, 288, 378–79
"Stand by Me" (King, Leiber, and Stoller),
    261
"Star-Spangled Banner, The," *309*
startle reflex, 16, 29
state-dependent memory retrieval, 52
statistical processing, 26
"Stay a Little Longer" (Tillis), 105
Steely Dan, 149, 155, 184, 243, 291
Stegemöller, Elizabeth, 136
Stein, Seymour, 205–6
Steinbeck, John, 251
Stevenson, Bryan, 166–67
Stewart, Rod, 287
"Still Learning How to Fly" (Crowell), 227
Stills, Stephen, 149, 203
stimulus generalization, 96–98
Sting, 18, 244
Stravinsky, Igor, 296, 311, 312, 315–16
stress
    displaced persons and, 145–46

mental health conditions and, 164
movement disorders and, 112
music therapy for mental health conditions and, 166
playing an instrument and, 112–13
trauma and, 140
See also PTSD; relaxation
striatum, *28,* 115, 336, 338, 340, 341
string theory, 319
stroke, 120, 170–72, 173, 178, 179, 351
stuttering, 105, 108, 109, 110–12, 113–14, 355
STX1A (syntaxin gene), 221
subconscious, 10
substance misuse, 164, 238, 241
substantia nigra, 128, 336, 342
subthalamic nucleus, 336
Sullenberger, "Sully," 37
"Summertime" (Gershwin), 96–97, 98, 186
Sunny Nights (Huxley), 266
Sun Ra and his Arkestra, 98
superior olivary nucleus, *17*
superior temporal gyrus, *28,* 88
"Superstar" (Carpenter), 286
"Superstition" (Wonder), 66
"Surprise" Symphony (Haydn), 100
Swan Lake (Tchaikovsky), 52, 162
"Sweet Georgia Brown" (Bernie and Pinkard), 278
Swift, Taylor, 256, 267
swing feel, 32
Symphony No. 2 (Mahler), 74
synapses, 29–30, 50, 342
syntax, 280–82, *281, 282,* 285–86, 288–89, 294–95
syntaxin gene, *STX1A,* 221

tactus, 342
See also rhythm
Talking Heads, 205
"Tangled Up in Blue" (Dylan), 64–65
Taupin, Bernie, 142
TBI (traumatic brain injury), 120, 144, 332, 351
T cells (T lymphocytes), 198–99
Tchaikovsky, Pyotr Ilyich, 162, 213, 219, 298, *299*
tempo
lullabies and, 250
musical memory development and, 56
music therapy for multiple sclerosis and, 120

music therapy for Parkinson's disease and, 134
music therapy for stuttering and, 114
pattern matching and, 60
playlist design and, 263
universality of music and, 310
Williams syndrome and, 215
temporal lobe, *28,* 342
"Tequila" (The Champs), 286
thaats, 310
thalamus, 16, *17,* 18, 115, 245, 248, 342
Thaut, Michael, 119–21, 134, 137, 331, 332
"Theme from *The Addams Family*" (Mizzy), 63, 214
Theory and Practice of Vocal Psychotherapy, The (Austin), 333
Therapeutic Music Capacities Model, 317–18
therapeutic singing, 332
Thinking, Fast and Slow (Kahneman), 95–96
"Third World Man" (Steely Dan), 184
"This Is the Song" (McKagan), 161
"This Land Is Your Land" (Guthrie) 298, *299*
Thompson, William Forde, 317–18
"Three Blind Mice," 63
Tibetan singing bowls, 146
tics. *See* Tourette syndrome
Tillis, Mel, 105, 112, 355
Tilson-Thomas, Michael, 74
timbre
amusias and, 70
attention and, 64–65
cultural variations in, 310
defined, 9–10, 342
frequencies and, 306
lullabies and, 250
musicality and, 203
musical meaning and, 286, 288, 304
musical memory development and, 56
musical preferences and, 34–35
sound perception and, 17
time-bound nature of music, 91, 320
time signature, 215, 250, 339
Tirovolas, Anna, 313–14
tonal color, 342
See also timbre
tonality, 17, 31
tone deafness. *See* amusias
Tourette syndrome, 114–16, 356
trance states, 9, 10
transcendence. *See* musical transcendence

transcranial brain stimulation, 135
trauma, 139–46
    brain impact of, 139–40
    displaced persons and, 145–46
    generational cycling in, 148
    music therapy for, 140–41, 142–46,
        149–55
    professional vs. amateur musicians and,
        238
    PTSD and, 140, 141–42, 143, 144, 145,
        152
    sexual abuse, 150–52
    songwriting and, 149–51
traumatic brain injury (TBI), 120, 144,
        332
triple-feel, 33
tritone, 31
tritones, 287
12-tone system, 290
"Twinkle, Twinkle, Little Star," 250
tympanic membrane (eardrum), 22–23, 34
Tyner, McCoy, 278

unconsciousness, 85
"Unity" (Buhl and Levitin), 149–51, 153
universality of music
    musical meaning and, 296–97, 303, 305,
        307, 308, 310
    transcendence and, 324
unwanted music, 29, 84, 253
"Up Where We Belong" (Nitzsche,
        Sainte-Marie, and Jennings), 148

variety, 263–66
vasopressin, 188, 189
ventral prefrontal cortex, 111
ventral striatum, 338, 340
ventral tegmental area (VTA), 141, 342
vermis, 111, *111*
visual cortex, 18, 342
vocal expression. *See* singing
Vocal Psychotherapy, 333
von Ehrenfels, Christian, 60–61
Voyager probe, 184
VR (virtual reality) technologies, 104
Vuust, Peter, 30–31

Wagner, Richard, 320, 349
Walden, Chris, 182
Wang, Paul, 211
Warnes, Jennifer, 148
Wearing, Clive, 71, 82–83

Webb, Jimmy, 302–3
Weeknd, The, 97
Weill, Kurt, 36–37, 38
Welch, Bob, 383
Well-Tempered Clavier (Bach), 232, 296
Wernicke-Korsakoff syndrome (WKS),
        364
Wernicke's area, 69–70, 88, 342–43
Wesseldijk, Laura, 162
Western classical music
    musical memory and, 182
    performance interpretation in, 32,
        311–15
    professional musicians in, 240
    timbre in, 35
    See also specific composers
"We Will Rock You" (Queen), 62
"When the Saints Go Marching In," 298,
        299
White, Benjamin, 60–61
white matter tracts, *28*, 30–31, 110, 343
"Wichita Lineman" (Webb), 302–3
Williams, John Cyprian Phipps, 208
Williams, Mary Lou, 32
Williams, Robin, 90
Williams, Victoria, 118
Williamson, Cris, 142, 272–73
Williams syndrome (WS), 11, 208–21
    cognitive profile and, 201, 207,
        209–10
    fMRI research and, 218–20
    gene mutation and, 208–9, 221, *222*
    intelligence and, 209–10, 213
    perseveration and, 213
    physical manifestations of, 210–11
    research methods and, 215–17
    rhythm and, 209, 215
    singing and, 201, 212–13
    sociability and, 210, 214–15, 221
    sound fascination and, 213–14, 218
    treatment persistence and, 193
William Tell Overture (*Lone Ranger* theme)
        (Rossini), 62, 65
Wills, Bob, 32
Will to Love, The (Young), 234
Wilson, Ann, 339
"Wind beneath My Wings" (Silbar and
        Henley), 278
Winehouse, Amy, 242, 256
"Wings Cut from Fabric" (Buhl and
        Levitin), 153
Wittgenstein, Ludwig, 98

Wonder, Stevie, 35, 66, 205, 243, 278
Wooten, Victor, 35, 90, 188, 192, 233, 244
workplace environment, 253–58, 374–75

"Yankee Doodle," 55
Yellowbird, Shane, 339
Yiddish, 295

Young, Neil, 161, 185, 234, 243, 266
YouTube, 231

Zac Brown Band, 203
Zanes, Warren, 234
Zappa, Frank, 33
Zhu Zaiyu, 380